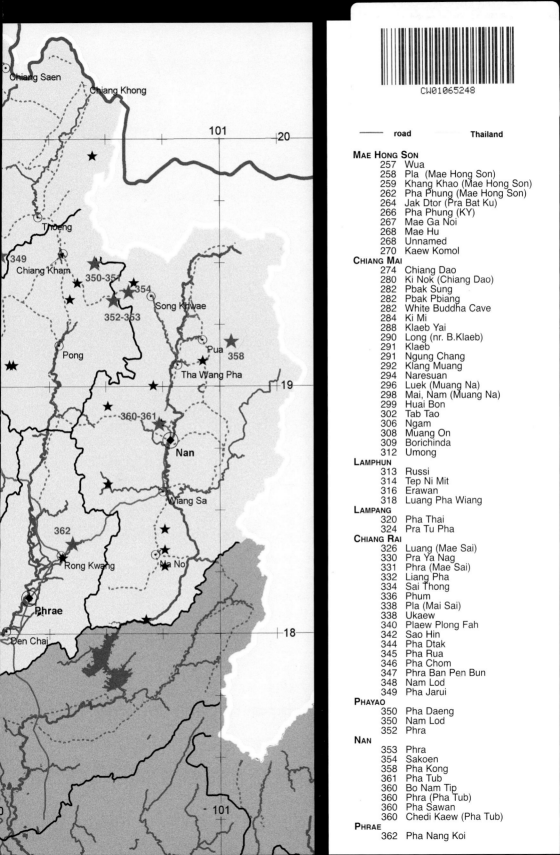

———— road   Thailand

Caves of Northern Thailand

CAVES OF
NORTHERN THAILAND

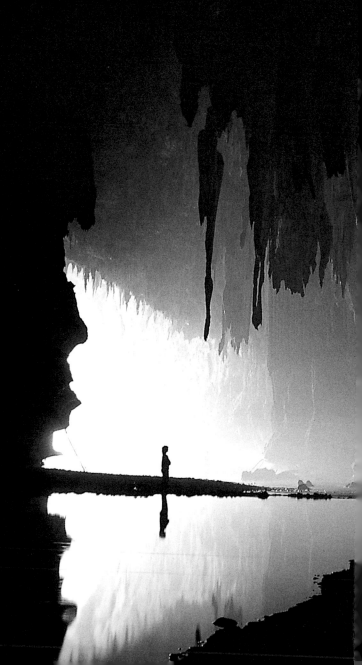

River Books

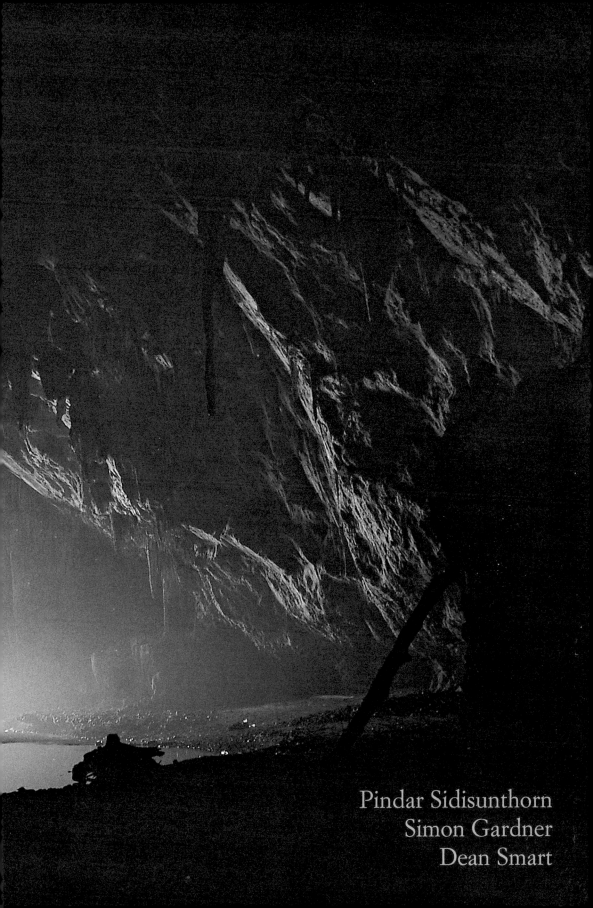

Pindar Sidisunthorn
Simon Gardner
Dean Smart

Previous page: Tham Lod, Pang Ma Pha. (© Dean Smart)

Front cover : Tham Jak Dtor, Mae Hong Son

Back cover : Tham Pet, Pang Ma Pha

 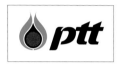

First published and distributed in Thailand in 2006 by
River Books Co., Ltd
396 Maharaj Road, Tatien,
Bangkok 10200 Thailand
Tel: (66) 2 225-4963, 2 225-0139, 2 622-1900
Fax: (66) 2 225-3861
Email: paisarn@riverbooksbk.com
Website: www.riverbooksbk.com

The background research of the book was funded by the Thailand Research
Fund (TRF). The printing of this edition is supported in part by the Thailand
Research Fund(TRF) and the PTT Public Company Limited.

Editor Narisa Chakrabongse
Design Suparat Sudcharoen
Production Paisarn Piemmettawat

ISBN 974 9863 13 5

Printed and bound in Thailand by Sirivatana Interprint Public Co., Ltd.

CONTENTS

PREFACE

In 1998-2000, Thai Research Fund (TRF) funded a project to conduct research on caves in Pang Ma Pha, Mae Hong Son Province. In order to disseminate the work to the general public, TRF requested the assistance of Pindar Sidisunthorn, Simon Gardner and Dean Smart to prepare a book entitled "Caves of Northern Thailand". The first two authors have many years of practical experience doing ecological and social research in Northern Thailand, and Dean Smart has spent over 10 years surveying caves throughout Thailand. The Pang Ma Pha project formed the basis for background information, with additional surveys of caves in other areas undertaken by the authors in 2003.

This is the first book ever published on the subject and includes chapters on geology, biology and archeology as well as conservation and cave management. The publication is richly illustrated with photos and drawings from over 100 caves throughout Northern Thailand, including some world-class caves which are of great interest for speleologists and archeologists. Many caves have been disturbed by human activities and in order to mimimize this impact it is vital to raise public awareness of the value of caves and the need to protect them. TRF hopes that this book will contribute to a better understanding of caves and help to preserve the cave resources of Thailand.

Prof. Dr. Piyawat Boonlong
Director, Thailand Research Fund

Summary of the Pang Ma Pha Cave Project 1998-2000

The project aimed to collect primary and secondary data relating to caves such as geology, hydrology, archaeology, forestry, wildlife, land use, and community. All data was organized as data base system in order to view relational data that could help classify caves and indicate cave transformation in the past, present and future. The research area covers Lang and Khong river basins in Amphoes Muang and Pang Ma Pha, Mae Hong Son Province.

There are around 170 caves comprised of 90 caves of geological and 80 caves of archaeological significance. The team sampled cave clusters representing well-distributed samples all over the river basins. These samples reflected geological and archaeological significance and were near villages, for example cave clusters of Muang Pham, Pa Daeng and Mae Lana. All researchers set up working groups to record specific data into data forms. However, they re-examined data accountability and reliability before transferring them into database system.

During 1998-2000 the research team explored around 46 caves. These caves respectively comprised 9 caves of geological significance and 37 caves plus 1 rock shelter of archaeological significance. Although geological and geomorphological features of the mountainous area are heterogeneous, karst topography is dominant in around half of the area. The physical domain reflects ecological complexity and abundant natural resources which have attracted people to settle down here since the pre-historic period. Nevertheless, current human activities such as agriculture, forestry or tourism have affected fragile cave resource and ecological systems. The database system of cave resources is useful for natural resource conservation or logical development of the area with minimal environmental damage.

Research Team: Dr. Sittipong Dilokwanich (Faculty of Environment and Resource Studies, Mahidol University); Dr. Kasem Kulparadit (Faculty of Environment and Resource Studies, Mahidol University); Dr. Chaiporn Siripornpibul (Division of Deep-well Drilling Development, Public Works Department); Dr. Rasmi Shoocongdej (Department of Archaeology, Faculty of Archaeology, Silpakorn University); John Spies (Independent Speleologist); Krit Chareonthong (Royal Forestry Department); Somsak Laoyipa (Royal Forestry Department).

ACKNOWLEDGEMENTS

FUNDING: production and printing of this book was provided by the Thailand Research Fund, without whom it would have been impossible to complete. We would especially like to thank Dr. Suchada Chinajitr for her assistance and patience throughout the process. In addition, the printing was supported in part by the PTT Public Company Limited.

RESEARCH: this book would not have been conceived without the background research of the Pang Ma Pha Caves Project and the Highland Archeology Project, both funded by the Thailand Research Fund. We are extremely grateful to all the researchers on these two projects and their assistants.

FIELDWORK: we would never have been able to find many of the caves without the help of John Spies, whose painstaking work over many years will form the foundation for speleological research in Pang Ma Pha for many years to come. Mana Kaewongwan was an excellent companion and an invaluable helper, whose patience and good humour kept us all going. We would especially like to thank all the local guides and assistants who helped us throughout the field work, Forestry Department officials for accomodation and support as well as many forest monks who gave freely of their time and knowledge. Many fellow cave surveyors have helped over the years, especially Nopparat Naksathit, Terry, Phil, Paul, Simon, Geu, Yan, Kui, Bill, Olly, Stu, Goose, George, Graham, Yut, Rob and Tony.

BAN THAM: during the course of preparing this book, we lived in Ban Tham Village, Pang Ma Pha for 8 months and we would like to thank all the people who made our stay there so memorable and worthwhile. In particular, Yot Chai and family, whose house we stayed in; Oi, Om and Pim who took such good care of our daughter; Palee, Bong, Cat, Nok, Jo, Gob, Hed for company and advice.

REVIEW: Robert Cunningham, Dr. Louis Deharveng, Dr. Sittipong Dilokwanich, John Dunkley, John Gardner, Paitoon Leksawad, Dr. Surapong Lerdthusnee, Dr. Rasmi Shoocongdej, Dr. Chaiporn Siripornpibul, M.L. Tosovan Tevakul, Dr. Chavalit Vittsyanon, Chavarit Kaokaew, Prayut Amonvatee

THAI-EDITION Duanpen Limsritrakul, Dr. Oy Kanjanavanit, Dr. Sittipong Dilokwanich, Weranuch Pintawanich, Duang Jai Rungrojcharoenkit, Prof. Padet Sidisunthorn, Pathai Sidisunthorn and Sunee Wongwaisayawan

SUPPORT: Department of National Parks, Wildlife and Flora Conservation staff, Chiang Mai National Museum, Princess Anthropology Centre, Siam Society Library

ILLUSTRATIONS: Pang Ma Pha Caves Project, Highland Archeology Project, Dr. Surapong Lerdthusnee, Kevin Kiernan (Nam Lang Map), Odel and Odel (Khao Kwian Map)

UNDERSTANDING: wife and son: Lek and A.J.

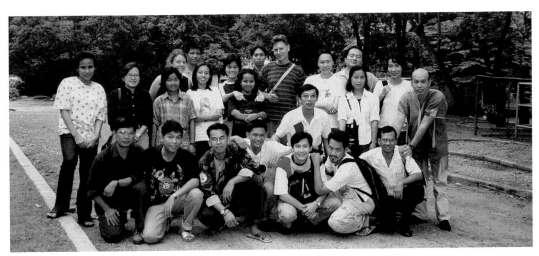

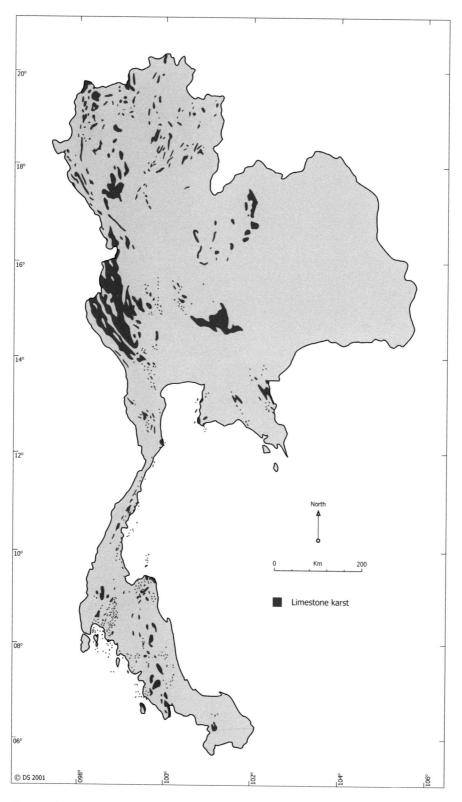

Limestone karst areas in Thailand.

© DS 2001

INTRODUCTION

Caves in Thailand

Around 4,000 caves are currently known in Thailand but experts estimate that this is probably less than half the real number. Another 4,000 caves are still waiting to be discovered. The most authoritative list to date is John Dunkley's "The Caves of Thailand" published in 1995 with a supplement in 1997. This invaluable reference lists over 2,000 caves from almost every province in Thailand, including 450 entries from Northern Thailand. The majority of the caves in this book are to be found on Dunkley's list.

Many new caves have been discovered since 1997, particularly in Western Thailand which has the largest area of limestone in the country and almost certainly houses the largest number of caves but has not yet been thoroughly surveyed.

Southern Thailand also has a considerable number of caves including some spectacular sea caves and the world-famous bird nest collecting sites. Most of the limestone in the southern region is in the form of isolated towers and the majority of caves are less than a few hundred metres long.

Eastern Thailand has a very limited area of limestone and consequently only a small number of caves. However, this region contains many sandstone rock overhangs which are of great archeological interest including some excellent prehistoric rock paintings and ancient religious sites.

Central Thailand is largely alluvial flood plain with only isolated pockets of limestone but even here there are caves in every province including extremely important cultural sites such as Tham Phothisat in Saraburi and the caves of the Khao Ngu range in Ratchaburi which house Thailand's oldest Buddhist artifacts dating from the 7th century AD.

World Record Caves (2004)

Longest Caves

Cave Name	Country	Length (km)
1 Mammoth	USA	579
2 Optimisticeskheskaya	Ukraine	214
3 Jewel	USA	207
4 Holloch	Switzerland	189
5 Lechuguilla	USA	180
6 Wind	USA	177
7 Fisher Ridge	USA	172
8 Siebenhengste Hohgant	Switzerland	145
9 Sistema Ox Bel Ha	Mexico	122
10 Ozernaya	Ukraine	122

Deepest Caves		Depth (m)
1 Veronja (Krubera)	Georgia	1710
3 Lamprechtsofen	Austria	1632
4 Mirolda / Lucien Bouclier	France	1626
2 Jean Bernard	France	1602
5 Torca del Cerro (del Cuevon)	Spain	1589

from Bob Gulden's website www.pipeline.com, Jan 29 2004

Thailand's Record Caves (2003)

Longest Caves

Cave Name	Province	Length (km)
1 Phra Wang Daeng	Phitsanulok	13.11
2 Mae Lana**	Mae Hong Son	12.60
3 Nam Lang**	Mae Hong Son	8.55
4 Takobi**	Tak	7.50
5 Luang**	Chiang Rai	6.22
6 Krachaeng**	Yala	5.51
7 Chiang Dao**	Chiang Mai	5.19
8 Pung Hung**	Mae Hong Son	4.44
9 Yai	Petchabun	4.15
10 Pha Mon **	Mae Hong Son	4.05

Deepest Caves		Depth (m)
1 Big House Cave *	Chiang Mai	190
2 Pha Phuak **	Mae Hong Son	166
3 Dead Robber Cave *	Chiang Mai	165
4 Turakit Maiset	Kanchanaburi	157
5 Pig's Play*	Chiang Mai	150

*** covered in this book   *other caves in Northern Thailand*

Caves in Northern Thailand

For the purposes of this book, Northern Thailand refers to the eight northernmost provinces of Thailand – Mae Hong Son, Chiang Mai, Lamphun, Lampang, Chiang Rai, Phayao, Nan and Phrae. This region is situated between 18 - 20° North and 98 -101° East, covering a total area of 88,371 km² which represents 17% of the whole of Thailand.

The area is divided by a series of mountain ranges which separate the four major tributaries of the Mae Nam Chao Phraya, Thailand's principal river. Most of Northern Thailand is part of the Chao Phraya watershed, with the exception of Mae Hong Son province which drains west into the Salween and Chiang Rai province which drains northeast into the Mekong. There are five peaks over 2,000 m including Doi Intanon (2,665m), the highest mountain in Thailand. The highest limestone mountain is Doi Chiang Dao (2,225m) which has a famous cave at its base.

An estimated 20% of the region is composed of limestone or other carbonate rocks, concentrated in Mae Hong Son and Chiang Mai provinces but with substantial outcrops throughout the region. The majority of the limestone is Permian in origin although there are also substantial areas of Ordovician, Silurian/Devonian and Triassic limestone. All of these areas have the potential to contain caves.

Cave Surveys

Northern Thailand contains more reported caves than any other region of Thailand but this is probably due to ease of access rather than higher cave densities. The only area in Northern Thailand which has been extensively surveyed for caves is Pang Ma Pha District in Mae Hong Son Province. Although Pang Ma Pha is only 1,200 km² (5% of Northern Thailand), a total of 176 caves have been discovered. The high density of caves in this district can be partly explained by the large proportion of limestone bedrock (47% total area) but is probably also the result of intensive survey efforts. A series of Australian and French expeditions in the 1980s produced the first systematic surveys of the most significant caves, including the (then) two longest caves in Thailand, Tham Mae Lana and Tham Nam Lang. Many caves owe their discovery to the exhaustive commitment of John Spies, a local resident and a vocal supporter of cave conservation. He recently joined a team of Thai scientists from Mahidol and Silpakorn Universities lead by Dr. Sittipong Dilokwanich. This multi-disciplinary team adopted a holistic approach to produce a database of all known caves covering archaeology, geology, hydrology, ecology and social aspects.

No other area in Northern Thailand has been surveyed in comparable detail, so it is not possible to judge if the density of caves in Pang Ma Pha is truly exceptional. Individual caves such as Tham Chiang Dao and Tham Luang (Mae Sai) have been intensively studied, particularly by a series of outstanding French expeditions who are the only group which has seriously attempted to study cave biology in Northern Thailand. Their efforts have resulted in the discovery of many new troglobitic species which are endemic to Northern Thailand – i.e found nowhere else in the world.

Caves have occasionally been mentioned in travellers' journals and other publications which can be an invaluable source of information, despite their non-technical background. Some of the early European travellers' accounts provide a fascinating insight into the importance of caves in the religious observances of the local people. Carl Bock's colourful account of his visit to Tham Tab Tao in 1884 is a particularly good example and his journals also mention other caves in Tak, Chiang Mai and Chiang Rai provinces.

Caves were a prominent feature of the Tourist Authority of Thailand's promotional theme for 2004, "Unseen Thailand" and have recently been included in several Thai Television documentaries as well as the BBC.

Following recent upsurge of visitor interest, the Royal Forestry Department now recognises the value of caves as tourist attractions and many national park pamphlets draw attention to caves in their area, although in most cases no detailed information is provided. Tham Chiang Dao is one of the few caves where a detailed guide to the history and legends associated with the cave is readily available in both Thai and English versions.

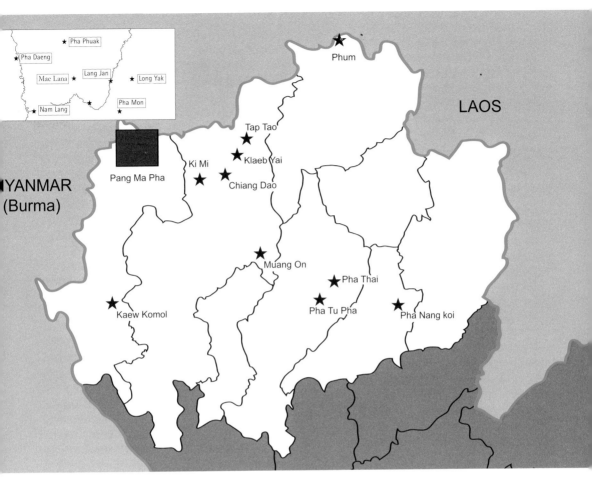

Unsurveyed Areas

Only a handful of all known caves in Northern Thailand have been comprehensively surveyed or mapped to any degree of accuracy and there are almost certainly many hundred caves which are still awaiting discovery. There are large areas that have never been surveyed in any detail, including such promising areas as Chiang Dao National Park, Doi Angkhaeng and Mae Ping National Park as well many other smaller areas scattered throughout the region. There are undoubtedly many opportunities for enterprising cavers to make exciting new discoveries, even without specialist equipment. The records for both the longest and the deepest caves in Thailand have recently been broken, and it seems likely that even longer and deeper cave systems are still waiting to be discovered.

Caves to Visit

The 100 caves selected for this book are only a small sample of the estimated 2,000 caves in Northern Thailand. We have included many of the most spectacular wild caves and the most famous temple caves, as well as a broad range of less well-known caves. There are representative examples of all of the major cave types found in the region – tourist caves, sporting caves, temple caves and archaeological sites. There are caves from all 8 provinces in Northern Thailand, although inevitably the samples are concentrated in areas which are more readily accessible and for which there is most prior information and survey details, even though this may not be an accurate overall picture of cave distribution in the region. Pang Ma Pha District is by far the best studied area and represents about 40% of the caves in this book but probably contains only 10% of the total number of caves in Northern Thailand. Chiang Dao and Mai Sai Districts also house many caves and are readily accessible. We have not included caves with extensive vertical sections or caves with extensive sections of CO_2 except for demonstration purposes. A large number of caves in Northern Thailand fall into these two categories but they can only be visited by experienced teams with specialist equipment.

Temple Caves

Tham Chiang Dao (Chiang Mai) is the most famous cave in Northern Thailand and has been an important pilgrimage site for many centuries. The cave houses a large collection of religious imagery, including the oldest historically – documented artifact of any cave in Northern Thailand. **Tham Phum** (Chiang Rai) is the most sacred cave in Northern Thailand, supposed to have been visited not only by the historical Buddha but also by six previous ones and predicted to be visited in the future by the Buddha Metteya. Records of donations go back over 650 years, which are the oldest written evidence of any cave in Northern Thailand. The cave is elegantly shaped like a vase and the placement of statues enhances the natural harmony of the surroundings. **Tham Tab Tao** (Chiang Mai) is another important temple cave which has been venerated for many hundreds of years. The outer day-lit chamber houses two gigantic golden Buddha statues surrounded by his followers, one seated and the other reclining. Another shrine hidden in the dark recesses of the vast inner chamber imparts a strong sense of the impermanence of our existence and is a spiritual experience in its own right. **Tham Muang On** (Chiang Mai) is another cave which is reputed to have been visited by the Buddha and has one of the most magnificent active stalagmites in Northern Thailand.

Many temple caves host festivals and are well worth visiting during these times. One of the most well known is the Songkran festival held in **Tham Pha Nang Koi** (Phrae). A large number of caves in Northern Thailand are still occupied by monks, either temporarily by wandering forest monks or permanently by resident monks. You will come across these caves throughout the region and they are well worth visiting as an insight into local culture and beliefs.

Show Caves (tourist caves)

These are caves which have been developed as tourist attractions and can be visited by anyone without the need for special equipment. The most important ones have lighting systems and a network of paths and are an excellent introduction to caves for the inexperienced. **Tham Lod** (Pang Ma Pha) receives over 100,000 visitors a year and is the mainstay of the local economy. A well-organised system of local guides with lanterns and rafts is available throughout the year and this is a very good example of the potential benefits of natural resources for local communities. The cave has been open to tourists since the 1980s and is on all the Mae Hong Son circuit tour itineraries. **Tham Pha Thai** (Lampang) is easily accessible close to the main Lampang-Phayao road and was recently declared a National Park.

The cave was visited by King Rama VII in 1926 and is managed by the National Parks Department who have recently installed an excellent trail and lighting system. **Tham Kaew Kamol** (Mae Hong Son) is one of the best decorated caves in Thailand with a magnificent display of crystals and is the only known example of a hydrothermal cave in the region. The cave was only opened to the public in 2002 after extensive infrastructure development including paths, lighting systems and a visitor centre.

Adventure Caves

Northern Thailand has a wide variety of sporting caves ranging from the extremely easy to the highly demanding in both physical and technical terms. Cavers of all backgrounds are sure to find plenty to suits their level of experience. Bear in mind that even an easy cave can be dangerous if you do not follow safety procedures – see page 162. Inexperienced cavers should never enter any cave without a guide. **Tham Long Yaow** (Pang Ma Pha) and **Tham Ki Mi** (Chiang Mai) are both suitable for beginners and contain some good examples of cave formations. Local guides are available from nearby villages. For those looking for a bit more of an adventure, **Tham Pha Daeng** and **Tham Bung Hung** (Pang Ma Pha) are fine stream caves with some challenging low sections and active speleothems. **Tham Seua** and **Tham Lom** (Tiger/Wind Caves) in Pang Ma Pha and **Tham Klaeb Yai** (Chiang Mai) are excellent examples of cave development, the latter is particularly interesting because it is a rare example of a paragenetic cave (see page 17). **Tham Pha Phuak** (Pang Ma Pha) has one of the most spectacular entrance chambers of any cave in Thailand and is definitely worth seeing, although the inner sections become progressively more difficult and dangerous. **Tham Mae Lana** (Pang Ma Pha) is the longest cave in Northern Thailand and one of the most beautiful and sporting but is not for the inexperienced. The upper levels contain outstanding cave formations which are extremely sensitive to accidental damage and should only be visited by fully conservation-conscious cavers. **Tham Nam Lang** (Pang Ma Pha) is a magnificent cave with some of the largest river passages in the world but is quite physically challenging with some swimming sections even in the dry season. The awesome entrance chamber alone is well worth a special trip and does not require any special experience or equipment. **Tham Pha Mon** (Pang Ma Pha) is one of the best decorated caves in Thailand but is extremely sensitive to accidental damage and can only withstand very limited visitor levels. The cave is currently managed by the Thai Military and is closed to the general public.

At the other end of the spectrum, vertical cavers and suicidal maniacs who enjoy nothing better than slow death by CO_2 poisoning will find plenty to amuse themselves. One of the attractions of serious adventure is discovering virgin passages which no one has ever visited before. There is plenty of scope for making new discoveries in Northern Thailand, but such caves are naturally not included in this book as you have to go and find them for yourself!

Archaeological Sites

The majority of archaeological sites are highly fragile and can never be replaced once damaged. Even well-intentioned visitors can unwittingly destroy apparently insignificant evidence which may be crucial to our understanding of the remote past. For this reason, we have not included access details for the most sensitive sites, even though these sites are an important part of Thailand's cultural heritage. Interested parties should contact Chiang Mai Museum or Silapakorn University for more information.

Prehistoric sites often have very little to interest even well-informed visitors as there is usually almost nothing left to see on the surface. The exception to this are the spectacular log coffin sites in Pang Ma Pha which are well-worth experiencing. Only a few sites are suitable for visitors as most of the best surviving sites are in remote locations with difficult access. The outflow entrance of **Tham Lod** has some coffins in a dry upper chamber which is probably the easiest place to see them, although these particular examples are in very poor condition and do not give visitors a true sense of their grandeur. **Tham Lang Jan** is close to the main road and is a much better place to experience this ancient culture as some of the coffins are in reasonable condition and still on their original support beams. **Ban Rai Overhang** is one of the most impressive and atmospheric sites with over 20 coffins, including the biggest yet found in the region. This is the best studied log coffin site in Pang Ma Pha and was intensively surveyed by a team of archeologists between 2000 and 2002. A signboard at the site explains excavation methods and the significance of the findings.

Several dozen sites in Northern Thailand contain examples of prehistoric rock art, although many of them are easily overlooked. The best place to see rock art is **Phra Tu Pha** (Lampang), which is readily accessible near highway 1 between Lampang and Phayao. This site houses a magnificent collection of over 1,800 paintings, including depictions of many different animals as well as a multitude of hand prints. The trail to **Ob Luang Overhang** (Chiang Mai) passes over a spectacular gorge and makes a pleasant trip although the paintings themselves are very faint. All the prehistoric artifacts have been removed to Chiang Mai Museum, which houses an excellent display and a model of the burial site.

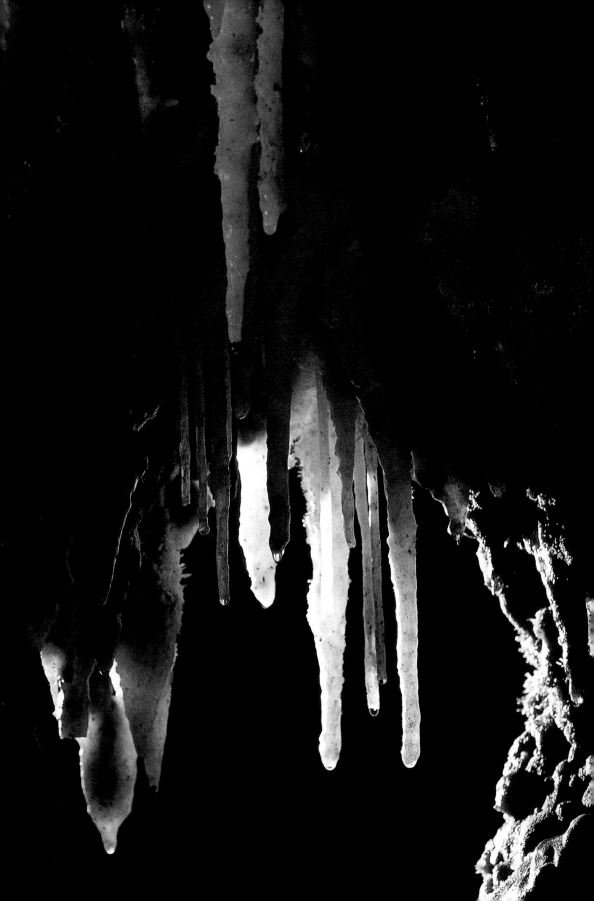

GEOLOGY

Caves are found in almost every situation on earth from the equator to the poles, from beneath the sea to high on the sides of mountains and from tropical climates to deserts. They form in virtually every rock type: sandstone, shale, gypsum, granite and basalt to name just a few. It is with limestone though that they realise their greatest fruition.

Limestone

Limestone is a sedimentary rock made up of the mineral calcium carbonate ($CaCO_3$). It is formed either from the accumulation of animal remains, whose shells and skeletons are made of $CaCO_3$, or, by precipitation of the mineral from water. Deposition takes place mostly in warm, shallow seas and the widespread occurrence of this environment, both today and in the past, has resulted in limestone being a relatively common rock. Marble is limestone that has been recrystallised by heat or pressure. Both limestone and marble are well suited to cave develop- ment. They are hard and have the strength to support the roof of a cavity. They also possess a very special property – they dissolve in acids.

The limestones of northern Thailand vary in both age and type. The oldest are marbles of Precambrian age found near the Phumiphol Dam. These outcrop as a thin band about 500 m wide running northward from near the dam wall. They have never been explored for caves.

Outcrops of impure, thinly bedded limestone of Ordovician age are widespread throughout southern Mae Hong Son and Chiang Mai. These form quite substantial areas and contain some significant caves including those of Mae La Noi. The vast area centred on the Mae Ping River in southern Chiang Mai was originally mapped as Ordovician. However, recent studies of fossils have found that, at least in part, the limestone is Devonian-Silurian in age. This raises the possibility that other areas believed to be Ordovician may in fact be younger. The next and most important cave bearing limestone in northern Thailand is the Permian. These limestones are pure, massively bedded and very strong. Numerous outcrops can be found all over the area. The most extensive are those of Pang Ma Pha where many large caves are located. The scenery formed in the Permian limestone is spectacular: precipitous towers, knife-like ridges, deep dolines and razor-sharp sculpturing are typical features. The third highest mountain in Thailand, Doi Chiang Dao (2,225 m), is made of Permian limestone. Limestone also occurs within Triassic age rocks of the Lampang/Phrae area. These rocks probably contain caves but data is lacking.

Caves and karst are formed in limestone by the solution in water. To understand them, we need to understand this process.

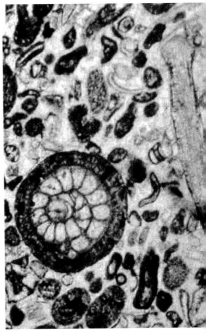

Microscopic shell remains within limestone. (© Pang Ma Pha Caves Project)

What is a Cave?

A cave is any natural underground cavity, without natural light, large enough for a human being to enter. Proto-caves are small cavities that have a realistic potential to grow and become caves. Micro-caves are all the other tiny cavities, cracks and pore spaces within the bedrock.

What is Karst?

Karst is a landscape formed by solution of the bedrock. It is often characterised by a particular suite of landforms including towers, pinnacles, blind valleys and dolines. The bedrock may be exposed over large areas and drainage is largely underground through caves and other cavities.

Opposite: Straws in Tham Pha Mon.

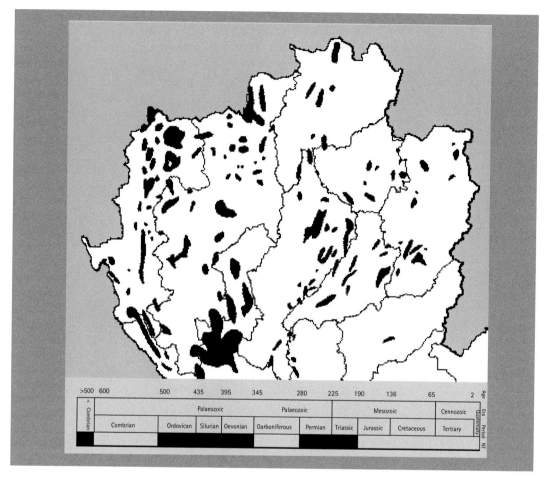

Age	>500	600		500	435	395	345		280	225	190	136		65		2	Age
Era	ʌ Cambrian			Palaeozoic				Palaeozoic			Mesozoic				Cennozoic		Era
Period	Cambrian			Ordovican	Silurian	Oevonian	Oarboniferous		Permian	Triassic	Jurassic	Cretaceous			Tertiary		Period
																	Quaternaly NT

Distribution and age of limestone rocks in Northern Thailand.

The Solution of Limestone

Calcium carbonate or calcite ($CaCO_3$) is not very soluble in pure water. The maximum that can dissolve in one litre is just 15 mg at $25°C$. By comparison, quartz (the main ingredient in sandstone) has a maximum of 12 mg. If nature relied on pure water to dissolve calcium carbonate, caves formed in limestone would be almost as rare as they are in sandstone or quartzite. Obviously, this is not the case and the reason why limestone contains many more caves is that it dissolves in acid. Nature produces many different kinds of acid, all of which are capable of dissolving limestone. By far the most important in the formation of caves and karst is carbonic acid. This acid is formed when water dissolves carbon dioxide gas. The water picks up the gas as it falls through the air as rain and, more importantly, as it soaks down through the soil. Here, respiration by plant roots and animals produces high concentrations. Measurements at Doi Chiang Dao and in the doline of Tham Hued have found soil carbon dioxide levels of 2.5% and 2.8%. Atmospheric carbon dioxide is just 0.05%.

The natural solution of limestone by carbonic acid is achieved via three chemical processes. The first two processes are happening wherever water, pure or natural, comes into contact

The Solution of Calcium Carbonate

In nature, water dissolves calcium carbonate (CaCO3) in three simultaneous, reversible processes. Each process is distinct yet related.

1) A small amount of $CaCO_3$ can be dissolved directly by the water. Solution is very slow and little CaCO3 can be dissolved. However, this process is important for providing the carbonate ions that are used in processes 2) and 3).

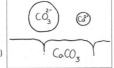

CaCO3 (solid) ⇔ Ca^{2+} + CO_3^{2-} (solution)
calcium carbonate calcium + carbonate

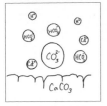

2) Water molecules can split up to make hydrogen ions (acid). This combines with the carbonate from 1) to produce bicarbonate which is much more soluble. However, the amount of acid available is very limited and solution is very slow.

H_2O ⇔ H^+ + OH^-
water ⇔ acid + hydroxide
Ca^{2+} + CO_3^{2-} + H+ ⇔ Ca^{2+} + HCO_3^-
calcium + carbonate + acid ⇔ calcium + bicarbonate

3) In nature, rainwater combines with carbon dioxide gas to form carbonic acid. Lots of hydrogen ions are now available to make bicarbonate and solution proceeds much more quickly.

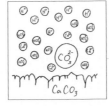

H_2O + CO_2 ⇔ H_2CO_3
water + carbon dioxide ⇔ carbonic acid
H_2CO_3 ⇔ H^+ + HCO_3^-
carbonic acid ⇔ acid + bicarbonate
Ca^{2+} + $CO3^{2-}$ + H + ⇔ Ca^{2+} + HCO_3^-
calcium + carbonate + acid ⇔ calcium + bicarbonate

The above processes are often summarised as the following:
$CaCO_3$ + CO_2 + H_2O ⇔ Ca^{2+} + 2 HCO_3^-
calcium carbonate + carbon dioxide + water ⇔ calcium + bicarbonate

with limestone. First the limestone physically dissolves into the water. This process is important for releasing the carbonate ions that are used in the other processes. The second process involves the break up of water molecules to produce hydrogen ions (acid). The hydrogen combines with the carbonate created by process one to form bicarbonate. This is much more soluble and dramatically increases the amount of limestone that can go into solution. However, the number of hydrogen ions produced by this method is very limited and solution is restricted. The third process is the formation of carbonic acid. This makes lots of hydrogen ions available to create bicarbonate and dramatically speeds up the rate in which the limestone dissolves.

Water will continue dissolving calcium carbonate until a maximum is reached. When this point is reached, the water is said to be saturated. Water that still has the capacity to dissolve more is called undersaturated. Above this point and it is supersaturated. The amount of calcium carbonate at which water becomes saturated varies due to a number of factors. These factors may increase the saturation point or lower it. They also affect the rate at which the solution process proceeds.

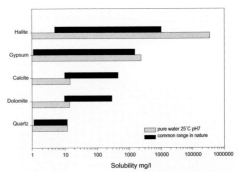

Mineral		Maximum solubility at 25˚C pH7,mg/l	Common range in nature, mg/l
Quartz	SiO_2	12	1 - 12
Dolomite	$CaCO_2$.$MgCO_3$	14	10 - 300
Calcite	$CaCO_3$	15	10 - 450
Gypsum	$CaSO_4$	2,400	0 - 1,500
Halite (rocksalt)	NaCl	360,000	5 - 10,000

Solubility of some common rock-forming minerals in water. Dolomite and calcite are unusual in that they are more soluble in nature than they are in the laboratory. Natural water contains acids that enhance the solution of these minerals.

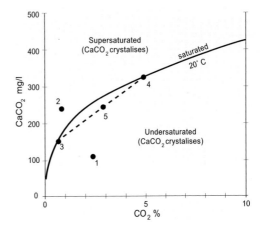

The line on this graph shows the maximum amount of calcium carbonate that can dissolve in water at different concentrations of carbon dioxide (20°C). Points plotting below the line are undersaturated, meaning the water has the capacity to dissolve more $CaCO_3$, e.g. (1) the stream sinking into Tham Hued - before the water has reached the cave and the limestone. Points plotting above the line are supersaturated and $CaCO_3$ crystallises out, e.g. (2) the stream emerging from Tham Khang Khao, Mae Hong Son - where the water has already been in contact with the limestone. The fictitious examples 3, 4 and 5 demonstrate the very important process of mixing corrosion. Number 3 is phreatic water at the water table. Number 4 is vadose water flowing down through the limestone above. Both waters are saturated with $CaCO_3$. When they meet and mix together, number 5 results. This new water is undersaturated and capable of dissolving more limestone.

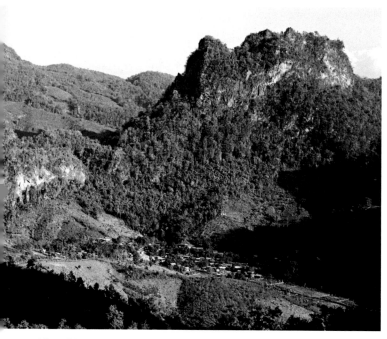

View of the Ban Luk Khao Lam doline and surrounding karst landscape, Pang Ma Pha.

Limestone Solution Rates

Measurements of limestone solution rates are rare for Thailand. It is possible though, to calculate roughly how much limestone is being dissolved away by looking at the calcium carbonate content of rivers leaving karst areas. The Mae Khlong River of western Thailand drains part of the largest area of karst in Thailand – the Western Karst Complex. Water measurements here reveal an average calcium carbonate content of about 130 parts per million. If we recalculate this figure for the total amount of water flowing down the river in an average year (about 11 billion cubic metres), we discover that nearly 4,000,000 tons of limestone are being carried out to sea each year. This may sound a lot, but it equates to the surface of the karst being dissolved away at a rate of less than 100 mm per thousand years, less than the rate in which the mountains are being uplifted.

The speed at which solution proceeds depends on a complex variety of factors. By far the most influential factor is the acidity of the water and this is controlled largely by the concentration of carbon dioxide gas. An increase in carbon dioxide from 0.03 % to 0.3 % raises the saturation point of the water from 55 to 150 mg/l. The water emerging from Tham Khang Khao (Mae Hong Son), for example, contains 1.0 % carbon dioxide and the calcium carbonate content is correspondingly high at 250 mg/l. In fact this water is slightly supersaturated.

Physical factors such as the amount of water available, its velocity, how turbulent the flow is, the shape of the cave passage and the length of time that the water can stay in contact with the rock also affect the rate of solution. A large volume of water obviously holds a greater amount of calcium carbonate than a small volume. So, areas with high rainfall experience more rapid solution. Fast flow rates can remove already saturated water from the area, bringing in fresh, acidic water to continue the action. However, very fast flow rates inhibit solution by reducing the time in which the water can be in contact with the rock. Passage shape is very important as it determines the surface area of the exposed limestone. For example, water filling a passage with a round cross-section will touch less limestone than if it is filling a more complex shape. Turbulent water (continually mixing) greatly increases the rate of solution by taking saturated water away from the limestone. Turbulence is a very important factor in the early stages of cave development (see later).

Temperature has an inverse relationship with solubility. The warmer it gets, the less calcium carbonate can go into solution. The warmth of tropical waters acts against the solution of limestone, although the water is often cooled as it enters a cave. The effect of temperature is not very significant anyway.

The presence of other minerals or metals greatly influence solution rates by increasing or decreasing the level of saturation. For example, sodium chloride (salt) dramatically increases the level. This partly explains why sea water can hold up to 2,000 mg/l calcium carbonate. Minerals that are chemically similar to calcium carbonate and dissolve more easily, e.g. gypsum, get in the way of solution and slow it down. Lead, copper and manganese from mine wastes and phosphate from fertiliser also depress solution. Cave sediments have indirect effects on solution. Thick floor deposits can shield the limestone from the water and concentrate solution onto the walls or roof. Sediments carried by streams can 'sandblast' hard crusts away, exposing the fresh limestone underneath that is easier to dissolve.

Underground, it is very likely that water from two different sources will join and mix together. Even if both waters are saturated with calcium carbonate and are unable to dissolve any more, mixing will result in a solution that is undersaturated and the water will become aggressive again. This effect is called mixing corrosion and is an important process in the development of caves at the water table.

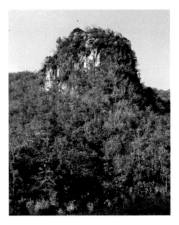

Tower behind Ban Tham Lod, Pang Ma Pha.

KARST LANDSCAPES

Karst is any landscape created by solution of the bedrock in water. It can form in limestone, dolomite, gypsum, rock salt and, if given enough time, sandstone and quartzite. The solution process gives karst many unique properties and characteristics that set it apart from all other landscapes. For example, karst has special hydrological properties in which most of the drainage flows underground through caves. The biodiversity of karst is unique as well and often contains numerous rare and endemic species. Karst is best regarded as being a system of complex interactions between the water, air, soil, rock, living organisms, energy and time.

The integrity of karst (and its caves) depends upon the preservation of these components and their interactions. A change to any one part can cause a disastrous chain-effect throughout the whole system.

Some 12% of the land surface of the Earth and around 18% of Thailand is karst. Virtually all of the karst in northern Thailand is formed in limestone, with only a few small areas of marble karst. The karst can usually be recognised by a particular suite of landforms. This suite includes dolines, towers, pinnacles and, of course, caves.

The word 'karst' is derived from 'Kras' – the name of a karst region straddling the border between Italy and Slovenia. The word 'kras' also means bare, stony ground in Slovenian.

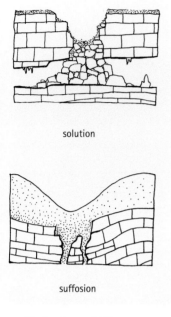

solution

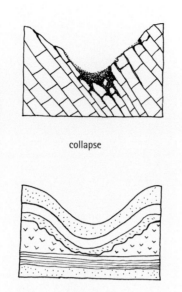

collapse

suffosion

subsidence

The four types of doline.

Dolines

Perhaps the most ubiquitous and characteristic landform of karst is the enclosed depression or doline. Dolines play a vital role in karst hydrology by directing surface drainage underground and into caves. Dolines may range in diameter from less than a metre to several kilometres and in depth from a few centimetres to hundreds of metres. There are four basic types of doline. **Solution dolines** are by far the most frequent in northern Thailand and can be seen throughout the area. They form at points where solution of the limestone surface is concentrated, such as along fractures. The limestone is removed more rapidly at these points than the surrounding area and eventually a depression forms. **Collapse dolines** form by collapse of a cave roof and can be quite spectacular features. Nam Bor Phi is a large collapse doline that is 100 m deep and 150 m wide with vertical walls. **Suffosion dolines** form in loose sediments that are overlying cavernous rocks. The sediments gradually trickle, or suffuse, down into the cavities in the rock leaving behind a hollow at the surface. **Subsidence dolines** form where there is a layer of soluble rock beneath the surface. As this rock dissolves away it leaves behind an empty space into which the overlying strata subside. This type of doline can be extremely large, but usually not very deep.

With continued growth, it is likely that several dolines will join together to become a compound doline, or **uvala.** The doline containing Tham Pung Hung is a uvala made up of at least 6 small dolines within the main depression. Another type of doline seen in Pang Ma Pha is the **polje.** These are large, flat-floored depressions with steep sides. Streams and marginal springs drain into poljes and the water drains out via sinks and underground routes. Poljes are often subject to seasonal flooding when the sinks become overpowered by the flow of water. The Nam Lang valley downstream of Soppong is a classic polje.

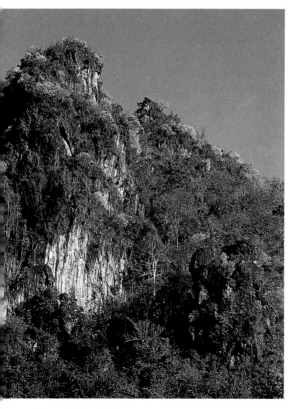

Towers in Pang Ma Pha.

Towers and Pinnacles

As dolines develop they capture an increasing proportion of
the drainage and so concentrate the solution process into
themselves. This leaves the areas in between with less solution and
a slower surface-lowering rate. Over time, these areas become high
points and may eventually develop into tall, steep sided mountains
called **towers**. Many of the towers in Pang Ma Pha are of this type.
Towers can also form as residual hills on plains where the
limestone bedrock is dissolved away evenly over a wide area, for
example southern China. Dipping beds of limestone within other,
softer rocks can create towers, as can blocks of limestone raised
by faulting. Once towers have been raised above the surrounding
area, solution and erosion around the base becomes important
procees for steepening the sides by undercutting. **Cones** are
similar to towers, but are more conical in shape. They develop in
weaker limestones that do not have the strength to support tall
cliffs and become towers.

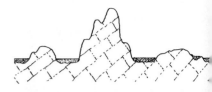

Corroded plain.

   Pinnacles are smaller than towers and develop initially beneath
a deep covering of soil in limestone with closely spaced fractures.
Solution is concentrated along the fractures, which are incised
deeply downwards. Subsequent soil erosion may then expose the
pinnacles and they become subject to above ground solution
processes and sharpened.

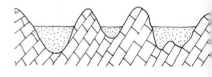

Valleys buried by sediment.

Karren

Covering the surface of most of the karst in northern Thailand is
a bewildering array of small-scale channels, ridges, pits and spikes.
These features are known by the collective name 'karren'. Karren
can be divided into two major groups – those that form on bare
rock and those that form under a covering of soil or vegetation.
Bare rock types are characteristically sharp and well defined. Some
common ones you are likely to see include: **rainpits** – circular pits
formed by falling water; **ripples** – horizontal waves formed by
rhythmically flowing water; **rillenkarren** – shallow, vertical
channels formed by running water and **spitzkarren** – residual
spikes. Sub-soil types are smooth and more rounded. They
include: **rundkarren** – rounded channels and **kavernosen karren** –
intricate mazes of tubes and cavities. **Kamenitza** are an
intermediate type. They are shallow, flat-bottomed basins formed
on bare rock surfaces beneath a covering patch of algae. The
transformation between bare rock and sub-soil karren at the soil
surface is often quite abrupt and can be used as an indicator of
soil erosion. Exposure of the sub-soil types demonstrates that soil
erosion is taking place and can even be used to gauge the rate at
which it is occurring. Most of the karst areas of northern Thailand
have sub-soil karren exposed around their bases, proving that
rapid soil erosion is happening in the area.

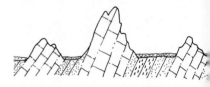

Interbedded with other rocks.

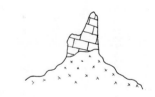

Pedestal on another rock.

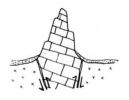

Block uplifted by faults.

Five modes of tower development.

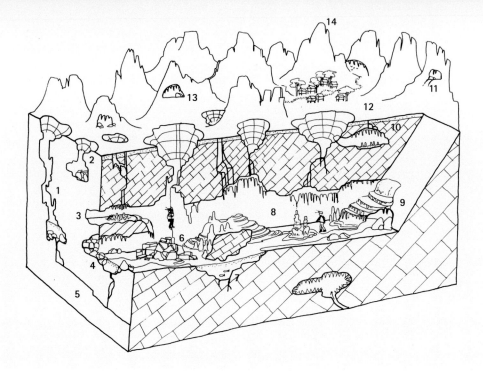

Typical features of a sub-tropical, mountain karst system in northern Thailand.

1. vadose cave
2. collapse doline
3. abandoned outflow
4. active outflow
5. phreatic protocave
6. breakdown
7. sumped passage
8. active stream cave
9. flowstone blockage
10. hydrothermal cave
11. karst window
12. solution doline
13. inactive cave
14. towers

THE BIRTH OF CAVES

In the past, it was believed that caves were created by catastrophic events like massive earth movements tearing open the Earth's crust or the floodwaters of Hell washing the ground away beneath our feet. These dramatic explanations could not be further from the truth. Nature takes its time when creating a cave – many thousands or even millions of years.

In the beginning, a cave is nothing more than a water-filled cavity in the limestone. At this stage it is called a '**micro-cave**'. The water flowing through the micro-cave barely seeps along and enlargement by solution of the sides is incredibly slow. Eventually, a network of tiny tubes develops called **anastomoses**.

The flow of water through these tubes is laminar. This means that the sides and centre of the flow move parallel to each other and there is no mixing. Solution is still very slow because only part of the water touches the limestone. Eventually, one of the tubes reaches the critical diameter of about 5 mm and the flow of water inside suddenly becomes irregular or turbulent. The sides and centre of the flow now mix together allowing more water to come into contact with the limestone. This dramatically speeds up the rate of solution and this tube begins to enlarge more rapidly than the others. As it gets bigger, the tube starts to steal the water flowing along its neighbours, increasing its water volume and flow velocity and enhancing the solution process even more. This tube has now grown into a '**proto-cave**'. Given sufficient time it will grow and develop into a true cave, one large enough for people to enter. The type of cave it becomes depends on its position relative to the water table. It can either be a 'phreatic' cave – formed at or beneath the water table, or a 'vadose' cave – formed above the water table.

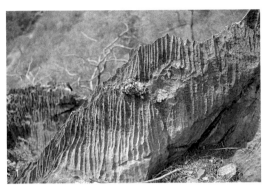

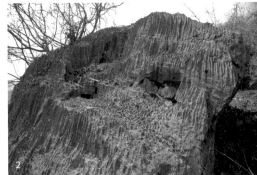

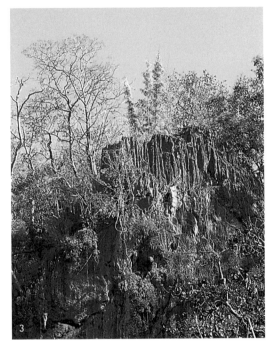

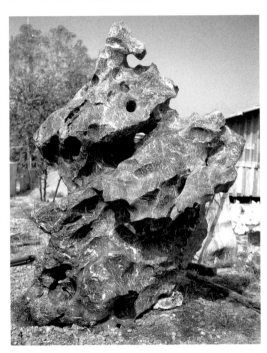

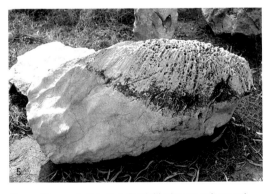

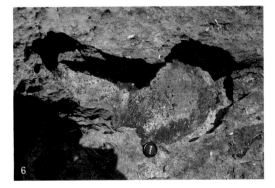

Types of karren: 1. sharply defined rillenkarren on bare rock 2. rillenkarren with rainpits and small spitzkarren on the central, flat area 3. large rillenkarren 4. sub-soil kavernosen karren on an excavated boulder

5. excavated boulder showing soil surface boundary (dark line) with sharp, bare-rock karren above and smooth, sub-soil karren beneath 6. Kamenitza. (© Dean Smart except photo 3)

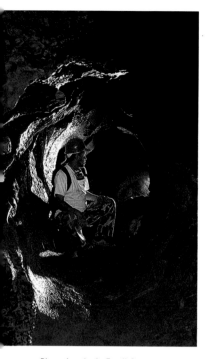

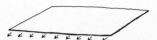

1) Water seeps slowly along a fracture in the rock. Solution in this micro-cave is very slow.

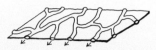

2) Eventually, the water dissolves out a network of tiny, interconnecting tubes called anastomoses.

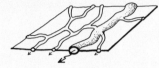

3) One tube reaches the critical diameter of about 5mm and the water flowing inside becomes turbulent. Solution dramatically speeds up and the tube begins to grow more rapidly.

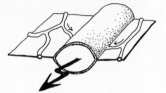

4) As it grows, the tube steals water from its neighbours, increasing its water volume and flow rate. Solution is enhanced still further. This tube is now a proto-cave and will continue growing into a proper cave given time.

Phreatic tube in Fossil Cave.

Phreatic Caves

Phreatic caves are characterised by being filled, or at least periodically filled with water. This results in the limestone being dissolved away evenly on all sides of the passage, producing smooth, uniform shapes. Water movement through phreatic caves is generally slow.

There are four types of phreatic cave. The first is **bathyphreatic.** In this type, the cave follows a deep, looping path far below the water table. It descends quickly and does not rise again until near to its spring. A famous spring in France, the Fontaine de Vaucluse, is a good example. Here the water rises from a depth of at least 315 m. The Nam Rue Takhean spring near Tham Tab Tao is probably a bathyphreatic cave as well, although exploration cannot proceed beyond 20 m depth due to a blockage of tree trunks. Bathyphreatic caves are permanently filled with water.

At the other end of the phreatic cave spectrum is a type commonly found in northern Thailand – the **epiphreatic** cave. These caves form at or just below the level of the water table and spend time in both the phreatic and vadose zones. The periods of development under phreatic conditions, though, have a much greater influence on the formation of the cave. Mixing corrosion due to percolation water joining with that in the cave is a particularly important process. This is evidenced by the many roof pockets seen in epiphreatic caves. These bell-shaped pockets are dissolved out at the point where the percolation water enters the cave and where the mixing occurs. Epiphreatic caves have a relatively flat profile compared to bathyphreatic caves. The route

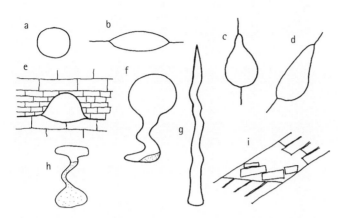

Cave passage cross-sections:
a. Phreatic tube in solid limestone
b. Phreatic along horizontal fracture
c. Phreatic along vertical fracture
d. Phreatic along inclined facture
e. Phreatic in dissimilar beds
f. Phreatic with vadose trench in floor – 'keyhole'
g. Vadose canyon
h. Paragenetic
i. Modified by breakdown in thin inclined bedding

they take between the sink and the spring is generally shorter and leads to faster flow rates. Examples in northern Thailand of caves with epiphreatic origins include: Tham Nam Lang, Tham Nam Mae Lana, Tham Susa, Tham Lod and Tham Chiang Dao.

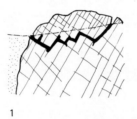
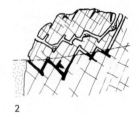
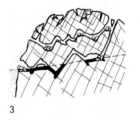
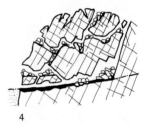

Epiphreatic caves often have passages developed on multiple levels, each level representing a former water table. These caves are important in that they can provide a record of valley deepening at the spring and an insight into regional erosion rates. Tham Chiang Dao has at least three levels of development.

Four types of phreatic cave.
1. Bathyphreatic.
2. Multiple loop.
3. Mixed multiple loop and epiphreatic.
4. Epiphreatic. (Ford and Williams 1989)

Given time, bathyphreatic caves evolve into epiphreatic caves passing through two intermediate stages on the way. These stages are: '**multiple loop phreatic**' caves – a series of deep downward loops with occasional returns to the water table; and '**mixed multiple loop and epiphreatic**' – a number of deep loops with epiphreatic sections in between.

Phreatic caves can be recognised by the smooth, uniform shape of their passages. They tend to be round or oval in cross-section and have smooth walls with few projections. **Scallops** – the heel-shaped hollows formed by swirling currents – are large, an indication that they formed in slow moving water. Bedrock roof pendants may hang from the ceiling, though these can also be paragenetic in origin.

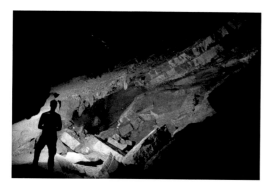

Thin inclined bedding with breakdown, Fossil Cave. (© Dean Smart)

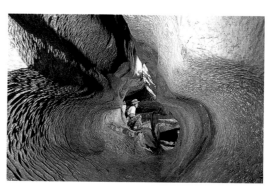

Passage with small scale scalloping, Tham Klaeb Yai.

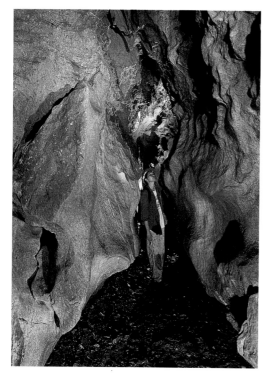

Tham Pung Hung is a master stream cave, draining many other caves probably including Pha Phuak.

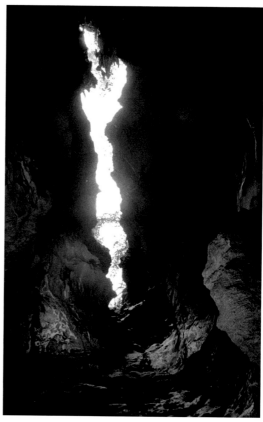

Vadose canyon, Tham Talu (Muang Phaem).

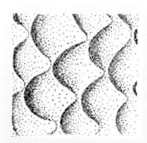

Scallops form when swirling currents dissolve away the limestone. The steep side indicates the upstream end and the sharp point indicates downstream. The large scallops in the photo on the right were formed by slow moving water. (Curl 1974)

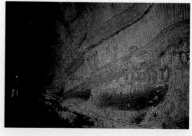

Vadose Caves

Above the water table, in the vadose zone, air can occupy some of the cavities in the rock. Water entering this zone flows a lot quicker and is more turbulent than beneath the water table. It also tends to follow steeply descending paths from the point of entry to the water table below. Waterfalls and vertical shafts are common features of vadose caves.

Many vadose caves develop directly from phreatic caves, especially epiphreatic. The phreatic cave forms in the normal way but is forced into the vadose zone by a lowering of the water table brought about by rapid erosion at the spring and/or mountain uplift. If water continues to enter the cave at the same point, the gradient between the sink and the newly created, lower spring will steepen and the stream will start to cut a vadose trench into the floor. This creates 'keyhole' shaped passages with an abandoned phreatic tube at the top and an active vadose canyon underneath.

The development of vadose caves happens relatively quickly. Rapid flow rates and the turbulent nature of the water increase the rate of solution. Also, the water often carries large quantities of sediment giving it abrasive power. The formation of rockmills – holes in the floor drilled out by spinning rocks – aids in the cutting of trenches and enlargement of the cave is helped by the collapse of the roof and walls. Fallen blocks are quickly removed by the stream. In the airspace of the cave, stalactites, stalagmites and other speleothems grow.

Vadose caves are distinguished by their steep profiles with waterfalls, tall, narrow passages and sharply defined features. The walls may show a sequence of horizontal incuts and projections. Rockmills may be present in the floor. Scallops are small and well defined indicating formation by fast flowing water. Vadose canyons often meander around tight bends.

The tall, narrow canyon passages of Tham Ban Luk Khao Lam are classic examples of vadose cave development. From the entrance, a series of tall, narrow canyons less than 2 m wide but up to 20 m high drop down five vertical shafts to a depth of 113 m. At the bottom of the cave is a short section of horizontal phreatic passages.

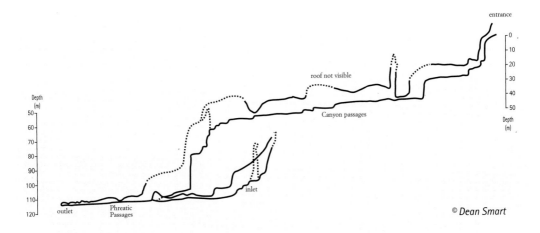

© Dean Smart

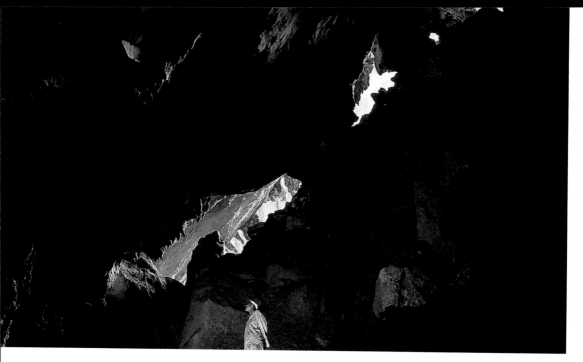

Tham Pha Phuak descends steeply downwards at 45° along the line of the bedding planes.

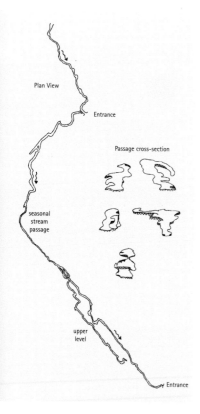

Tham Klaeb Yai. (survey by Assoc. Pyreneenne de Speleologie, 1986).

Paragenetic Caves

Paragenetic caves look superficially like vadose caves. They have tall, narrow passages that meander, walls with horizontal incuts and projections and sharply defined features. However, paragenetic caves are phreatic in origin. The similarity they have with vadose caves is due to an unusual method of development – they grow upwards and not downwards.

Upward growth begins when the floor of a phreatic cave is buried under sediment. This shields the floor from the water and its solutional activities. Solution is instead concentrated onto the ceiling and walls, which expand upwards and outwards. If sediment deposition keeps paces with the upward migration of the ceiling, the cave will keep growing upward until the water table is reached. If sedimentation stops, the cave will grow laterally. Subsequent stream action may remove much of the sediment from within the cave.

Telling the difference between a paragenetic cave and a vadose cave is not always easy, but there are a few clues that can help. Paragenetic caves contain much more sediment, patches of which are often stuck to the walls and near the ceiling. The passages may also have lots of unusual projections and a completely flat ceiling. There will be no key-hole shaped cross-sections. Paragenetic caves meander more at ceiling level, whereas vadose caves meander more at floor level. Finally, the horizontal incuts and projections on the walls of a paragenetic cave are always in pairs. An incut always opposes an incut and a projection always opposes a projection. In vadose caves, they are often, but not always, offset from each other.

Northern Thailand has several fine examples of paragenetic caves. Tham Klaeb Yai is particularly good, showing all the classic characteristics.

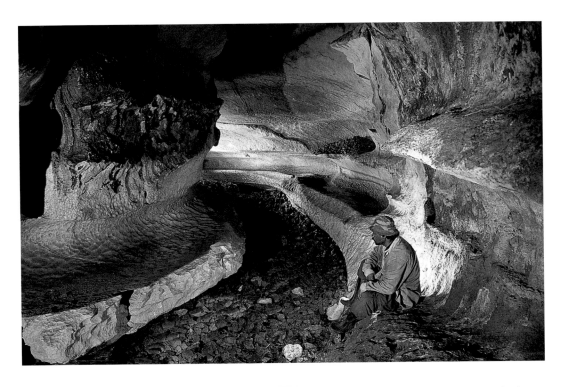

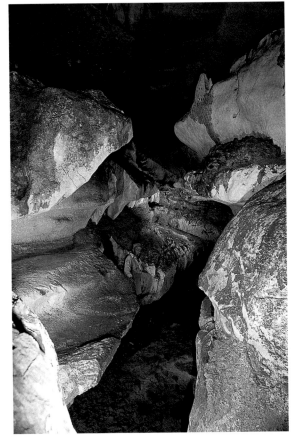

Top and Left: Tham Klaeb Yai is a good example of a cave formed by paragenesis. The passages contain lots of sand and gravel, some of which is cemented to form bridges. The passages also have odd shaped cross-sections.

Cave Shapes

Developing caves take on shapes that are in part governed by the geology of the surrounding rock. Therefore, by looking at the shape of caves it is possible to learn something about the local geology.

Faults and joints affect cave shape in a number of ways. In limestone with closely spaced faults and joints, the water is able to follow many different routes and a maze cave results. The phreatic cave Tham Khao Kwian is an example of a maze cave formed on three sets of closely spaced fractures orientated NW-SE, N-S and SW-NE. The cave passages follow these orientations almost exclusively. Development of some of the passages in the Tham Suea/Tham Lom system has been influenced by two sets of small, vertical faults filled with pink calcite. The two sets are perpendicular to each other at NW-SE and SW-NE. The main passage alternates between each set, turning 90° bends as it switches orientation. Where perpendicular faults cross the main passage, lateral enlargement has occurred creating a series of small chambers. Vadose caves tend to follow vertical fractures during their steep descents. The tall, narrow canyons of Tham Ban Luk Khao Lam follows two sets of perpendicular faults.

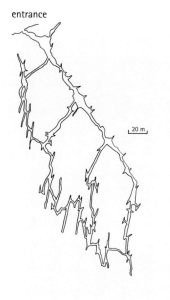

entrance

20 m

Above: *Plan view of Tham Khao Kwian, a phreatic maze cave formed on three sets of fractures oriented NW-SE, N-S and SW-NE. (Odel)*
Right: *Plan view of Tham Nam Lang The bends of the main passage are the result of the river being forced to flow around the noses of plunging folds. (Kiernan 1991)*

Below: *Fossil Cave is formed in thinly bedded limestone dipping to the North. The bedding has influenced the cross-sectional shapes of the passages.*

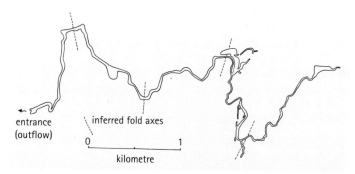

entrance
(outflow)

\ inferred fold axes

0   1

kilometre

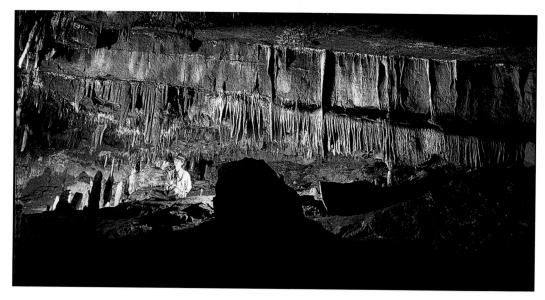

Tham Nam Lang is a large river cave with long, sweeping bends. The bends are the result of the river flowing around the noses of broad, plunging folds in the limestone. Lack of a major fault in the area meant that the river could not follow a straight course and punch through the folds. Instead the river followed the next easiest path – the bedding planes of the limestone. Unfortunately, the low gradient of the cave could not provide enough pressure to push the water up and over the folds and the river was forced to take a longer, more meandering route around the noses. Another cave displaying shape influenced by bedding planes is Fossil Cave. The ceiling of the stream passage in this cave slopes to the north and the floor is littered with breakdown blocks. This is the result of the cave being formed in thinly bedded limestone, which dips northwards.

Caves Formed at Rock Boundaries

The contact between limestone and another rock type is a place where caves commonly develop. Tham Pang Kham is formed between a limestone roof and a non-soluble metamorphic rock floor. The metamorphic rock can be clearly seen outcropping at the outflow end of the cave. Another interesting aspect of this cave is that it is perched about 120 m above the floor of the valley at the outflow. The small stream flowing through the cave has not been able to erode down through the resistant metamorphic rock and keep pace with the erosion of the valley outside.

The inclination of the guiding fracture has a strong influence on the shape of passage cross-sections. Horizontal fractures produce broad, flat passages and vertical fractures give 'tear-drop' shapes. Inclined fractures produce intermediate forms. Differences in the solubility of the limestone beds produces irregular forms.

Inactive Caves

Eventually, the water table drops beneath the cave and even the vadose waters find an alternative route. All of the water drains out leaving the cave high and dry. The growth stage in its life has come to an end and the cave becomes 'inactive'. Inactive caves are very common in northern Thailand and can be found virtually everywhere. Many have been turned into temples and some are important archaeological sites.

Inactive caves can form out of the upper levels of epiphreatic caves, for example, Tham Tab Tao and the upper cave of Tham Pha Daeng (Phayao). Or, they may be remnants of ancient cave passages and cavities broken into by the downward progression of surface erosion. Tham Muang On is an ancient, phreatic cavity that has been breached in this way. However they are formed, inactive caves have several things in common aside from a lack of flowing water. They tend to be shorter than active caves ending abruptly in collapses or sediment and speleothem fills. Some exit on the other side of the hill. Most were originally phreatic caves although their shape is often heavily modified by collapse of the roof and walls. Layered sediments usually cover the floor. Inactive caves are by no means dead. Processes such as speleothem

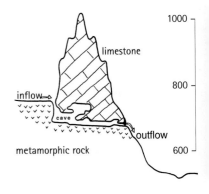

Tham Pang Kham is formed at the contact between the limestone and a non-soluble metamorphic rock. The cave is perched 120 m above the valley floor because the small stream flowing through the cave has not been able to erode down through the resistant floor. (Kiernan)

growth and sediment deposition will be ongoing. There will undoubtedly be a healthy ecosystem inside. People may also come along and take shelter or make use of the caves resources.

Many thousands or perhaps millions of years after becoming inactive, surface erosion finally reaches the cave. Cracks in the rock above the cave weaken the cave ceiling and cause it to collapse. Daylight enters for the first time and the cave becomes a part of the surface environment. It has reached the end of its long life. The large collapse doline Nam Bor Phi is a cave that has reached this stage.

Other Methods of Cave Formation

The formation of caves is not only limited to solution of limestone by carbonic acid. Nature can create many different kinds of acid and all are capable of dissolving limestone. Caves are also found in other rock types, some of which are insoluble. These caves form in other ways.

Hydrothermal caves (caves formed by solution with hot water) are known to exist all over the world. The water comes from deep within the Earth's crust (where it is heated up) and rises to the surface via a fault or similar fracture. During its upward journey, the water may encounter limestone and begin dissolving it. Caves formed by hydrothermal water have a maze-like character. They may also be lined with crystals. Jewel Cave in the USA is a three-dimensional maze cave with nearly 208 km of passages, often lined with large calcite crystals. The hydrothermal cave Satorkopuszta in Hungary is a maze of small, round chambers with roof pockets called cupolas. Mazes in two dimensions form when the hot, rising water becomes trapped within a thin bed of limestone, for

Tham Sakoen, an inactive cave perched high above the valley floor, showing extensive breakdown.

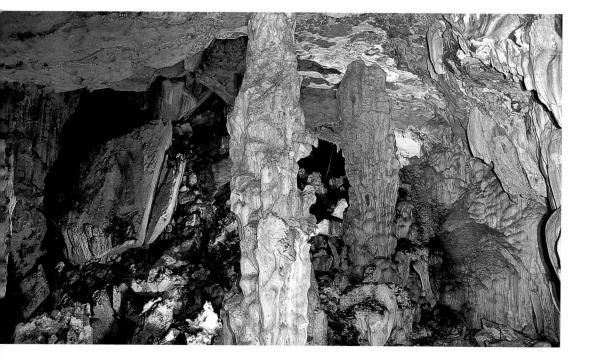

example, Devis Hole Mine Caverns, England. A few examples of hydrothermal caves are known in northern Thailand. All occur in Ordovician limestone near Mae La Noi and were discovered during mining operations. Tham Kaew Komol is the most extensive cave found to date with around 200 m of development. The cave descends steeply through a series of small, irregular shaped chambers with roof cupola. The chambers are lined with crystals such as calcite nailheads and spar and aragonite needles.

The limestone caves of Carlsbad Caverns and Lechuguilla in the state of New Mexico, USA, have been formed by solution with sulphuric acid instead of carbonic acid. Sulphuric acid is particularly potent and can dissolve large quantities of limestone very quickly. Very extensive and large caves can be produced. Lechuguilla is one of the longest caves in the world with over 176 km of passages. Carlsbad Cavern's Big Room has a volume in excess of 1,000,000 m₃. A characteristic by-product of limestone solution in sulphuric acid is the mineral gypsum and this is found in large quantities throughout the caves. Sulphuric acid can also be produced through the actions of bacteria. The bacteria oxidise hydrogen sulphide gas for energy and the acid is a by-product (see next chapter). This process is modifying the lower passages of Movile Cave in Romania.

Landscapes that look like karst, but have formed under different processes are called **pseudokarst**. A pseudokarst landscape, such as the sandstone mountains of northeast Thailand, may possess every feature that true karst has, but it forms more through physical erosion than solution. Pseudokarst caves form where water flows widens weaknesses in the rock by abrading the walls. Rock shelters and shallow caves are sandblasted out by strong winds. Pseudokarst caves can be found in virtually any rock type, but they are most common in sandstone, conglomerate and granite.

It is possible for caves and karst to form in quartzite by solution, but it takes a very long time. The ancient quartzites of Venezuela have been exposed for hundreds, possibly thousands of millions of years. During that time, a true solutional karst has developed complete with large caves, dolines and towers.

Caves can also form in lava flows during volcanic eruptions. These **lava 'tubes'** are created when the exposed surface of a flow cools down and hardens more rapidly than the hot, central part. A solid roof forms over a tube in which the lava continues to flow. When the volcano stops erupting, the lava drains out leaving an empty tube behind. Some very long lava tubes are known. Kazamura in Hawaii is a 65.5 km long tube formed in basalt. The Western USA, Kenya and Korea also have extensive lava tube caves. In northern Thailand, old basalt flows occur around Lampang, but no tubes have yet been reported.

Perhaps the most bizarre cave forming method of all is found at Kitum Cave, on Kenya's Mount Elgon. The cave was originally formed by solution of a layer of volcanic ash. Since the cave became inactive, it has been extended and reshaped by elephants literally eating away the rock in their search for mineral salts.

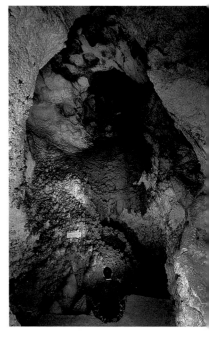

Tham Kaew Komol is a cave formed by hydrothermal water. It is a series of small irregular chambers lined with calcite and aragonite crystals.

SPELEOTHEMS

Stalactites (the ones that hang down), stalagmites (the ones that go up) and all other secondary mineral deposits found in caves are collectively known as speleothems. An astonishing variety of shapes, size, colour, mineralogy and mode of origin is displayed.

Over 200 minerals have been found in caves and some of these occur nowhere else. By far the most common speleothem mineral is calcite ($CaCO_3$). Calcite is deposited in caves by saturated percolation water dripping off the ceiling and trickling down the walls. The process is the reverse of that which dissolves limestone and involves the loss of carbon dioxide gas rather than taking it up. As the water enters the cave, it encounters the cave atmosphere which contains less carbon dioxide. An imbalance between the water and the air occurs which the water corrects by releasing carbon dioxide. This lowers the saturation point of the water and calcite is forced out of solution and into solid, crystal form. Direct evaporation of the water accounts for very little speleothem growth. Aragonite is chemically equivalent to calcite, but has a different crystal structure. Variations in the water chemistry with respect to carbon dioxide, calcium and magnesium determine whether calcite or aragonite is deposited. In Carlsbad Caverns, USA, an aragonite stalactite is growing directly above a calcite stalagmite. The change in water chemistry after depositing the aragonite is sufficient to change the mineral that crystallises out. Over time, aragonite converts to calcite.

Gypsum ($CaSO_4$) is the third most common speleothem mineral, although it is rarely seen in Thailand. An exception is Tham Pha Daeng (Phayao), which has gypsum growing out of a sediment-filled fracture in the wall.

Speleothem types found in caves in Northern Thailand.
1. *anthodites*
2. *helicites*
3. *column*
4. *bacon*
5. *shield*
6. *popcorn*
7. *vulcanites*
8. *stalagmite*
9. *straws*
10. *stalactites*
11. *drapery*
12. *'nom tham' stalactites*
13. *flowstone*
14. *pendulite*
15. *pool spar*
16. *gour pools*
17. *condulites*
18. *mud stalagmites*
19. *cave pearls*
20. *canopy*
21. *hydrothermal crystals*
22. *cave clouds*
(adapted from Hill and Forti 1986)

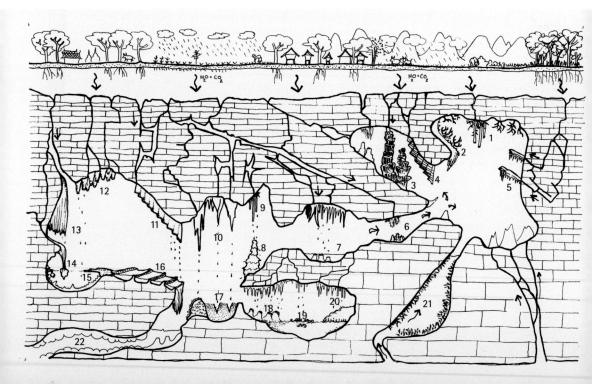

Pure calcite is colourless, but most calcite speleothems are coloured. These colours come from impurities. White or opaque speleothems form because of tiny bubbles of air or water trapped within the crystal structure. More exotic colours are due to traces of other minerals. Copper minerals produce blues and greens (for example in Tham Pha Mon), lead and zinc produce pale yellows and iron gives red and black colours. Black may also come from manganese. Creams and pale browns are due to organic acids being leeched from the soil above the cave. Virtually any colour is possible if the appropriate mineral is present. Discolouration of speleothems may come from sediment being deposited during floods or from people touching them.

Stalactites, Stalagmites and Columns

The growth of a stalactite begins with the development of a straw – a thin, hollow tube about 5 mm in diameter. Water flows down the inside of straws and deposits calcite only at the end. Because of this, straws grow quite quickly at between 0.2 and 20 mm per year. Straw stalactites are composed of a single crystal which usually breaks under its own weight before it reaches any great length, although specimens 6 m long are known in Australia. Slight changes in water flow or the drip point create straws with bends and spirals. If the central canal of a straw becomes blocked, the water will leak out and flow down the outside. Deposition now takes place on the sides as well as the end and a stalactite shaped like a carrot forms. Stalactites lengthen more slowly than straws (<0.1 to 3 mm per year) because their width is expanding as well as their length. Inside a stalactite, the crystals are arranged in concentric circles or 'rings'. These rings are similar to the annual growth rings of a tree and indicate alternating periods of rapid and slow growth. Christmas cave has a fine display of straws and stalactites.

Growing upwards from the floor of the cave are stalagmites. Stalagmites are generally fatter than their corresponding stalactites due to the splashing effect of the falling drips. They also grow more slowly and gain height at just <0.005 to 0.7 mm per year. Variations in the amount of splashing and deposition creates a stalagmite with wide and narrow layers. An extreme form of this consisting of a stack of thin layers is called 'pile of plates'. Tham Nam Lang has good examples. Inevitably, stalactites and stalagmites meet in the middle and join together to form a column. Columns are common features of many caves. Tham Lod contains a fine 21 m high example. Many of the columns found in caves of northern Thailand are broken or cracked across the middle. These breakages are most likely caused by settling of the cave floor due to earthquakes.

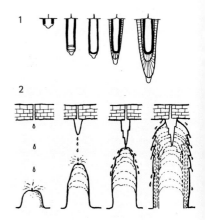

The growth of speleothems:
1. *development of a stalactite from a straw. (Prinz 1908)*
2. *growth of a column. (Gams 1981)*

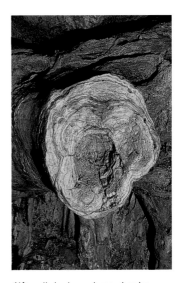

Naturally broken column, showing concentric growth rings.

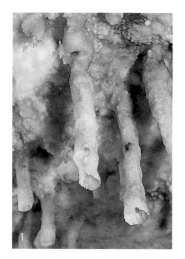
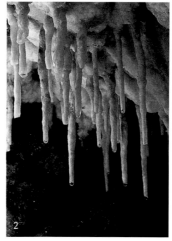
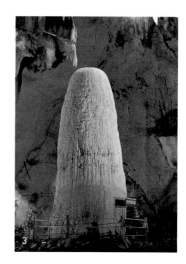
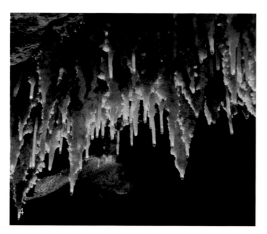
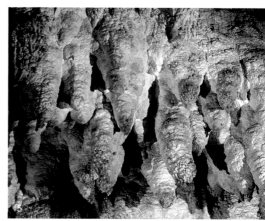
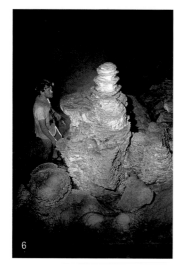
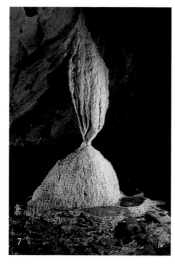
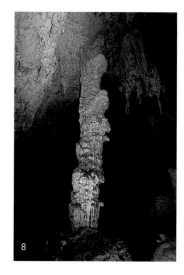

Examples of stalactites and stalagmites: 1. and 2. straws;
3. stalagmite 7 m, Tham Muang On; 4. straws and stalactites,
Tham Pha Mon; 5. rounded stalactites, Tham Pha Daeng;

6. "pile of plates" stalagmite, Tham Nam Lang (© Dean Smart);
7. stalactites and stalagmites just joining to form a column,
Tham Pang Kham; 8. column 7 m high, Tham Lod.

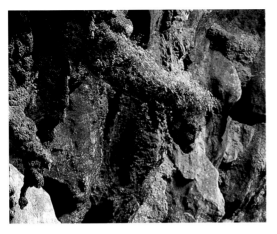
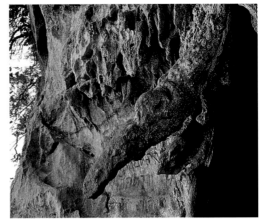

Two examples of eucladioliths.

Eucladioliths

Stalactites in cave entrances often bend towards the daylight. This is not because of the wind, but due to the action of green plants living on the side facing the light. These plants obtain nutrients via photosynthesis and take some of the carbon dioxide they need from the water running down the stalactite. This causes calcite to be deposited faster on the side with the plants and the speleothem grows in the direction of the light. This form of speleothem is called a eucladiolith and almost every large cave entrance has them.

Flowstone

Flowing water deposits thin sheets of mineral on the walls and floor. This is 'flowstone'. Flowstone has no shape of its own, it simply moulds to the shape of the cave it is covering. Some of the largest speleothems in northern Thailand are flowstone, such as those of Tham Nam Lang and Tham Pha Mon.

1

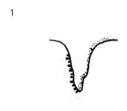

2

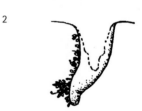

3

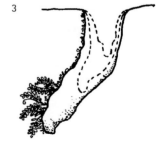

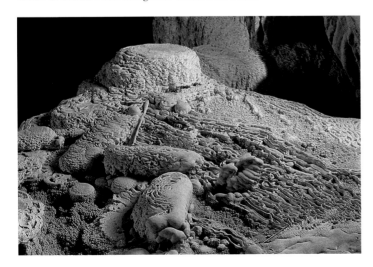

Flowstone in Christmas cave.

The growth of a eucladiolith:
1. *talactite starts developing*
2. *mosses, algae and small plants grow on the side with daylight*
3. *the plants draw carbon dioxide out of the water running down the stalactite, causing faster deposition on that side and bending the eucladiolith towards the light*

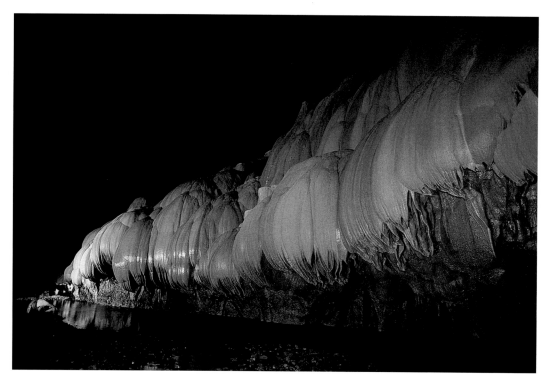

Tham Nam Lang. This flowstone is 100 m long, one of the largest speleothems in Thailand.

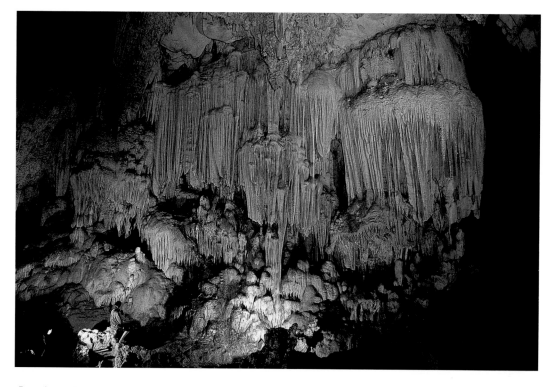

Extensive section of flowstone in Tham Naresuan.

Canopies and False Floors

Canopies are commonly seen in caves of Northern Thailand. These are thick flowstones that grew over the top of a sediment floor. Subsequent removal of the sediment by a stream leaves the flowstone hanging in the air, looking like a canopy. Pebbles and other sediments can usually be found stuck to the flat underside of canopies.

Similar to canopies are false floors. These are thin flowstone sheets stretched across the passage that were deposited as layers within a sedimentary sequence. Re-excavation of the sediment left them suspended. False floors differ from canopies in that they were once an integral part of the floor and not just a deposit on the top.

Below: False floor, Tham Nam Rin.

Bottom: Canopy with flowstone fringe, Tham Pha Mon.

Left: Flowstone, Tham Pha Mon.

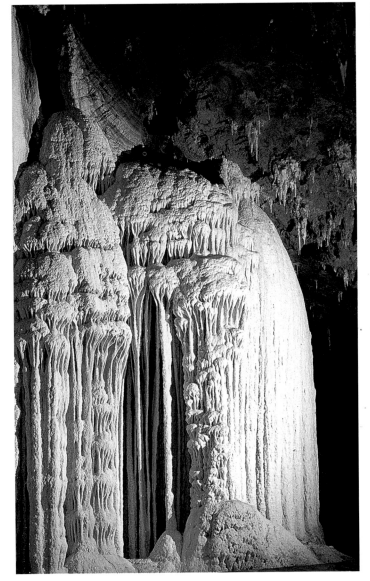

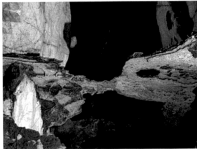

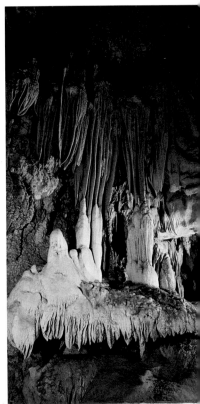

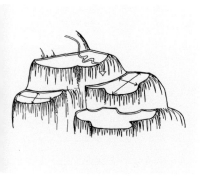

Tiered gour pools. The pool length is about double the width.

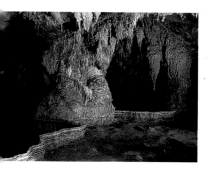

Rimstone dam in Tham Pha Mon.

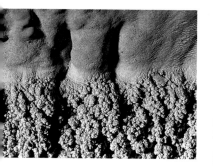

Cave coral and high water mark.

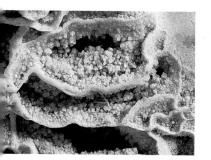

Pool spar.

Gour Pools (rimstone dams)

Streams flowing across sloping floors deposit a type of speleothem called gour pools or rimstone dams. These are calcite dams that grow on obstructions, such as pebbles or ripples on the floor. They often build up into tiers with small waterfalls in between. The size of gour pools is related to the steepness of the slope on which they form. Large pools grow on shallow gradients and small pools grow on steeper slopes. Gour pools also show a remarkable consistency in their dimensions – the length is nearly always double the width. The reason for this is unclear. Very often, gour pools are lined with knobbly projections called cave coral, or crystals known as pool spar. Gour pools are also found outside of caves at surface waterfalls such as Wang Kaew Falls, Lampang.

Erratic Forms

Sometimes a straw stalactite will grow horizontally, straight out from the wall. It may be twisted and bent and may be branching like a small tree. These are helictites and they grow into these strange shapes because of tiny imperfections in the crystal structure. These imperfections build up as the helictite grows and cause it to bend. The bends are not caused by wind currents. Anthodites are similar to helictites in that they grow horizontally and bend and twist, but they do not have a central tube and are not hollow. Aragonite needles and the small bristles growing on other speleothems are also classed as anthodites. The aragonite anthodites of Tham Mae Lana are excellent.

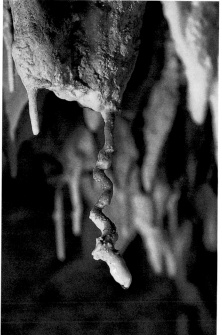

Spiral straw in Tham Pha Mon.

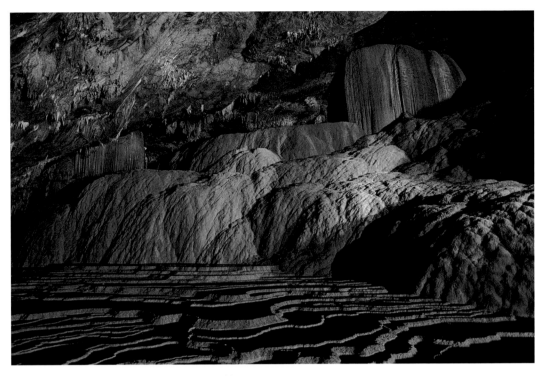

Fine series of gour pools and flowstone in Tham Pha Mon.

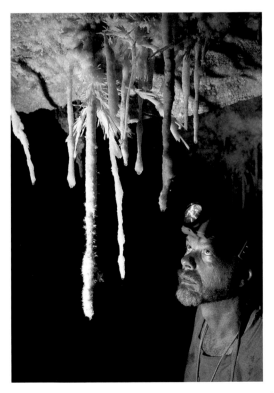

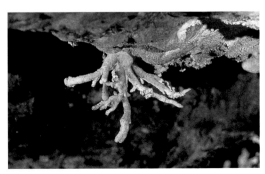

Erratic forms in Tham Long Yaow.

Aragonite anthodites and straws in Tham Mae Lana. (© Dean Smart)   *Erratic forms in Tham Long Yaow.*

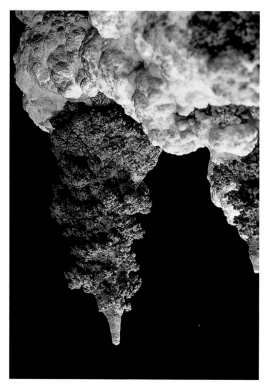

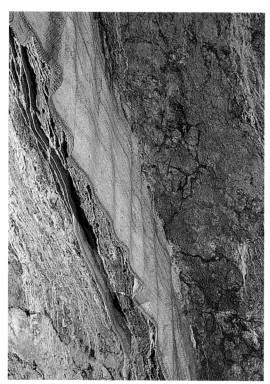

Popcorn growing on a stalactite in Tham Pet.

Right: Curtain in Tham Kai Mook.

Below: Calcite coated pebbles in tiny gour pools, Tham Boong Hoong.

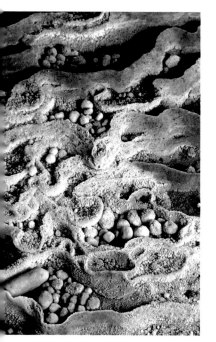

Popcorn

Popcorn are unusual in that they grow by direct evaporation of the water and not by loss of carbon dioxide gas. They are the small, irregular, coral-like projections growing in draughty places, very often on other speleothems. They can occur individually or in large clusters. Popcorn looks similar to cave coral, which only grows in water and never in the air. Tham Lod and Tham Pakarang have good popcorn.

Curtains (drapery)

On sloping cave ceilings or walls, curtains or draperies can develop. These grow where the water runs down the rock in a line instead of dripping off. Crystals are deposited along the line and successive layers build up into a curtain. Curtains with light and dark bands are called 'bacon'. The bands are similar to the rings of a stalactite.

Cave Pearls (oolites)

Solid balls of calcite deposited around grains of sand or other material are called cave pearls or oolites. Often they grow in small clusters or 'nests'. The calcite builds up in successive layers around the grain, which is prevented from being cemented to the floor or to another pearl by being continually agitated by dripping water. Cave pearls can grow quite large. The largest single pearl ever found weighed over 1 kg. Sometimes, rounded cube shaped pearls and even hexagonal pearls form. There is a fine balance between the amount of dripping water being too little or too much. Too

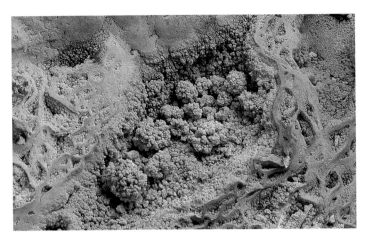

Irregularly shaped cave pearls known locally as 'noi na' - 'custard apple' pearls.

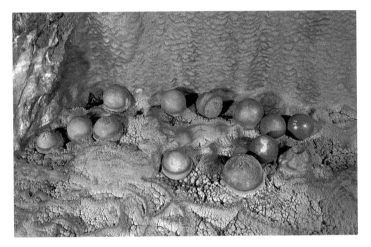

True cave pearls in Tham Nam Lang. (© Dean Smart)

little water and the pearls may cement together. Too much water and they may be bounced out of the nest. However, nature finds this balance often enough and cave pearls are commonly found throughout the world. Strangely though, not many caves in northern Thailand have them. Tham Mae Lana is exceptional in that it contains large numbers.

Other Speleothems Types

Shields form when a stream, slightly under pressure, enters a cave via a thin, horizontal crack. Crystals are deposited on both sides of the crack and build up in an outwards and upwards direction. Two parallel plates form with a thin crack sandwiched between (the extension of the wall crack) and very often stalactites hang from the underside. Shields may grow to be over 1 m in diameter.

An unusual speleothem that occurs in Tham Pha Mon is the **mud stalagmite**. These are cone-shaped stalagmites made from mud mixed with calcite. They resemble small volcanoes, which has led to the name 'volcanites' sometimes being used.

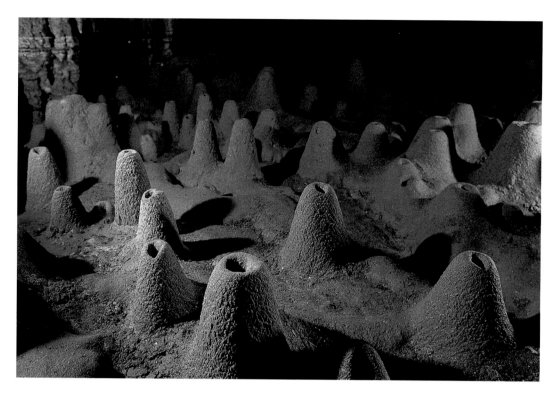

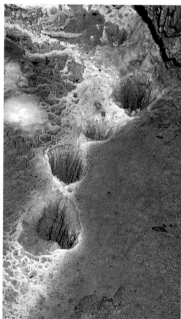

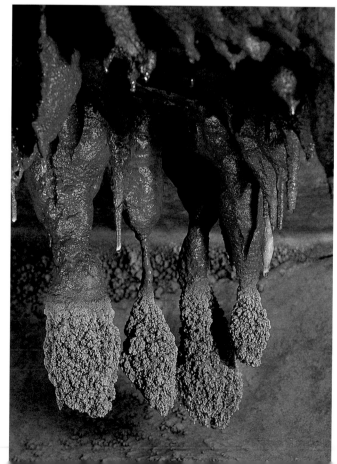

Top: *Mud stalagmites, Tham Pha Mon.*
Above: *Conulites, Tham Long Yaow.*
Right: *Pendulites, Tham Luek (Muang Na).*

Mud stalagmites form in pools of saturated water beneath drips carrying sediment from the surface. This sediment builds up into cones and the calcite in the pool water cements it together. The holes in the top are drilled out by dripping water after the pool has drained.

Pendulites are club-shaped stalactites hanging down into pools of calcite saturated water. The 'club' end is formed from cave coral growing on the end of the stalactite.

Conulites are calcite-lined holes drilled into sediment floors by dripping water. Sometimes the calcite lining extends above the top of the hole. Shallow, plate-like conulites are called 'bird baths' and form by splashing drips.

Cave Sediments

Nowhere else on Earth is such a wide variety of sediments found in one place as in caves. Caves owe this uniqueness to the fact that cave sediments are derived from two distinct sources – inside the cave and outside.

The majority of cave sediments come from outside and are transported in by streams. The streams erode the sediment out of the catchment area and move it downstream towards the cave. By the time it reaches the cave, the sediment has been ground into cobbles, pebbles, sand, silt and mud. Inside the cave, sediment is deposited at every opportunity and the type of sediment laid down depends upon the water carrying it. Fast moving, perennial

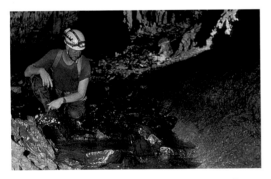

Black manganese oxide coated cobbles. (© Dean Smart)

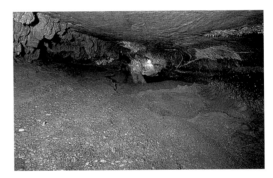

Tham Kaeng Khao (Mae Hong Son) the entire cave is floored with a diverse sediment train, including coarse gravels.

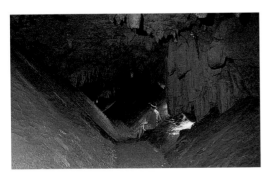

Large mud banks line the seasonal stream passage of Tham Luang (Mae Sai).

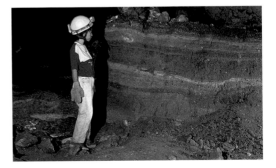

Layered cave sediments. (© Dean Smart)

streams deposit long trails of sand and gravel, such as that seen along the entire length of Tham Nam Lang. Slow moving phreatic waters drop silts and clays. Mud is predominantly deposited by ponded water. Most of the sedimentary structures of surface environments are also found underground, for example: braided channels, point bars, cut-and-fill and ponded backwaters. Cyclic deposition produces sequences of graded beds, starting with pebbles at the base and changing gradually to sand or silt at the top. Then the next bed starts again with gravel. These are characteristic of flooding events.

Pebbles in cave streams often have a dark brown or black coating of **manganese** oxide. The manganese is picked up by the water during solution of impure limestone and is redeposited as manganese oxide where the water becomes turbulent and full of oxygen. If these manganese rich waters are ponded and fill the cave completely, the roof and walls turn black as well. Bacteria may also play a role in this process.

Large quantities of fine sediments occur as silt or mud banks in caves. Percolation water entering from the roof introduces these sediments from the soil above. In the phreatic zone, these fine grains are suspended in water and travel long distances before settling to the bottom. Large sequences of thin layers accumulate and can sometimes fill the cave to the roof. When the mud dries out, it shrinks and triple-$120°$ cracks form in the surface.

Breccia and other sediments

A deposit peculiar to caves is known as a cave breccia. This is a jumbled mixture of angular blocks of breakdown and speleothems set in fine clay. Quite often bones are mixed in. Each component of the breccia originally came from a different source. They were mixed together due to slumping or sliding of the floor. Sediments of biological origin may accumulate into vast deposits in certain caves. Huge piles of guano from bats and birds form beneath large colonies. The skeletons of dead animals can also accumulate in caves and at the bottom of shafts. In time, these biological sediments become mineralised and may gain economic importance.

Tham Pha Thai contains an unusual cave sediment – a layer of volcanic ash. How the ash got into the cave is not known, but it has been dated using the potassium/argon technique. The ash is 9.36 (±0.47) million years old. If the ash proves to be a primary deposit and not more recently washed in, this would make Tham Pha Thai at least this old.

Cave Breakdown

Pieces of rock that fall off the walls and roof of the cave are called breakdown. The blocks are angular and deposited in a random fashion on the cave floor. Breakdown blocks can range in size from a few centimetres to larger than a house. The blocks pile up beneath the cavity they are falling from. Many of the larger underground chambers have been created with the aid of breakdown, such as those of Tham Sakoen and Tham Tab Tao.

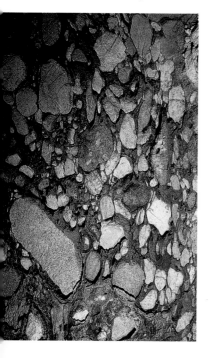

Breccia in Tham Pha Kong.

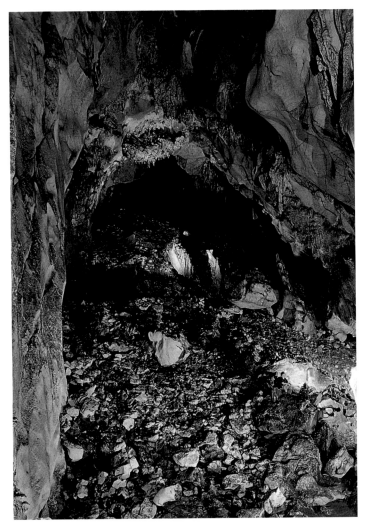

Massive rock pile in Tham Tab Tao.

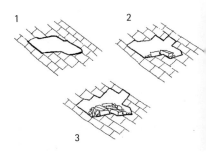

Stages of cave breakdown
1. unmodified cave passage
2. slight modification
3. heavy modification and breakdown pile

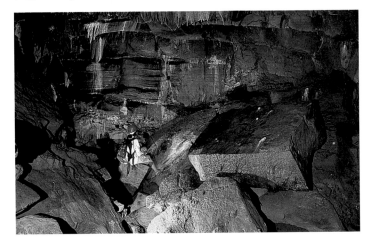

Slab break down in Fossil Cave.

Boxwork

Boxwork is a series of hard, flake-like projections sticking out of the cave wall. It often forms square patterns and resembles boxes (hence the name). The projections are an insoluble mineral, usually chert (silica), and are extensions of mineral-filled cracks in the limestone. Boxwork develops as a result of the surrounding limestone being dissolved away faster than the mineral, leaving it standing proud. Boxwork is not a true speleothem because the mineral was deposited before the cave existed.

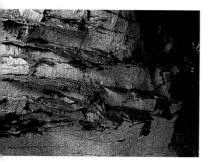

Chert boxwork in Tham Erawan, Lamphun.

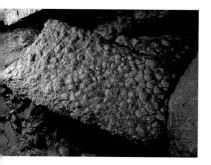

Detached slab full of large brachiopod fossils, Fossil Cave.

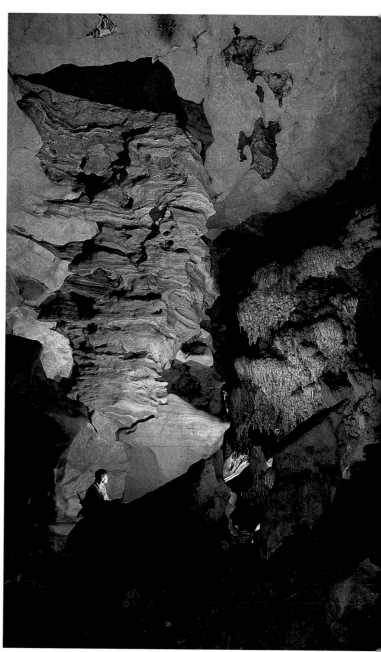

Redissolved flowstone, Tham Ki Mi.

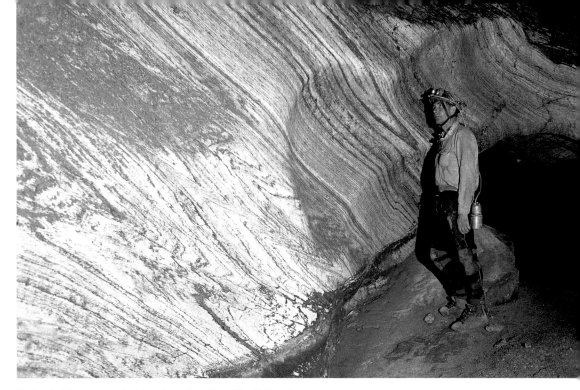

Redissolved flowstone wall with scallops, Tham Yang Wee. (© Dean Smart)

Redissolved Flowstone

Most speleothems are deposited in air-filled caves and their presence in a phreatic cave is an indication that the cave has entered the vadose zone. However, quite a few caves have flowstones that appear to be cut through. The internal structure is visible and large scallops are present on the outside. They also appear to be an integral part of the cave wall. These are flowstones that have been 'redissolved' by renewed, phreatic cave-forming processes. They provide strong evidence for multiple episodes of cave development in the phreatic zone due to rising and falling water tables. In some caves, at least four of these episodes can be recognised. What causes these large-scale fluctuations in the water table is a mystery though.

Fossils

Limestone is commonly made up of the skeletal and shell remains of animals. Wave action usually breaks most of these remains up during deposition and renders them unrecognisable. However, it is still fairly easy to find complete fossils within the limestone walls of caves. Caves are good places to look for fossils as they provide convenient access points to fresh rock exposures. The solution process may also selectively dissolve away the limestone surrounding fossils and leave them standing proud of the wall. Some remarkably well preserved fossils can be seen. The aptly named Fossil Cave in Pang Ma Pha contains a bed of numerous, large brachiopod fossils and in many caves it is quite easy to find crinoid (sea lily), coral and gastropod fossils.

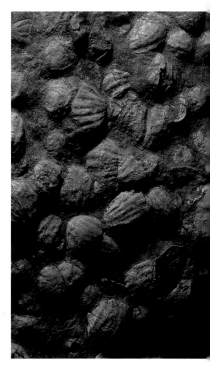

Fossil brachiopod shells, Fossil Cave.

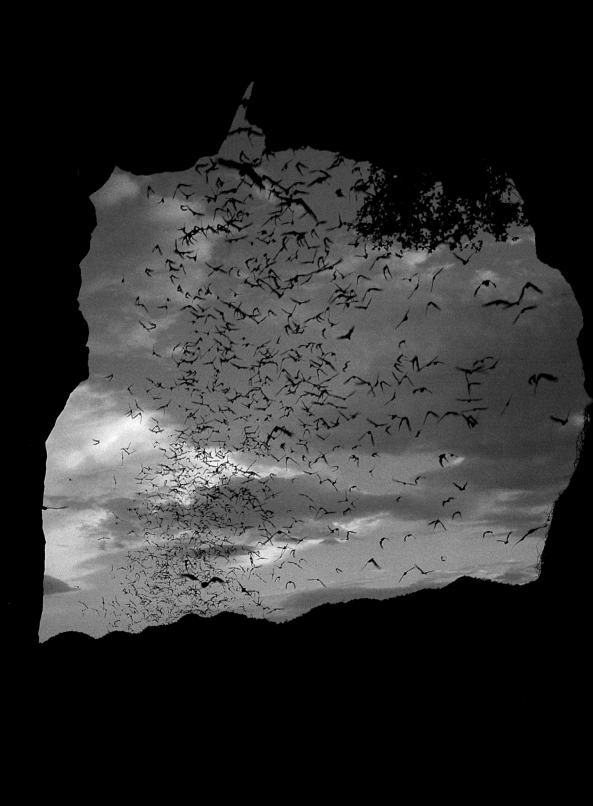

CAVE LIFE AND ENVIRONMENT

The cave environment is a world very different to ours. It lacks one of the most fundamental aspects of life on the surface – sunlight. This means that caves cannot support green plants (the basis of most food chains on this planet) and they receive no direct warmth from the sun. It also renders visual navigation and communication impossible. There is only one other environment on earth where it is truly dark – the bottom of the deepest oceans.

The cave environment is also permanently humid and remains a constant temperature. Energy supply may be very limited. Or, all of the energy may arrive in one large load and then cease for several months. The atmosphere might be abnormal making respiration difficult or even deadly. The animals that choose to live under these conditions must adapt themselves in order to cope. No adaptation means no survival. Those that make the required changes turn into strange looking aliens from another world. They have no eyes and their skin becomes transparent. They feel their way around with extraordinarily long, sensitive feelers. They also grow slowly, reproduce less and live longer lives.

These are the true cave dwellers – fantastic creatures that can survive on the meanest of diets in one of the harshest of environments. These are the denizens of the dark.

Darkness

The most striking difference between the cave environment and the surface is that there is no light in a cave. No light at all. Not even the faintest glimmer. It is so absolutely dark that it is impossible to see anything, even things placed a few centimetres in front of your face. This lack of light is a huge barrier for surface animals that rely on vision for moving around, locating food and mates, avoiding predators and communicating with each other. It also upsets the timing of certain rhythmic behaviours, such as sleeping and waking, which are triggered by the daily cycle of light and dark. In order to be successful underground an animal must abandon its dependence on light – not an easy accomplishment.

Stability

Another important aspect of the cave environment is its stability. Obviously caves remain dark for a very long time – often for millions of years. They also possess a climate that is remarkably constant. The temperature, humidity and atmosphere remain quite stable, even though large and rapid changes may be taking place outside. The insulating properties of the surrounding rock and limited environmental exchange with the surface protect the cave from daily and seasonal variations. Even severe, worldwide climatic changes such as ice ages are buffered. Cave fauna benefit enormously from this stability as it affords them with protection and increases their chance of survival through changing surface

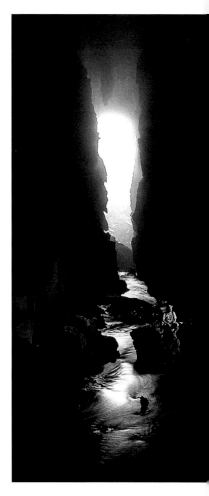

Tham Nam Lang is a large river cave which supports many cave animals.

Opposite: Stream of bats leaving cave at dusk. (© Dean Smart)

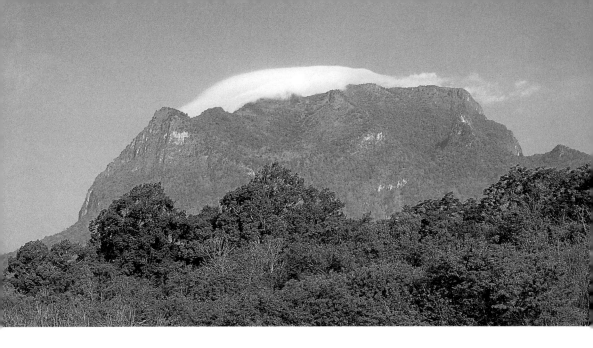

Doi Chiang Dao, Thailand's 3rd highest mountain (2,225 m). Cave P1 is situated behind the band of cloud.

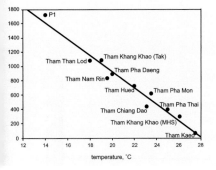

Air temperature versus altitude for selected caves of N. and W. Thailand.

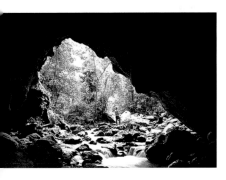

Tham Nam Lod, Chiang Rai, a short through cave with an active stream.

conditions. Outside relatives on the other hand are often faced with a daunting prospect – adapt or die. So, what is the cave climate like? How hot is it? How dry? Let's look at the conditions under which the cave fauna of northern Thailand live.

Temperature

The temperature of a cave matches the long-term average for the surrounding area. This is determined by latitude, altitude and the conditions of the local area. The area covered by northern Thailand is too small for noticeable changes in latitude to affect cave temperatures. However, differences can be detected due to altitude variation, which ranges from below 300 m at Phumiphol Dam up to 2,565 m on the summit of Doi Inthanon. As the altitude increases, cave temperature decreases. For example, Tham Khang Khao (Mae Hong Son), situated at an elevation of 300 m, is a cosy 26.1° C. Tham Pha Mon is a little higher at 620 m and is 23.5° C. The highest cave in northern Thailand where the temperature has been measured is P1 cave, high on the flanks of Doi Chiang Dao at 1,720 m. This cave is a chilly 14° C.

The local environment surrounding a cave can also have a significant effect on temperature. For example, caves located in the bottom of deep valleys tend to be cooler than those in bare limestone towers. Caves in mountain sides facing south are slightly warmer than those facing north. Some of the local factors affecting cave temperature are the type and quality of the forest cover, the amount and strength of sunlight reaching the area, humidity, aspect and biological activity. The presence or absence of flowing water also affects cave temperatures. Stream caves average about 1° C cooler than dry caves. This is because the water entering caves tends to be colder than the air (around 1 to 2° C cooler) and the temperature of the air is suppressed. However, with increasing depth into the cave, this difference decreases until the air and water temperatures become the same.

Humidity

Relative humidity tends to be high in most caves, between 90 and 100%. It is kept high by the continual presence of water in streams, trickling down the walls and dripping off speleothems. Restricted air movement also helps by preventing drier, outside air from entering the cave. Where draughts do blow through caves, the relative humidity changes in response to different air temperatures and water vapour content. Cool air cannot physically hold as much water vapour as warm air. It requires a smaller amount of water vapour to reach 100% relative humidity. So when warm daytime air enters a cave and is cooled, the relative humidity quickly rises to 100%. During the night, the opposite happens. The outside air entering the cave is colder and, as it warms up, the relative humidity drops. Generally speaking, large windy caves alternate between being like a tropical rain forest during the day and like a desert at night.

Air Movement

Draughts and winds in caves are caused by two main effects – 'breathing' and by the 'chimney' effect. **Breathing** occurs in caves with one entrance. It happens when the air inside the cave moves in and out due to barometric pressure changes outside. As the pressure outside rises, air is pushed into the cave and when the pressure drops, air is sucked out. Gradual pressure changes induce only weak breathing effects because the air pressure inside the cave has time to balance with that outside. Dramatic changes in pressure like those created by thunderstorms force air in and out very quickly. Very strong winds can be produced, especially in large volume caves with small entrances.

Chimney effects occur in caves with two entrances located at different altitudes. The effect produces air movement in two directions depending on the time of day. During the daytime, the air inside the cave is cooler than the air outside. This makes it heavier and causes it to sink towards the bottom entrance. At night the opposite occurs. The cave air is now warmer and it rises, like the smoke in a chimney, out of the top entrance. The upper and lower entrances of Tham Lom (Wind Cave) are separated by more than 100 m altitude and a powerful chimney effect blows through the cave, providing inspiration for its name.

Atmosphere

A phenomenon called 'bad air' occurs in many caves in northern Thailand. This is a situation where the atmosphere contains abnormal amounts of the gases carbon dioxide (CO_2) and oxygen (O_2).

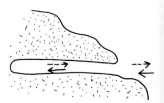

Breathing – air moves in and out of a single entrance when the barometric pressure outside changes.

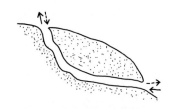

Chimney effect – air moves up and down through a cave with two entrances depending on the time of day.

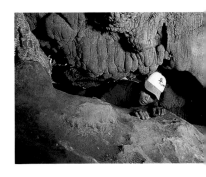

A chimney effect is very noticeable in the narrow passages of Tham Seua.

Tham Hued has one of the highest recorded CO$_2$ levels in Thailand.

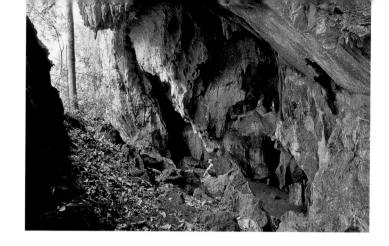

Types of Cave Animal.

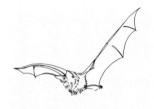

Trogloxenes use caves temporarily, for example: bats, porcupines and crickets.

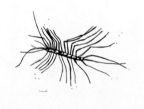

Troglophiles live their entire lives inside caves but are also found outside. For example, Scutigera centipedes live in caves and also in the forest.

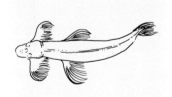

Troglobites live exclusively in caves and nowhere else. Very often these animals show adaptations to life in caves. For example, Cryptotora thamicola, the waterfall climbing fish, has no eyes and no skin colour.

Opposite: Leaf litter and decaying roots often supply energy to cave.

Normal air contains about 0.05% CO$_2$ and 21% O$_2$. Bad air is said to occur when CO$_2$ exceeds 0.5% and/or there is less than 18% O$_2$. High CO$_2$ levels are particularly acute in Tham Hued and in the side passages of Tham Pha Thai where 5 and 6.5% have been measured. This is more than 100 times the normal atmospheric level. There have been fewer O$_2$ measurements to date, but initial findings have shown that O$_2$ levels decrease at the same rate as CO$_2$ increases, at least in the caves studied. This suggests that some of the bad air found in the caves has a biological origin, coming from the decay of organic debris and the respiration of roots and animals. Abnormal cave atmospheres can also be caused by hydrothermal activity. Gases like carbon dioxide, hydrogen sulphide and even carbon monoxide can rise from deep inside the earth and find their way into caves.

Contrary to popular belief, carbon dioxide does not sink to the bottom of caves or into floor hollows. The fact that bad air is often concentrated in these places is because the floor is site of production. This is where the organic material and the organisms responsible for producing bad air are located. Bad air is being produced in all caves, but in most cases it is continually being removed by air currents and replaced with fresh air from outside. In poorly ventilated caves and places with little air movement, such as dead end passages and floor hollows, the bad air accumulates and concentrates. If it is not flushed out, it may reach highly dangerous levels (see Cave Exploration chapter).

Energy Supply

Life is sustained on the surface of the planet by sunlight and green plants. Every living thing ultimately depends on the energy provided by the sun's rays and the ability of green plants to photosynthesise them into nutrients. However, beneath the surface, in caves, there is no sunlight and there are no green plants. So how do the animals living in caves gain their energy? What do they eat?

The energy supply for most cave ecosystems comes from outside and is derived from sunlight and green plants. It enters caves in various ways and in several forms. Streams carry energy into caves in the form of leaves, branches and other vegetation. Trogloxenes, such as bats and porcupines, also provide energy by

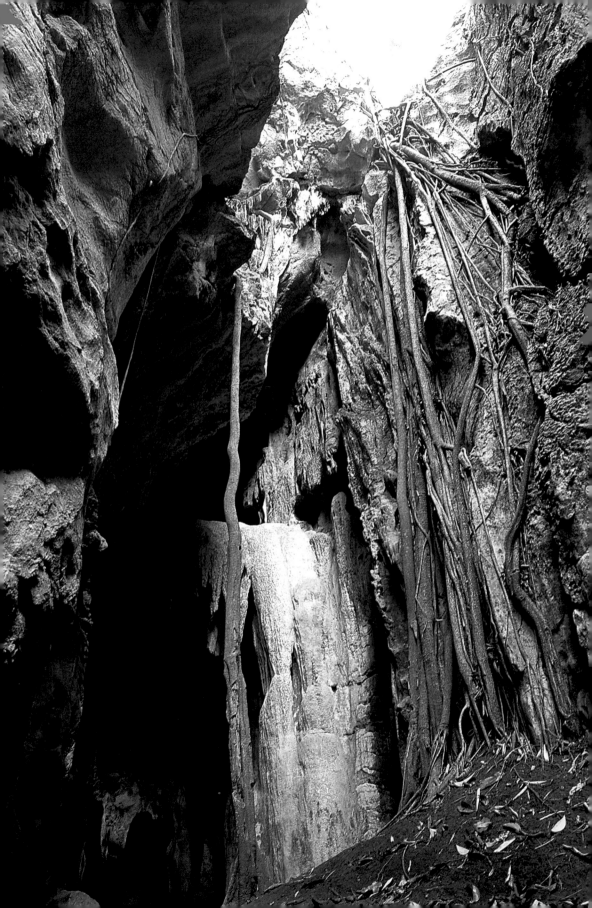

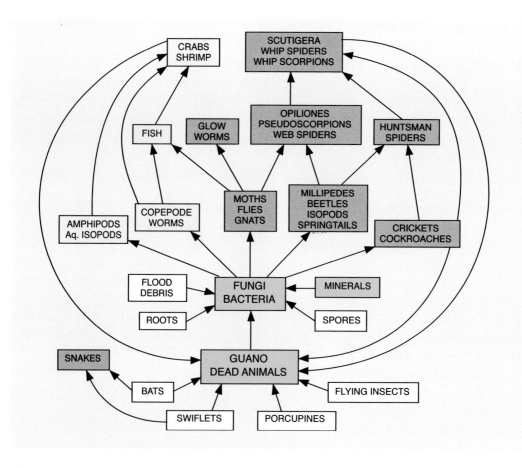

Generalized tropical cave food web.

Cave communities and energy supply.

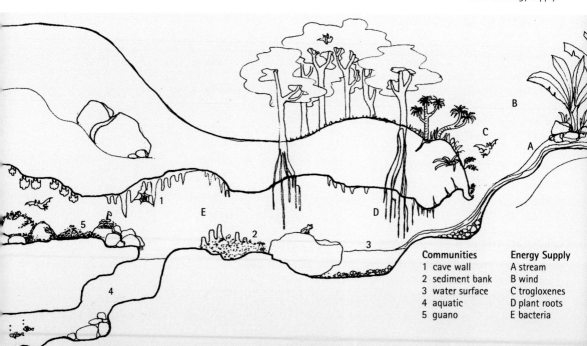

Communities	Energy Supply
1 cave wall	A stream
2 sediment bank	B wind
3 water surface	C trogloxenes
4 aquatic	D plant roots
5 guano	E bacteria

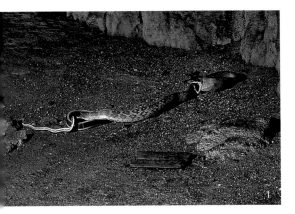

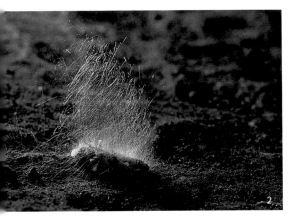

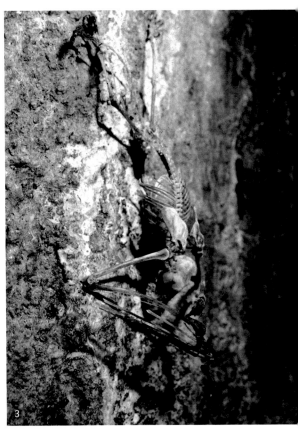

feeding outside and returning to the cave where they drop their faeces, shed parasites or die. Flying insects, pollen and spores are carried into caves on air currents. The roots of trees and other plants often penetrate cave ceilings and walls and provide another energy source. Inattentive surface animals may wander into caves by mistake or fall down surface shafts. If these animals do not find a way out, they will quickly die from starvation, or their injuries, and their carcasses will become a source of energy.

Some of the energy entering caves from outside can be eaten directly by the animals living there. The carcasses of dead animals are scavenged by beetles, crickets, crabs and shrimp. Spiders and 'glow-worms' prey on flying insects. Other forms of energy, such as plant remains and bat faeces are not so readily available. Bacteria and fungi need to break down these materials first and absorb the nutrients trapped inside. Then, animals like springtails, millipedes and isopods graze upon the bacteria and fungi.

Not all of the energy in a cave ecosystem comes from outside though. Some is produced within the cave itself by special bacteria called **chemo-autotrophs**. These bacteria are common underground and produce energy by converting iron, sulphur, nitrogen and manganese minerals. For example, *Ferrobacillus ferroxidans* obtains energy by oxidising iron compounds. *Nitrobacter* does it by

1. Cave racer snake eating a bat.
2. Fungus growing on bat guano.
3. Bat carcass. (© Dean Smart)

converting ammonia into nitric acid. Another bacterium, *Desulphovibrio,* turns sulphates into sulphides and oxidises organic material. The by-products created by the action of *Desulphovibrio* are carbon dioxide and hydrogen sulphide gas. Chemo-autotrophic bacteria provide an energy source that is independent of the surface. In some cases, this source is quite significant and may form the majority of the energy available. There are even a few, extremely rare examples around the world where cave ecosystems are powered entirely by bacterial energy. One such example is the unique Movile Cave located near the Black Sea coast of Romania. This small cave, only 250 m long, was completely sealed from the outside world until 1986 when excavation of a shaft accidentally broke into one of its passages. Inside the cave, the explorers found an amazing world of troglobitic animals including worms, snails, isopods, spiders, centipedes and insects – 48 different species in total with 34 being endemic. The ecosystems energy comes solely from the action of chemo-autotrophic bacteria, which oxidise sulphides in the slightly thermal, sulphur-rich groundwater.

Movile Cave receives virtually no energy input from outside. Today, an airtight door protects the cave and access is strictly limited to researchers only. It is recognized as a UNESCO World Heritage Site. To date, no cave ecosystems based on chemo-autotrophic bacteria have been found in northern Thailand. However, there is much hydrothermal activity and 'sealed' caves such as Tham Kaeo Komol are occasionally found by quarrying and mining operations. It is possible that a Movile Cave type ecosystem does exist here and is just waiting to be discovered.

The total amount of energy supplied to a cave ecosystem, whether from outside or through chemo-autotrophic bacteria, is normally quite low compared to most surface habitats. In a dry cave with a single entrance and no bats, the total amount of available energy can be extremely low. Another characteristic of the energy supply to a cave is that it can be very sporadic, arriving at particular times of the year or only in certain places within the cave. A stream cave for example would receive a huge amount of energy during the rainy season, but only a trickle for the rest of the year. Localities receiving bat guano regularly move as the bats search around for their most favoured roosting conditions. The bats may find these somewhere else in the same cave or in a different cave entirely. When the bats do move, obviously their supply of guano moves with them.

However energy is supplied to a cave – a constant trickle or an intermittent mass – is of little consequence to the animals living there. What matters most is that the supply of energy remains predictable. Survival depends on a similar amount of energy arriving in the same place and at roughly the same time as it did the last time (and the many times before that). Any deviations from this, such as a sudden reduction or increase in the amount of energy, a change in timing or location – can have disastrous effects on the ecosystem. This, together with the generally low levels of energy involved, is why cave ecosystems are amongst the most fragile on earth.

White millipede, Glyphillus *sp.*
(© Dean Smart)

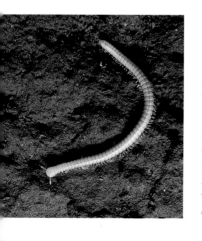

Adaptation

Animals move into caves for a variety of reasons.
Some do it to escape changing climates or harsh
living conditions on the surface. Others go looking
for shelter. Some are opportunists seeking alternative
food sources and others wander in accidentally.
Whatever the reason for going underground, most
animals find that the cave environment is a harsh
place to live. Those most likely to survive are the
ones coming from habitats with some similarities to
the cave environment, for example soil and leaf litter.
These animals are 'pre-adapted' to a certain degree
and this helps them overcome the initial difficulties.
However, caves present some unique challenges and
these must be overcome before true success can be achieved.
This is done through physical and behavioural adaptation.

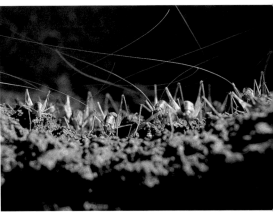

*Crickets have long feelers which are
indispensible in the darkness of a cave.
(© Dean Smart)*

  In the dark of a cave it is impossible to see anything, so
features like eyes and colour are useless. Over time, cave animals
lose these features becoming blind and without skin pigments.
The troglobitic fish *Cryptotora thamicola* has already lost these
features. It has no eyes, not even an optic nerve, and its skin is
completely depigmented. It appears pink because the red blood
inside its body shows through.

  Cave animals still need to be able to find their way around,
locate food and mates and avoid predators. They do this by
developing feelers and other organs to sense what is around them.
The feelers grown by cave animals are extra long and extra
sensitive. Those belonging to cave crickets, for example, are many
times longer than their bodies and are constantly waving around
gathering information. Troglobitic fish enhance their ability to
detect chemical changes and movement in the water by having a
greater number of sensory pits, especially on the head.

  Non-aquatic cave animals face a serious problem in caves
where relative humidity levels often reach 100%. At this
maximum level, the water content of the air is higher than that of
an animals' body fluids and moisture absorbs into the animal until
it eventually drowns. Cave animals develop special techniques for
keeping water out and for getting rid of excess water in the body.
The hard casings of some invertebrates have waxy, waterproof
coatings. Other animals filter the water out of their bodies with
incredible efficiency. Only the true cave dwellers – the **troglobites**
– possess such adaptations and they are the only ones that can
occupy places where relative humidity stays at 100%. In actual
fact, the troglobites seem to prefer such places and are more
commonly found living in wall cracks and floor hollows than in
the large, windy main passages of a cave. A curious outcome of
living in 100% relative humidity is that many troglobites find it
difficult to distinguish between terrestrial and aquatic habitats.
Animals normally associated with dry land such as beetles, isopods
and millipedes are often seen walking around and feeding on the

How do animals adapt to life in a cave?
lose their ability to see
lose their skin colour
grow long feelers or other sensory organs
develop waxy waterproof coatings
grow more slowly and live longer
mate less frequently
produce fewer offspring
develop efficient food storage systems
are less aggressive towards each other
react less if they are disturbed

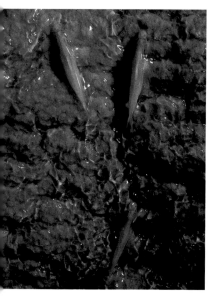

Cryptotora thamicola, a troglobitic blind fish endemic to N. Thailand. (© Dean Smart)

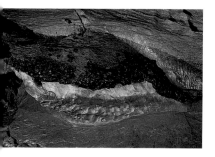

Black coating of iron and manganese oxides, produced by chemo-autotrophic bacteria.

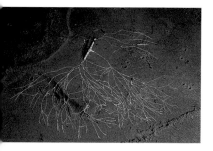

Fungal hyphae radiating from a piece of rotten wood.

bottom of streams and pools. Likewise, troglobitic fish are able to leave their watery habitats and move around on the damp cave wall and floor. *Cryptotora thamicola* is often seen doing this and it does so apparently in absolute comfort.

Adaptation to a poor food supply is where cave animals excel over their surface relatives. They can survive on the tiniest amounts of food for long periods of time. This does not mean that they are always on the brink of starving to death though. On the contrary, cave animals are strong and healthy. They are able to survive by saving energy at every opportunity. They do this by growing more slowly, living longer lives, having larger bodies and taking longer to reach maturity. They also mate less frequently and produce fewer offspring and eggs. Finding food is made more efficient through having specially adapted feelers and sense organs. At times when food is abundant, energy is stored in the body as fat to be used later when food becomes scarce. Cave animals tend to be less aggressive towards each other than surface animals and they react less if they are disturbed. Finally, growing eyes and having coloured skin requires energy. Troglobites save by not having them. Troglobitic fish also save by not growing scales.

Once adaptation to the cave environment is complete, the animals have little choice but to stay there. The changes they have undertaken preclude an existence elsewhere and return to the surface is out of the question. But if they prove successful in the underground world, why would they want to go anywhere else?

CAVE LIFE

A surprisingly large and diverse number of animals have made the move into caves in northern Thailand. Some of these are true troglobites showing the special adaptations needed for life underground. Others are occasional visitors seeking shelter and protection and some are able to live happily in both worlds, both above and below ground.

The animals living in caves often show remarkable levels of endemism. Entire species are very often limited to just one cave. Sometimes they are confined to just one small area in just one cave. This makes them particularly vulnerable to extinction, especially the troglobites. So, when looking for cave animals, please try to avoid disturbing them and their habitats.

Bacteria and Fungi

Caves are full of bacteria. More than 10 million of them live in every gram of sediment. Of course, it is impossible to see individual bacteria without a very powerful microscope, but we can see the results of their work as they go about decaying organic debris and converting minerals. One such bacterial product is the black, sometimes sticky coating on cobbles and walls of stream caves. This coating is a mixture of manganese and iron oxides, some of which are produced by chemo-autotrophic bacteria living on the surface of the rock. Another cave deposit associated with bacteria is 'moonmilk'. This is a soft, white substance that feels

The Animal Kingdom, showing groups found in caves

Phylum	Class	Order	English Name
Mollusca	Gastropoda		snails, slugs
Arthropoda	Crustacea	Decapoda	crabs, shrimp
		Isopoda	Isopods
		Amphipoda	amphipods
	Chilopoda		centipedes
	Diplopoda		millipedes
	Arachnida	Araneae	spiders
		Scorpiones	scorpions
		Amblypygid	whip spiders
		Uropygid	whip scorpions
		Opillones	opiliones, harvestmen
	Ellipura	Collembola	springtails
	Insecta	Diptera	flies, gnats, glow-worms
		Dictyoptera	cockroaches
		Orthoptera	crickets, grasshoppers
		Coleoptera	beetles
		Hymenoptera	bees, wasps, ants
		Lepidoptera	moths
Chordata	Teleostei		fish
	Reptilia		snakes, lizards, turtles
	Amphibia		frogs, toads
	Aves		birds
	Mammalia		bats, porcupines, serow

like cream-cheese when it is wet and like talcum powder when it is dry. Bacteria and blue-green algae are often found in moonmilk suggesting that may they play a role in its creation.

Fungi flourish in the cool, damp environment of caves. They are carried into caves as spores by the wind and by unsuspecting animals or people. Once there, if some of the spores happen to land on a suitable food source, the fungus will start to grow. Fungi sustain themselves by decaying organic material such as faeces and the remains of dead animals and plants. Long, white threads called hyphae spread out over the food and secrete enzymes to cause the decay. The products of the decay are then absorbed back into the fungus. Different fungi decay different things. For example, those found on bamboo are not the same as those found on wood and the white hairs that sprout from the bodies of dead moths are different again. Fungi in caves only very rarely grow fruiting bodies – the above ground part of surface fungi that makes and disperses the spores, the 'mushroom'. It takes a lot of energy for a fungus to produce a fruiting body and there is usually not enough available in the cave ecosystem to allow it.

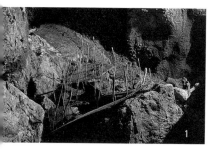

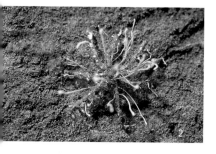

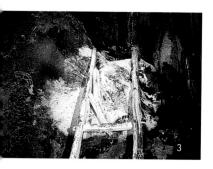

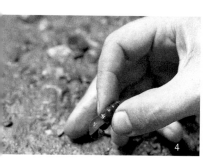

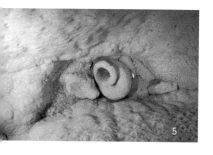

Snails and other Molluscs

Snails need calcium carbonate to build their shells and so it is no coincidence that they are very common in limestone habitats. Most live on the surface: in the soil, on plants, on cliffs and in water. They can often be found in caves, especially those with streams. None of the cave snails found in northern Thailand are troglobitic though.

Snails are a very diverse and important group of animals and thousands of species have been described. Some snails are herbivores and some are predatory. Others are not fussy and will eat anything that is available. There are aquatic snails and terrestrial snails. Shell sizes range from microscopic to several centimetres in diameter. The shells of some species are smooth and plain. Those belonging to some 'micro' snails are twisted and spiralled into fantastic shapes. This great variety is partly due to many snail species having very limited ranges, perhaps a single mountain. Some species have been found to be extremely restricted – living in just one small basin on the side of a single mountain. Their diversity and, often, limited range makes snail species especially prone to extinction through disturbance and habitat loss.

Spiders and other Arachnids

When you go into a cave, the first animal you are likely to see is a spider. These predators are by far the most abundant, large invertebrates found in caves. One reason for this is that most surface spiders are already well suited to the cave environment and require little adaptation in order to survive there. This is particularly true for the web-building spiders, which are very common in caves. To find prey, the web-builders can simply follow the vibrations made by a trapped insect and need not use their eyes. In fact, some species are already blind. Webs constructed by cave-dwelling spiders are not stretched across the open cave passage as you would expect. Rather they are pressed flat against the wall. This arrangement is designed to catch crawling prey like crickets, beetles and isopods, which are usually more common in caves than flying insects. Other web-building spiders live in burrows with trip lines radiating out from the entrance. They sit and wait in the burrow for prey to walk past and touch one of the lines. When this happens, the spider rushes out and catches it. Free-roaming, hunting spiders are less common underground than web-builders, but they can grow to enormous sizes. 'Huntsman' spiders (Heteropoda) have a leg span of up to 20 cm. The cave huntsman have smaller eyes than their surface relatives and probably hunt by touch using their long legs as feelers. Apart from the free-roaming spiders found near cave entrances, which are probably trogloxenes, most spiders found in caves are troglophiles. Only a few troglobitic species have been found in northern Thailand to date. One lives in Tham Chiang Dao. It is a small, blind web-builder of the genus Spermophora that is waiting to be formally named.

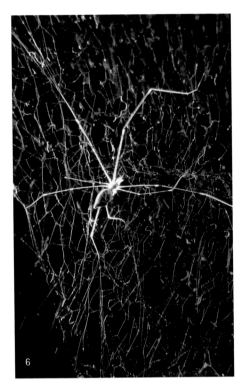

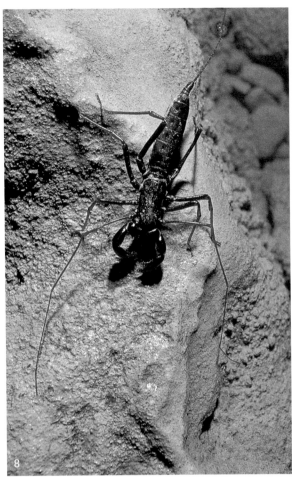

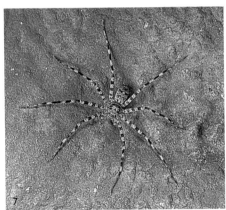

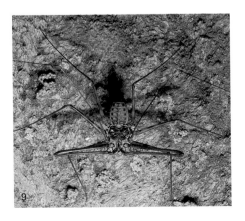

Oppesite page:

1.-3. Many types of fungi can survive in caves. (2 © Dean Smart)

4. Tiny mollusc living in a stream inside a cave.

5. Large, calcite-coated snail shell. (© Dean Smart)

This page:

6. Web-building spider, Pholcida. (© Dean Smart)

7. Free-roaming huntsman spider, Heteropoda.

8. Uropygid (whip scorpion). (© Dean Smart)

9. Amblypygid (whip spider). (© Dean Smart)

Another type of arachnid commonly encountered underground is the Amblypygid, or 'whip spider'. These magnificent creatures have round, flattened bodies up to 4 cm long and long spindly legs. They can move quite rapidly in any direction – even sideways. At the front of the body there is a pair of arms loaded with ferocious spikes. These are used to grab prey (usually crickets), tear it into little pieces and position it ready for eating. Prey is detected using the extra-long front pair of legs which have been modified to act like feelers. Amblypygids are not dangerous to people.

Uropygids, or 'whip scorpions', are no less impressive. These 6 cm long arachnids are usually black and have heavily armoured bodies with a long thin tail. On the head are two powerful arms that are used to grab prey and rip it apart. The tail has no sting, but some species can defend themselves by spraying acid.

Large seething masses of small animals with long legs are often seen in shady cave entrances. These are opiliones ('Harvestmen' in the UK, 'Daddy Long Legs' in the US). Most opiliones have bodies 0.5 to 1 cm long and legs up to 10 cm in length, though some giant tropical species can have leg spans of more than 30 cm The legs of opiliones seem to fall off very easily. This is actually an act of self-defence and is done voluntarily. The idea is to fool a predator into thinking it has caught the opilione, when actually it has caught nothing more than a detached leg and the opilione makes its escape. Unfortunately, broken off legs never grow back. Most opilione species are predators with a few being scavengers. Tham Klaeb Yai is the type locality for a genus and species of blind opilione called *Fangensis leclerci*.

Crabs, Shrimp and other Crustaceans

Crabs and shrimp, like snails, require calcium minerals to construct their exoskeletons and are especially common in limestone areas. They are versatile creatures and can readily cope with the underground environment. Their food is flexible. They can filter sediment, scavenge dead animals and eat flood debris. Some are predators, catching and eating other invertebrates. Large individuals may even catch and eat fish. Some cave populations are probably permanent, though most crabs and shrimp found in caves are trogloxenes. No troglobitic species have been recorded from northern Thailand. The shrimp species *Macrobrachium sirindhorn* was initially discovered in Tham Pha Mon, although it was later found to be a surface dweller as well. *Potamon doichiangdao* and *Potamon namlang* are two surface crab species described from the Doi Chiang Dao and Pang Ma Pha karst areas.

Amphipods and isopods are small crustaceans with seven pairs of legs. The easiest way to tell them apart is to look at their body shape. Isopods have broad, flat bodies whereas amphipods are more slender. These animals are among the most diverse of all to be found in subterranean habitats and there are more than 1,200 cave species known worldwide with many troglobites. Aquatic

White isopod. (© Dean Smart)

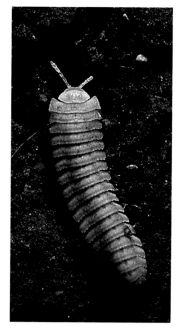

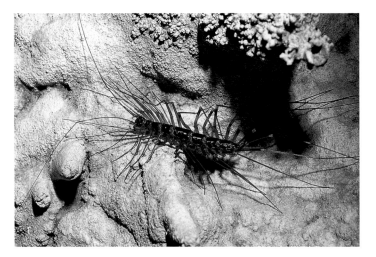

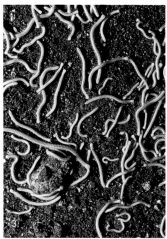

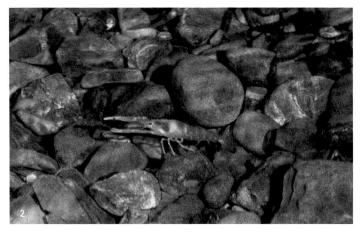

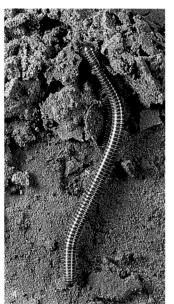

species are the most common. They look rather like small, white shrimp and live in streams, pools and the groundwater. Some live in the permanently water filled cracks above caves. Occasionally, heavy rain flushes them out into the cave where they congregate in small inlet pools. Small wood lice are commonly seen running around on sediment banks in caves. These are terrestrial isopods. Both troglobitic (white) and non-cave adapted (grey) species can be found. They feed on fungi growing on decaying wood or guano and defend themselves by rolling into a ball. Tham Khang Khao (Mae Hong Son) is the type locality for two species of isopod. The first, *Exalloniscus thailandensis,* is a terrestrial species with eyes and lives in the dry upper level of the cave. The second is aquatic and a true troglobite. Its name is *Indoniscus deharvengi.*

Centipedes and Millipedes

Cave centipedes (Scutigera) are spectacular animals about 10 cm long. They can run unbelievably quickly on their long, arched legs, earning them the nickname 'Hairy Mary'. Scutigera are fierce predators, hunting around the cave for any unfortunate cricket that gets in their way. The long antennae detect the prey, which is

1. Centipede (Scutigera) in Tham Nam Rin.
2. Shrimp in Tham Pha Daeng. (Pang Ma Pha).
3. Noodle millipedes on guano. (© Dean Smart)
4 Millipede in Tham Long Yaow.

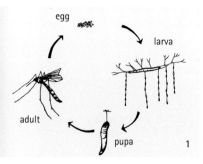

1. Life cycle of glow-worm.

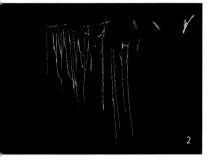

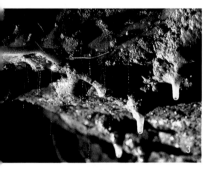

2-3. Glow-worm larvae and the sticky threads with which they catch their prey.

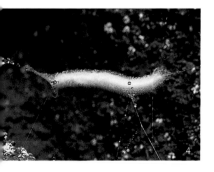

4. Sometimes the larvae become prey themselves, as with this one which has been overtaken by a fungus.

quickly killed by being injected with poison and then eaten. The large eyes of Scutigera are only able to see light and dark and they rely on their antennae to feel their way around, even outside of caves where they also live.

Small, white millipedes are often seen crawling around on guano deposits. These animals can be so numerous that they resemble noodles thrown on the floor, earning them the nickname 'Noodle Millipedes'. Their diet is very broad and they eat virtually anything they find: fungi, bacteria, dead animals, guano, leaves, wood, paper, etc. Millipedes move too slowly to run away from predators. When threatened, they roll up into a ball to protect themselves. Some can also secrete unpleasant chemicals to deter would-be attackers. A few millipede species are blind, such as *Dyomerothrix gremialis* of Tham Chiang Dao.

Springtails

Springtails, or Collembola, are a primitive group of animals related to insects and are commonly found in caves. They are very tiny creatures, only a few millimetres long and live in sediment, under flood debris or on the cave wall. They have a spring-like appendage on the tail that enables them to jump long distances. If you find a tiny, white creature that can jump like a flea, it is probably a springtail. Food consists of bacteria, fungi and other animals even smaller than themselves. On the surface, springtails live in soil and other places that closely resemble the cave environment. Many are pre-adapted to life underground allowing them to colonise caves with ease. Consequently, multitudes of cave species are known including many troglobites. Tham Chiang Dao accommodates at least 17 different species, eight of which are endemic to that cave. The cave-adapted genus *Troglopedetes* is also interesting. To date, 12 different species have been found living in 11 different caves of northern and western Thailand. With the exception of Tham Chiang Dao, which has two species, all of the other caves contain only 1 species. *Troglopedetes* is a good example of how cave animals evolve into different species after being isolated from each other. It also demonstrates the level of endemism that troglobites achieve with entire species limited to just one cave.

Insects

Small moths, flies and gnats abound in caves with lots of organic debris and guano. Spiders, opiliones, glow-worms and other small predators eat them. They can also be an important food source for cave fish.

Bunches of thin, transparent threads are often seen hanging from cave ceilings, especially in moist, windy caves. These are home to 'glow-worms' – the worm-like larvae of a type of gnat. The worms are about 2 cm long and nearly transparent. One glow-worm inhabits each bunch of threads and it can usually be seen crawling along the one horizontal strand that links the hanging threads together. The threads are sticky and used to trap flying insects for food. The name 'glow-worm' comes from similar larvae found in the famous Waitomo Caves of New Zealand. These larvae produce light to attract prey. For some unknown reason, the glow-worms of northern Thailand do not seem to glow. Perhaps they should be renamed 'dull-worms'?

Cockroaches are scavengers and will eat almost anything. Guano is a perfect food source for them and they find ideal living conditions in caves with lots of bats. They are also excellent opportunists and will move into previously unpopulated caves to eat the rubbish left by people. Unfortunately, once inside the cave they will eat everything they can, depriving the existing cave fauna of food and pushing them out. However, not all cockroaches are a symbol of filth and pollution. Most species are in fact very clean living. Troglobitic cockroaches exist though none are known in northern Thailand.

Every cave in northern Thailand seems to have lots of crickets. These ubiquitous creatures, commonly called 'cave' or 'camel' crickets, live in large groups near to patches of guano. They move around by jumping from one place to another using their long hind legs like springs. Jumping is a highly unusual form of motion for an animal that lives in the dark as it can easily lead to a collision with the cave wall or something much worse like a predator. Cave crickets avoid collisions by having enormous antennae many times longer than their bodies. These antennae constantly wave around, feeling for obstacles and predators and searching for food. Most cave crickets feed on the fungus and bacteria living in guano. A few species are carnivores and prey on birds' eggs and chicks. Even though cave crickets are able to find sufficient food in caves, the majority are actually trogloxenes. On dark, damp nights when the outside temperature is similar to that of the cave, they go out and feed in the forest. Upon returning to the cave, the crickets spread droppings, die or fall victim to predators thus supplying the ecosystem with energy. Some cave ecosystems rely almost entirely on this energy source, particularly dry caves with no wind and few bats. Huntsman spiders, whip spiders, whip scorpions and centipedes are the main predators of cave crickets.

There are more species of beetle on this planet than anything else. Caves also contain many different types and they play a wide variety of roles in the ecosystem. However, most beetles found in tropical caves are not cave-restricted and troglobites are quite rare. An exciting exception to this was the 1986 discovery of a new genus of beetle called *Ozaenaphaenops*. These totally blind beetles, found living in caves of northern Thailand, were the first true troglobites of their family to be found in the tropics. Only two species are known so far. One is endemic to Tham Pha Mon and the other is endemic to Tham Ukaew. (Tham Kou Khan)

Bees, wasps, ants, termites and other colonial insects often build their nests in or near cave entrances. They are never found far

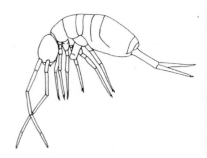

Troglopedetes longicornis is a troglobitic springtail found only in Tham Pha Mon. The long, forked tail is used for jumping. *(from Deharveng and Gouze 1993)*

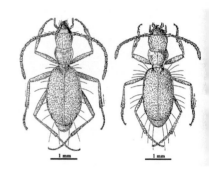

Blind, troglobitic beetles of the genus Ozaenaphaenops - O. leclerci *from* Tham Ukaew (left), O. deharvengi *from* Tham Pha Mon (right). *(Courtesy of Deharveng)*

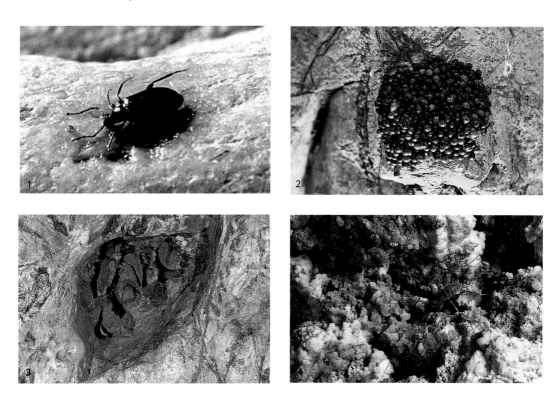

1. *Cave beetle, possibly Ozaenaphaenops. (© Dean Smart)*
2. *Colony of ladybird beetles.*
3. *Wasp nest.*
4. *Cricket on calcite crystals.*

underground though. Occasionally, a group of bright green grasshoppers might be found hibernating in a crack in the cave wall. Large moths also use cave entrances as places to hibernate.

Fish (Contributed by Dr. Chavalit Vidthayanon)

Troglobitic fish are found throughout the world in both temperate and tropical regions. The majority of the 90 species currently known live in freshwater streams, although some species have been found in estuarine and sea caves. Thailand supports nine species including three new species described in 2003, placing Thailand as the third most diverse country in the world after China (15 spp.) and Mexico (10 spp.).

Two species of troglobitic fish are found in northern Thailand and both are endemic to caves of Pang Ma Pha District. The first species, *Schistura oedipus,* is a loach type fish up to 8 cm long. It is pinkish white or pinkish yellow and the eyes have degenerated into black dots. It inhabits slow moving pools and its main food source appears to be small moths that fall onto the surface of the water. It has been found in five apparently unconnected caves and has an estimated population size of 10,000-100,000 individuals. The populations in the five caves are genetically distinct and exhibit differing levels of eye reduction.

The second species in the area is the remarkable *Cryptotora thamicola* which was only described in 1988 and is currently the only known species of its genus in the world. This tiny fish, only 4 cm long, lives in rapidly flowing water and is only found in two caves. It clings to the rock like a lizard and does not swim like a normal fish. It can do this because it has hundreds of tiny hooks on the underside of its fins. The fishes' body and fins are also shaped in such a way that the flowing water pushes the fish down, into the rock, helping it 'stick' even more. *Cryptotora thamicola* is completely blind and has no eyes at all, not even an optic nerve. The skin is totally transparent and without scales. The internal organs and muscles are clearly visible through the skin and the fish only appears pink because of its blood showing through from beneath.

Cryptotora thamicola has one of the lowest total population sizes of any cave fish in the world with only a few hundred individuals. The genetic diversity of the known populations is also very low, further increasing its risk of extinction. The species is being threatened by over-collecting, habitat disturbance and water-borne pollution. More research is urgently needed on the demography, biology and ecology of both species of troglobitic fish in Northern Thailand to enable better conservation of these unique animals.

1. Cryptotora thamicola. (© Dean Smart)
2. Troglobitic Schistura. (© Dean Smart)
3. Brook Carp Neolissochilus stracheyi.

Normal surface fish often enter caves for temporary visits and can sometimes be found many kilometres from the nearest entrance. Some of these may be troglophilic, though mostly they are trogloxenes. Many caves owe their name to the shoals of fish found near their entrances. One of the most impressive sites is the small spring at Tham Pla, Mae Hong Son which supports hundreds of large Brook Carp *(Neolissochilus stracheyi)*. These fish grow to an impressive size, averaging 80 cm in length and weighing more than 5 kg. Specimens more than 1 m long are sometimes seen. The fish grow so big because of the abundant food supply from tourists and because the local people consider them to be sacred and very rarely eat them.

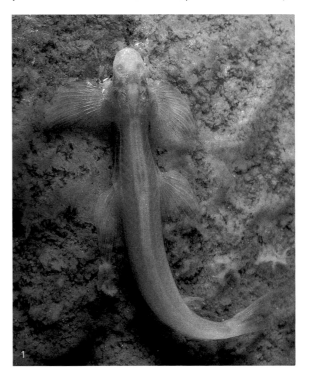

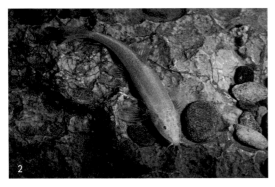

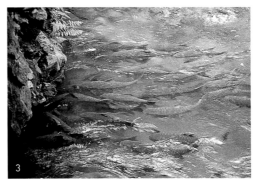

Reptiles and Amphibians

Snakes are quite often seen in the entrance area of caves. This is not because they are common in caves, but because they are easier to see than out in the forest where their camouflage hides them. Snakes enter caves for various reasons: some may wander in by mistake, some seek temporary shelter from the dry, blazing sun of the hot season and others may follow their favourite prey inside, such as frogs, bats or other snakes. The most ubiquitous of snakes found in caves of northern Thailand is the Cave Racer Snake, *Elaphe taeniura.* This beautiful, harmless snake grows up to 2.5 m in length and is easily recognised by its pale colour, black and white striped tail and dark eye stripe. Some individuals probably live permanently in caves where they catch and eat their preferred food – bats. The bats are killed by constriction before being eaten. Cave Racer Snakes are excellent rock climbers and smoothly slither all over the walls and roof of the cave. Occasionally they hang from stalactites on the ceiling and pick flying bats out of the air.

Geckos and other lizards often inhabit cave entrances. They can usually be found in ceiling crevices. Some species are very pale in colour and appear superficially to be cave adapted. However, a closer look will reveal that they have large eyes and are in fact trogloxenes.

Frogs and toads are also frequently found just inside cave entrances, sometimes in large numbers. Most are temporary visitors, escaping the heat of the dry season by sheltering in the humid atmosphere of the cave. Others are permanent residents, such as the huge, dark grey toads that sit majestically on wall ledges and boulders. These are called 'Cave Toads' *(Bufo asper)* even though they are not restricted to caves. Cave Toads reach impressive sizes with some individuals exceeding 20 cm in length and weighing more than 1 kg. They have voracious appetites and will eat any small animal that wanders within range of their long, sticky tongue. Cave Toads, like all other toads and frogs, use their eyesight to catch prey and move around. Whilst the Cave Toad's eyes are extremely sensitive and it can see in very low levels of light, it cannot see in the dark. This is one of the main reasons why frogs and toads are never found in the dark zone of caves.

Bats

When asked to think of an animal that lives in caves, most people immediately think of bats. Even though all bats are trogloxenes and do not rely on caves for survival, they are so closely associated with the underground world that it would be unfair not to say more about them.

Bats are fascinating, valuable animals. Unfortunately, they have been maligned and persecuted for centuries by people who mistakenly believe them to be harbingers of death and evil. Many bats and even entire colonies have been exterminated because of

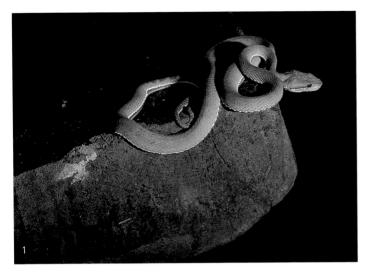

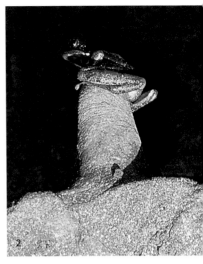

1. *Pope's Pit Viper* Trimeresurus popeorum.
   (© Dean Smart)
2. *Frog on stalagmite.*
3. *Cave Racer Snake* Elaphe taeniura.
   (© Dean Smart)
4. *Soft-shelled turtle Amyda cartilaginea.*

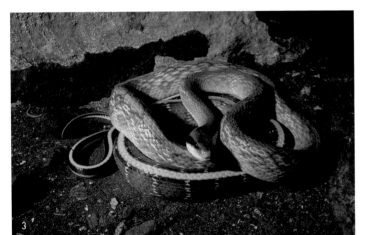

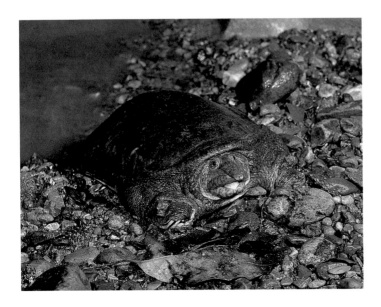

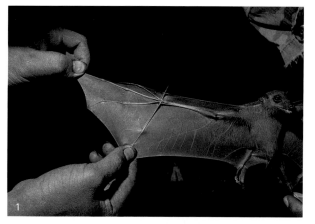

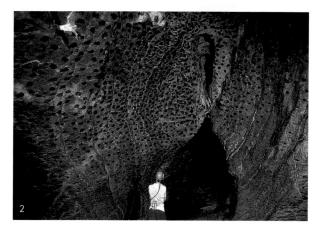

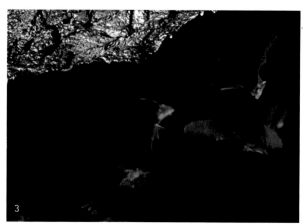

1. The Dawn Bat, Eonycteris spelaea, has a simple echolocation system. (© Dean Smart)
2. Roosting bats leave marks on the roof and walls of caves, Tham Pha Daeng (Phayao).
3. Bats in flight, Tham Khang Khao (Khun Yuam, Mae Hong Son).

these falsehoods. Bats deserve to be treated better as they are actually of great benefit to us. They perform important roles such as pest control, flower pollination and seed dispersal. Large colonies produce economic quantities of guano (their faeces) that can be sold for fertiliser. In some areas bats are eaten to supplement poor diets.

Bats are the only mammals with the ability to truly fly. Their wings are made from a thin membrane of skin stretched over elongated arms and fingers and attached at the base to the ankles. The design provides lift, forward motion and great manoeuvrability in the air. It works in a similar way to the wings of a bird where the air pressure on the top side is less than that on the underside.

Bats are nocturnal, meaning that they are only active at night. During the daytime they roost in quiet, sheltered places such as under the eaves of buildings, in hollow trees or inside caves.

In temperate climates bats survive the winter months by going into hibernation. The bats of Thailand, especially those living at higher altitudes, also go into hibernation for short periods. This helps them live through the frequent cold snaps of the November to February cold season. During hibernation, the bats reduce their activity to a minimum and lower the rate at which their body functions. Stored body fat provides them with energy. When the cold spell is over, the bats revive themselves and resume normal activities.

In Thailand as a whole there are more than 100 species of bat known to use caves and many of these can be found in caves in the north. Bat species into two main groups – the Microchiroptera (small, mostly insect eating bats) and the Megachiroptera (large, 'fruit' bats). The insect eating bats are much more abundant than the fruit eaters and have more species (99 compared to 18).

Most of the **Microchiroptera** hunt flying insects over fields and water or around treetops and street lamps. A few species will eat lizards, frogs, fish and even other bats. Prey is detected using a sophisticated echolocation system, or sonar. This marvellous innovation enables bats to 'see' in the dark using sound instead of light. The sound is a series of high-pitched clicks (inaudible to humans) sent out by the bat in a forward direction. Any obstacle or insect in the path ahead reflects the sound back to the bat, which then decides whether to take evasive action or attack. Some species have developed complex leafy-shaped noses that act like megaphones and enhance the sound they send out. All species have oversized ears that can hear the faintest of echoes. The system is so effective that some bats are able to detect objects as small as 0.07 mm in diameter and others can 'see' for several metres. Insect – eating bats provide a highly efficient and free pest control service. Each night, they go out and eat literally tons of insects. For example, a single colony of Wrinkle-Lipped Bats (*Chaerephon plicata*) with 250,000 individuals will consume around 2 tons nightly. This equates to millions of insects. Scaled up to cover the whole of northern Thailand, the number of insects and potential pests being removed each and every night increases to billions.

The fruit bats, or **Megachiroptera**, eat overripe fruit, nectar and pollen. They have fox-like faces with large eyes and small ears. Most do not have the ability to echolocate. Instead they rely on their eyesight and keen sense of smell to navigate and find food. Normally, these bats roost in trees and are only rarely found in caves. Species of the genus *Rousettus*, however, have a crude echolocation system in which the tongue is flicked to produce sound. This allows them to travel deep underground. Another cave-dwelling Megachiroptera is the Dawn Bat, *Eonycteris spelaea*. Fruit bats are important pollinators of plants. In some cases they are the only pollinator. For example, the durian fruit and the 'Sato' vegetable *Parkia speciosa* both rely on bats for this service. Bat pollinated flowers often have design features aimed specifically at bats. They tend to be large and hang 'upside down' with the opening at the bottom. They also release a strong odour at night. Fruit bats also help spread seeds around. The seeds are swallowed together with the flesh of the fruit and pass through the bat's digestive system. By the time the seeds are released in the bat's faeces, the bat may have flown many kilometres away from the original tree.

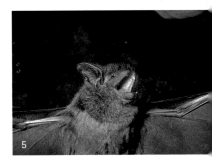

4. Hibernating Microchiropteran bat – Rhinolophus luctus. (© Dean Smart)
5. Insect-eating bat, Ia io. (© Dean Smart)

Other Mammals

The black and white quills of porcupines can often be found lying on the floor of caves. Porcupines are nocturnal. They sleep during the day deep inside caves where they find safety from predators. At night they come out to rummage around in the forest for food. Aided by their numerous, long whiskers, porcupines find their way around in the dark with ease and sometimes venture several hundred metres into caves. Their guano can be an important source of energy in some caves. Two species of porcupine inhabit caves of northern Thailand: *Hystrix hodgsoni* (large with long quills and a short tail) and *Atherurus macrourus* (small with a longer tail).

Piles of round, pellet-shaped droppings found in cave entrances belong to an animal called a Serow, *Capricornis sumatraensis*. This is a large goat-like animal that lives on craggy limestone mountains and uses the entrances of caves for shelter. Serow were quite common once, but hunting and habitat loss have reduced their numbers to the point where they are now rarely seen. Usually, their dung is the only sign that they are around.

1. *Porcupine footprints.*
2. *Canarium nuts, probably brought into the cave and eaten by a porcupine.*
3. *Porcupine quills. (© Dean Smart)*
4. *Bird nest made of mud and pebbles.*

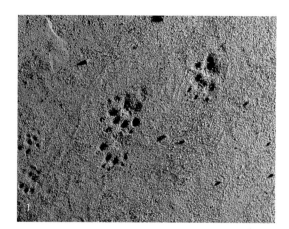

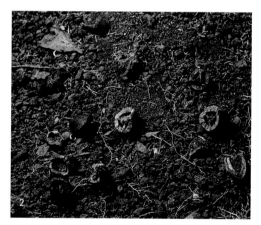

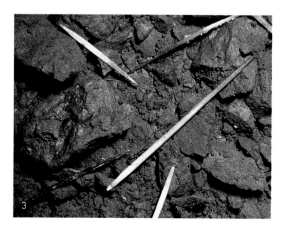

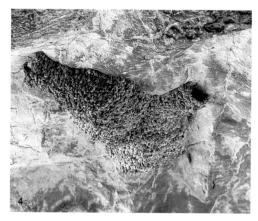

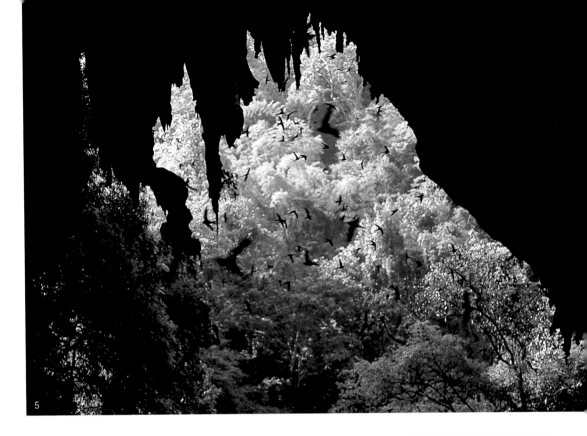

5

Birds

Owls, raptors and other birds sometimes nest on ledges in the entrance of caves. Most birds, however, are not closely associated with caves. The exception is a group of swiftlets that, because of their short legs, can only perch on vertical cliffs and cave walls. The group includes the Edible-Nest Swiftlet, *Aerodramus fuciphagus*, which builds the nests used to make the famous bird's nest soup. Tham Lot is home to a large, permanent colony of Himalayan Swiftlets, *Aerodramus brevirostris*. The birds leave the cave during the day to feed and return at sundown to roost. The evening fly-in is a spectacular sight. First, the swiftlets gather in a large swirling mass high in the sky above the downstream entrance. Then, long streams of birds peel off and spiral down into the cave. To navigate in the dark of the cave the swiftlets use a simple echolocation system, which is audible to the human ear as a series of rapid clicks.

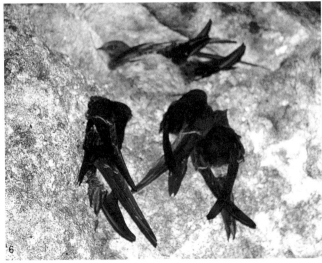

6

5. and 6. Himalayan Swiflets, Aerodramus brevirostris in Tham Lod. (© 6 Pang Ma Pha Cave Project)

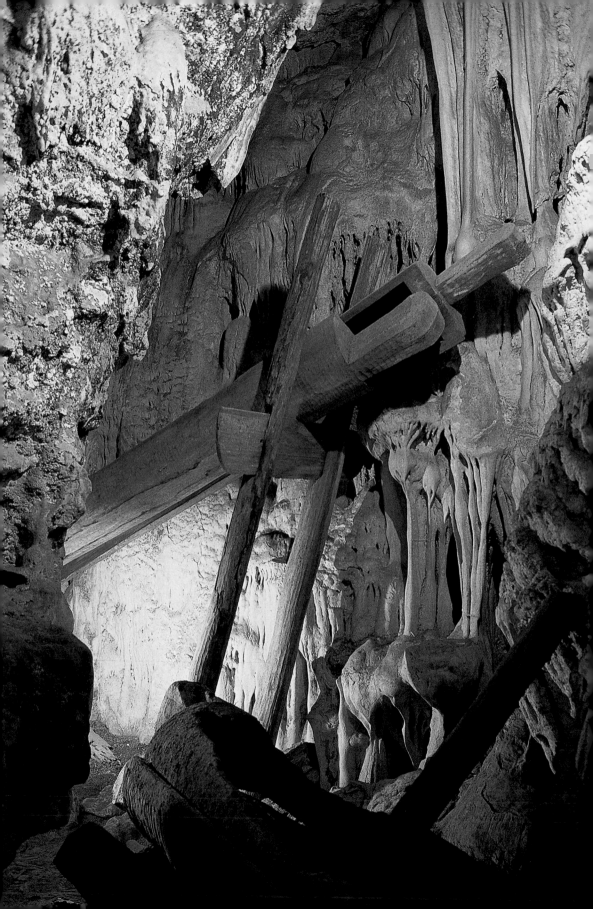

CAVE ARCHAEOLOGY

Caves and rock overhangs can provide us with an insight into the lives of our human ancestors that have used them as habitation or ceremonial sites since prehistoric times. Throughout the world, many of the most significant discoveries in archaeology and palaeontology have been made in caves. The natural conditions of low humidity, constant temperature and high alkalinity in many caves help to protect this vital evidence from the ravages of the changing external environment.

The study of cave archaeological sites in Northern Thailand is only just beginning and our current knowledge of this immense subject is very limited. Only a tiny fraction of all known caves in the region have been surveyed by archaeologists and there are almost certainly many important sites still waiting to be discovered. Archaeologists have already found many interesting and rich sites in a wide variety of locations scattered throughout the region. Current evidence indicates that caves and rock overhangs in Northern Thailand have been used by people as habitation sites, burial grounds and rock painting sites for at least 10,000 years. Evidence from several sites in Lampang Province has recently come to light which suggests that caves in the region may also have been used by *Homo erectus*, a human ancestor, as much as 700,000 years ago.

Visitors to caves and overhangs in Northern Thailand who look carefully have a good chance of finding evidence of prehistoric human occupation. Flake stone tools made from river pebbles and animal bones or shells which show signs of having been eaten as food are common in many sites throughout the region. Fragments of ancient pottery, rock paintings or even log coffins can also be found in certain areas. This archaeological evidence provides vital clues to help us understand the life style and beliefs of prehistoric people who used to live in the area and were part of the process which created the natural and human environment which we live in today.

Although many of these archaeological sites have been known for a long time, much of the evidence has not been properly recorded and it is only recently that any systematic surveys have been conducted. Very few sites have any effective protective measures, with the result that a great deal of this priceless heritage has already been disturbed or even destroyed completely. Unscrupulous collectors, acts of vandalism, mining activities and the development of caves as temples or tourist attractions have all taken their toll. In some cases, immediate protection measures need to be brought into effect to prevent the complete destruction of vital archaeological evidence.

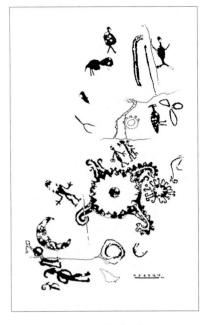

Rock paintings in Ban Rai Overhang, probably about 3,000 years old. (© Pang Ma Pha Cave Project)

Opposite : Teak log coffins in Tham Jabo, 1,200-2,200 years old.

Archaeological Timescales for Caves in N. Thailand		
Geological period	Artifact period	years B.P.
Late Holocene	Historical period	1,400 – current
Late Holocene	Metal Age	2,500 – 1,400
Middle Holocene	Metal Age/Neolithic	7,500 – 2,500
Early Holocene	Mesolithic	10,000 – 7,500
Late Pleistocene	Paleolithic	40,000 – 10,000
	(no artifacts)	250,000 – 40,000
Middle Pleistocene	(no artifacts)	730,000 – 250,000
Early Pleistocene	(no artifacts)	1,800,000 – 730,000

Chronology

The age of archaeological evidence found in caves varies from a few hundred to several hundred thousand years old. Archaeologists need to establish a system of dating in order to construct a time sequence of events and to compare the time sequence of different sites.

There are many ways of dating, but one of the most widespread and popular systems uses the type of archaeological artifact found to divide time into four main periods: Palaeolithic (Old Stone Age), Mesolithic (Middle Stone Age), Neolithic (New Stone Age) and Metal Age (sometimes further subdivided into Bronze Age and Iron Age). However, this system is often difficult to apply in the context of Northern Thailand because the same type of stone tools were used over extremely long time periods with very little change or development of style.

Another popular system of chronology is to use major geological events such as Ice Ages to define time periods. On this scale the entire history of the human race is confined to the last two periods – Pleistocene and Holocene. For the purposes of this book a combination of geological and artifact-based systems has been used to make sense of the complex history of cave use in Northern Thailand.

Paleoenvironment

The physical environment of Southeast Asia has been subject to many major changes in both climate and landscape. Geological evidence indicates that a series of global Ice Ages between 1,000,000 and 10,000 years ago resulted in drier and colder conditions that at present. The large volume of water locked up in the ice caps resulted in a lowering of sea levels worldwide which exposed the Sunda-Sahul continental shelf and connected the Southeast Asian mainland to Western Indonesia. The average sea level over the last 140,000 years was 65 m below current levels and at the height of the last Ice Age 18,000 years ago sea levels were more than 100 m below those today. As the climate became warmer and wetter, the melting of the ice resulted in rising sea levels which reached their highest levels for over 100,000 years. At the peak, 6,000 years ago, the sea was 6 m above present-day levels and the whole of Central Thailand was flooded as far as Lopburi, 100 km inland from the current sea shore. Since that time the sea level has been gradually receding and probably reached current levels about 1,500 years ago.

Although the current sea level is relatively stable, there are still continuous changes to the landscape as a result of ongoing geological processes. The Chao Phraya River deposits large quantities of silt annually, expanding the land area in the delta by a few metres every year. In Prachuap Kiri Khan, Petchaburi and Chonburi the coastline is retreating inland due to coastal erosion by the sea. These ongoing changes have had an effect on prehistoric human habitation, subsistence and migration patterns, which is reflected in the location of archaeological sites and the type of artifacts found in them. The flood plains of Central Thailand that used to be under water probably contain rich evidence of early human communities which are now buried beneath the alluvial sediment. The area which is now the gulf of Thailand may contain even older sites which are now lost to the sea. In this context, the mountainous regions of North Thailand are particularly important for archaeologists. These regions contain innumerable caves and overhangs hidden in the karst landscape which are all over 200 m above sea level and have probably been relatively unaffected by the dramatic changes experienced in the rest of Thailand over the past 140,000 years. The insight which such sites can provide into the life and culture of hunter-gatherer communities who used to live in the area is particularly important because there are no written records or other sources of evidence relating to these vanished cultures.

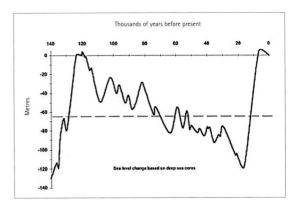

Sea level change from 140,000 years BP to the present, based on deep sea cores.

Sea shore line in Southeast Asia at the present compared to the average over the past 140,000 years.

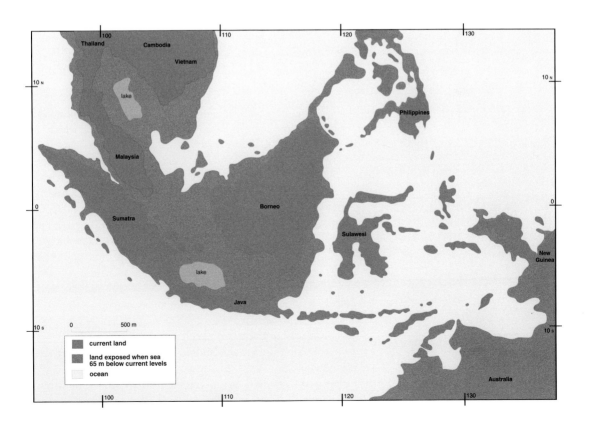

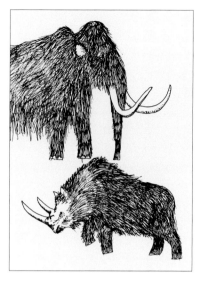

Woolly mammoth and Woolly rhinoceros - 2 extinct animals found in Europe during the Holocene.

Extinct Animals

Humans and their ancestors are not the only animals which have made use of caves in the recent past. A wide variety of animals have used caves as habitation sites, hiding places, breeding grounds and feeding areas. Many caves contains signs of this past use which can provide palaeontologists with vital clues about the evolution, distribution or migration patterns of these animals and to reconstruct the past environment as well as to understand the present.

Carnivorous animals such as bears, wild dogs, cats and birds of prey left the remains of their prey in the caves. Many used caves as shelter and died there. The excavation of cave floor deposits can lead to the discovery of extinct species previously unknown to science. Caves in South Africa have revealed evidence of several extinct animals from the early Pleistocene, for example *Notochoerus* (a giant pig), *Libytherium* (a relative of giraffes with antler-like horns) and *Hipparion* (a three-toed horse). Isolated habitats such as small islands are particularly rich sites for the evolution of distinct local species, many of which have since become extinct. In the Mediterranean, the islands of Malta, Sicily and Cyprus are known to have supported unique species such as *Choeropsis* (pigmy hippos) which have only become extinct in historical times. A wide variety of extinct animals are known from caves in mainland Europe, such as *Anancus* (mastodon) and *Homotherium* (sabre-toothed cat). Many of these extinct animals provide compelling evidence of dramatic climate change. In the Early Pleistocene, when Europe had a warmer climate than at present, extinct animals such as *Dicerorrhinus* (narrow-nosed rhinoceros) and *Palaeoloxodon* (straight-tusked elephant) lived in the area. In the colder Holocene, these animals were replaced by more hardy species such as *Mammuthus* (woolly mammoth) and *Coelodonta* (woolly rhinoceros). These species can be identified not only from the bones and other remains found in the caves but also from the rock paintings left by early humans.

Caves in Thailand have also provided rich evidence for the existence of many species which are no longer found in the area. An ancestor of orang utans, *Koratpithecus chiangmuanensis,* was found in a coal mine pit in Phayao province, dated 10-13 million years old. Giant pandas, now restricted to China, used to be found in Northern Thailand during the Middle Pleistocene when the climate was considerably cooler than is present. During warmer periods other species which are now only found in Africa, such as hyena and hippos have also been found, as well as extinct animals such as *Homotherium* (sabre-toothed cat).

Ancestors of Modern Humans

"Where Do We Come From?" is a question that we all ask ourselves. The discovery of fossils and stone tools of early humans in caves and rock shelters helps us to understand the evolution and development of our own ancestors. An increasing number of fossils are coming to light from around the world which may belong to our closest relatives or even to our direct ancestors. A bewildering variety of scientific names have been ascribed to these fossils, although there is as yet no general agreement amongst anthropologists as to the relationship between these species and ourselves. The most widespread species so far discovered is *Homo erectus,* which most experts believe is a direct ancestor of modern people. *Homo erectus* lived about 500,000-1,500,000 years ago and walked upright like ourselves, but had a thicker skull with a heavier jaw and smaller brain cavity. The oldest generally accepted evidence of *Homo erectus* has been found in Africa, but by 500,000 years ago they were widespread throughout Europe and Asia. Excavations in France show that *Homo erectus* in Europe learnt how to make stone and bone tools, hunted both large and small animals, built camps and covered themselves with animal skins. In China, several sites have been found including the famous 'Peking Man' site in Zhoukoudian Caves near Peking. *Homo erectus* bones at this site are probably about 500,000 years old and are associated with flake stone tools and a camp fire. In Indonesia, several sites on the island of Java are believed to be 800,000 years old.

In Southeast Asia the direct evidence for *Homo erectus* has been tantalizingly inconclusive, despite the high likelihood that they did formerly live in the area since they are known to have inhabited both China and Indonesia over 500,000 years ago. However, a major breakthrough was made recently with the

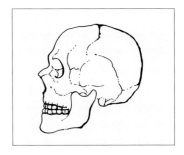

Homo sapiens sapiens, *modern human.*

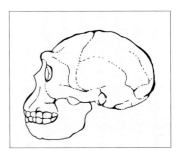

Homo erectus.

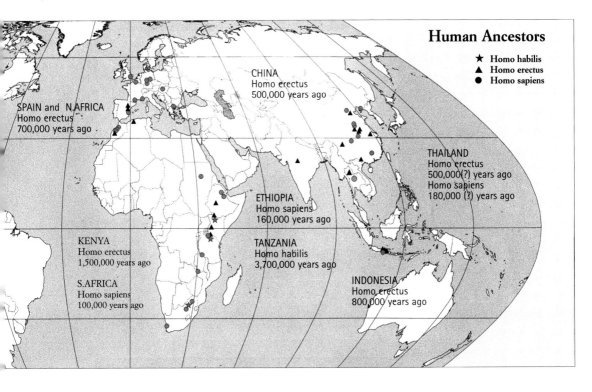

Human Ancestors

★ Homo habilis
▲ Homo erectus
● Homo sapiens

SPAIN and N.AFRICA
Homo erectus
700,000 years ago

CHINA
Homo erectus
500,000 years ago

THAILAND
Homo erectus
500,000(?) years ago
Homo sapiens
180,000 (?) years ago

ETHIOPIA
Homo sapiens
160,000 years ago

KENYA
Homo erectus
1,500,000 years ago

TANZANIA
Homo habilis
3,700,000 years ago

INDONESIA
Homo erectus
800,000 years ago

S.AFRICA
Homo sapiens
100,000 years ago

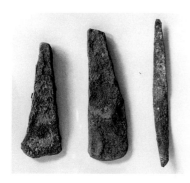

Iron tools from coffin sites in Pang Ma Pha.
(Highland Archaeology Project)

discovery of part of a *Homo erectus* skull in a limestone cave in Amphoe Kho Ka, Lampang Province. No direct dating has yet been carried out on this sample but the age has been provisionally estimated at 500,000 years by comparison with samples from China and Indonesia. At Khao Pha Nam, also in Lampang Province, river pebbles which bear a distinct resemblance to flake stone tools were found in association with charred hippo bones in sediments between layers of basalt dated 700,000 BP. If these are indeed stone tools then this is the oldest known evidence of human ancestors in Thailand.

A curious gap exists in the fossil record between these *Homo erectus* sites and the emergence of *Homo sapiens,* our own species. Remains dated 100,000 years old from a cave in South Africa are one of the earliest generally accepted pieces of evidence for *Homo sapiens* worldwide. Although older findings have been claimed from many parts of the world, the results are inconclusive. In Thailand, a single human tooth dated 180,000 BP has been reported from what might be a hyena lair in Vimannakin Cave, Chaiyaphum Province. Since this is the only evidence for humans or our ancestors in Thailand between 500,000 and 40,000 years ago this isolated finding is open to doubt.

Modern *Homo sapiens sapiens*, our subspecies, was dispersed throughout Africa, Europe, Asia and Australia by 40,000 years ago. Lang Rongrien Overhang in Krabi Province, dated 40,000 BP, is the earliest modern *Homo sapiens* site so far recorded in Thailand, although no human bones were found at this site. The oldest known modern human bones, dated 25,000 BP, were discovered in Mo Kaew Cave, also in Krabi Province.

Prehistory of Caves in Northern Thailand

Archaeological evidence of modern humans from all over the Old World shows the change and development of human culture over the last 40,000 years. Although this evidence is spread over a long timescale, a common sequence can be detected reflecting the pattern of the use of caves and overhangs in different periods.

Flake stone tools from Tham Liang Pha, Mae Sai.

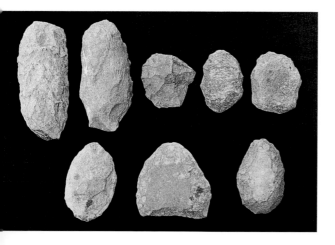

In Northern Thailand the earliest evidence of human use of caves is usually in the form of flake stone tools and animal bones or shells which show signs of having been eaten. Occasionally, the remains of camp fires have also been found which suggests that the caves were used as habitation sites or at least as temporary shelter. The lack of artifacts such as pottery or any evidence of domesticated plants and animals indicates that these early inhabitants were probably hunter-gatherers who used the caves and overhangs as temporary shelters or resting places. These habitation sites have been found in almost every province of

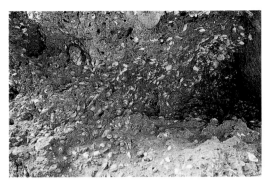
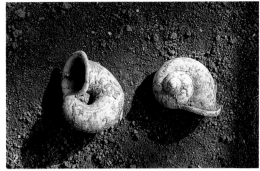
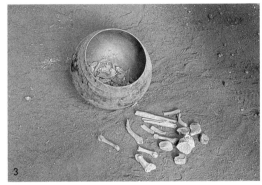
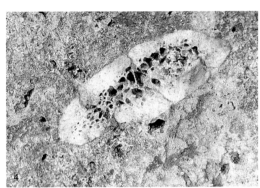

Northern Thailand, mostly dated from the late Pleistocene to the Middle Holocene, 12,500-2,500 years ago. Caves and overhangs were probably used as habitation sites as recently as the Late Holocene, only 1,000 years ago. Indeed, they are still used by hunters & illegal immigrants even today (see page 141).

From the Middle Holocene onwards, caves in Northern Thailand started to be used as ceremonial sites associated with burial rites or other rituals. Although some caves continued to be used as habitation sites, it seems unlikely that the same caves were used for both habitation and ritual purposes. Concurrently, the most spectacular examples of the use of caves as burial sites are the impressive log coffins found in Pang Ma Pha and adjacent regions which date from the Middle to Late Holocene, 2,300-1,300 years ago. Numerous rock paintings throughout the region also point to the use of caves and overhangs as sacred sites. The presence of polished stone tools, fragments of pottery, seeds of domesticated plants and metal tools in many of these ceremonial sites indicate that they were probably built by people who practised some form of agriculture. Whether these people were descendents of older hunter-gatherer groups or were immigrants from neigh bouring areas is still a matter of debate.

In the past few years two much older burial sites have been found in overhangs of Pang Ma Pha District, dated 9,000-12,500 years old. These sites point to the possibility that caves and overhangs in Northern Thailand have been used as ceremonial sites over a far longer timescale than is generally accepted.

1. and 2. Shell midden from Tham Liang Pha, Mae Sai.
3. and 4. Mammal bones including bovines and wild boar fromTham Huai Bon, Fang.

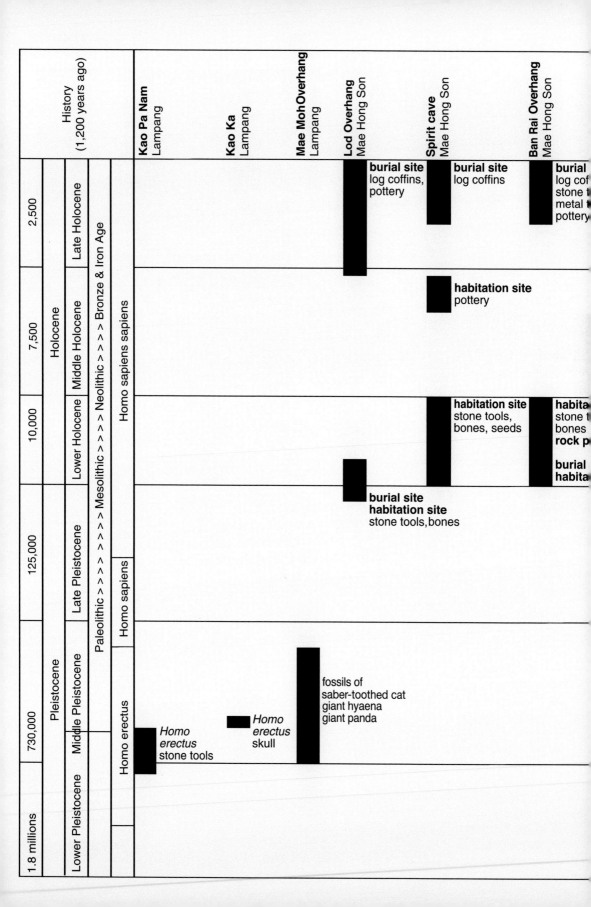

1.8 millions	730,000	125,000	10,000	7,500	2,500	History (1,200 years ago)	Kao Pa Nam Lampang	Kao Ka Lampang	Mae MohOverhang Lampang	Lod Overhang Mae Hong Son	Spirit cave Mae Hong Son	Ban Rai Overhang Mae Hong Son
	Pleistocene			Holocene								
Lower Pleistocene	Middle Pleistocene	Late Pleistocene	Lower Holocene	Middle Holocene	Late Holocene							
										burial site log coffins, pottery	**burial site** log coffins	**burial** log cof stone t metal t pottery
											habitation site pottery	
											habitation site stone tools, bones, seeds	**habita** stone t bones **rock p**
												burial **habita**
										burial site **habitation site** stone tools, bones		
Homo erectus		Homo sapiens		Homo sapiens sapiens								
		Paleolithic > > > > > > > Mesolithic > > > Neolithic > > > Bronze & Iron Age										
	Homo erectus stone tools											
		Homo erectus skull							fossils of saber-toothed cat giant hyaena giant panda			

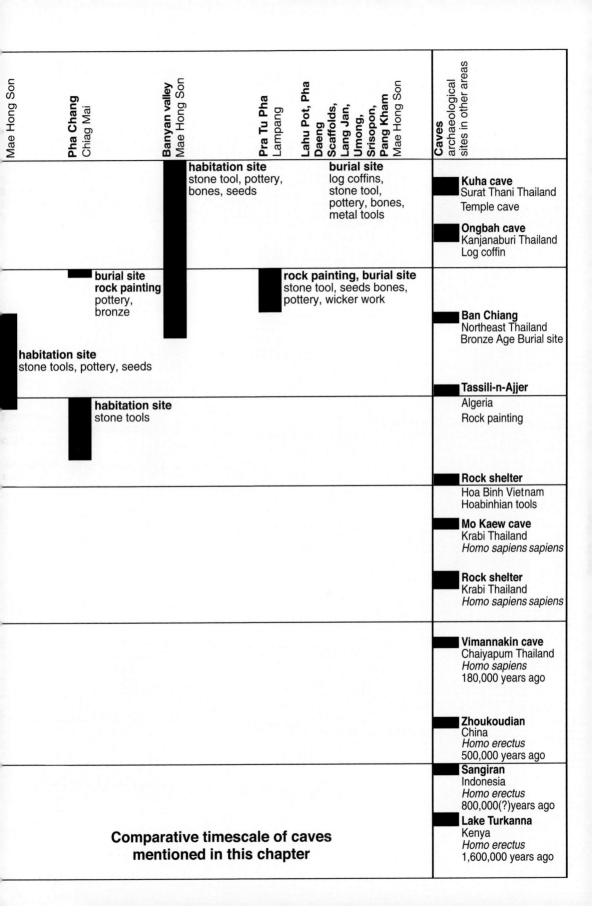

Mae Hong Son	Pha Chang Chiag Mai	Banyan valley Mae Hong Son	Pra Tu Pha Lampang	Lahu Pot, Pha Daeng Scaffolds, Lang Jan, Umong, Srisopon, Pang Kham Mae Hong Son	Caves archaeological sites in other areas
		habitation site stone tool, pottery, bones, seeds		**burial site** log coffins, stone tool, pottery, bones, metal tools	**Kuha cave** Surat Thani Thailand Temple cave
					Ongbah cave Kanjanaburi Thailand Log coffin
	burial site **rock painting** pottery, bronze		**rock painting, burial site** stone tool, seeds bones, pottery, wicker work		**Ban Chiang** Northeast Thailand Bronze Age Burial site
habitation site stone tools, pottery, seeds					**Tassili-n-Ajjer** Algeria Rock painting
	habitation site stone tools				
					Rock shelter Hoa Binh Vietnam Hoabinhian tools
					Mo Kaew cave Krabi Thailand *Homo sapiens sapiens*
					Rock shelter Krabi Thailand *Homo sapiens sapiens*
					Vimannakin cave Chaiyapum Thailand *Homo sapiens* 180,000 years ago
					Zhoukoudian China *Homo erectus* 500,000 years ago
					Sangiran Indonesia *Homo erectus* 800,000(?)years ago
					Lake Turkanna Kenya *Homo erectus* 1,600,000 years ago

Comparative timescale of caves mentioned in this chapter

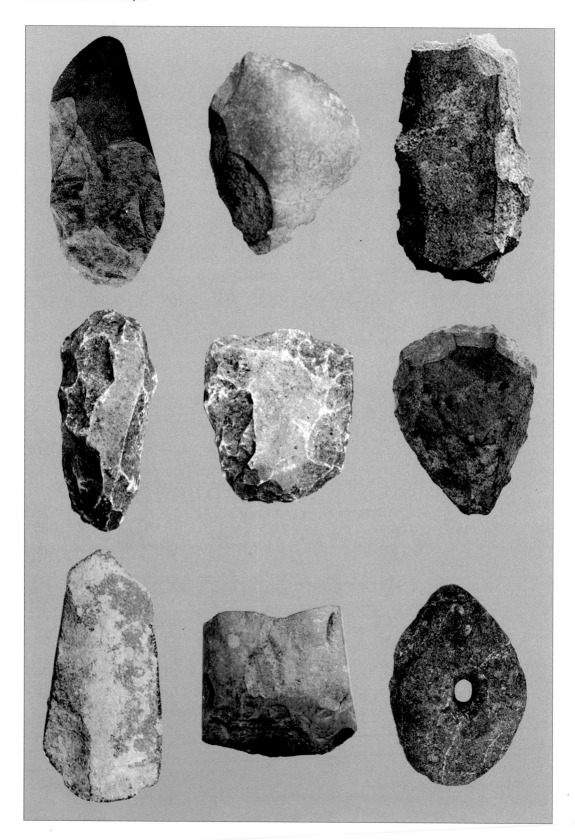

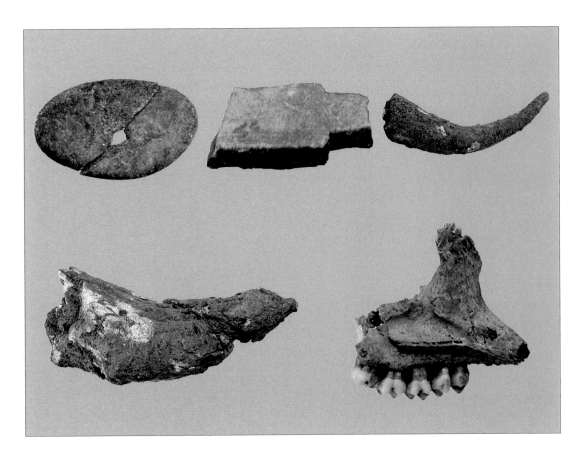

Hoabinhian Tools

The word 'Hoabinhian' was first used to describe prehistoric stone tools found during archaeological excavations in Hoa Bin, North Vietnam. These tools were made by chipping flakes from river pebbles and are typically found in association with shells or animal bones, which suggests that they were made by hunter-gatherer groups. The same style of tool was used over a period of at least 11,500 years during the Late Pleistocene and Early Holocene 20,000-8,500 BP. Similar flake stone tools have since been found in many locations throughout Southeast Asia including Northern Thailand.

Chester Gorman described the flake stone tools found in Spirit cave as 'Hoabinhian', based on similarities of style with tools from Hoa Bin. He used the term 'Hoabinhian Techno Complex' to refer to a pattern of technology, especially stone tool-making technology, resulting from the adaption of the prehistoric peoples of Southeast Asia to their environment. He stressed that the similarities between different areas could be due to a common response to similar environmental conditions, rather than to cultural interaction between the different groups of prehistoric people. Many researchers prefer to use the term **sumatralith** to describe this type of stone tool because it does not imply any cultural similarities with other areas.

Facing page and this page:
Prehistoric stone tools and bones found in caves and overhangs in northern Thailand.
(© all photos of Highland Archaeology Project)

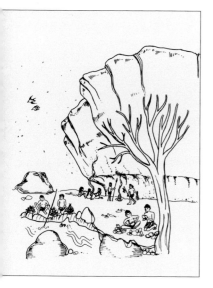

Caves as Habitation Sites

Caves and overhangs have been used as habitation sites by humans and our ancestors over a period of many hundred thousand years. Long before modern *Homo sapiens* appeared on the scene, caves were being used by *Homo erectus* and archaic *Homo sapiens* groups as shelter and hiding places. They usually choose to live under overhangs or close to cave entrances rather than deeper inside caves without natural light where there was the possibility of unknown dangers lurking within. Most of the sites are located not too far from permanent water sources, typically raised slightly above the level of the plain where there would have been less risk of flooding or contracting malaria and other diseases.

In Northern Thailand, the evidence indicating prehistoric habitation sites is usually in the form of flake stone tools, animal bones and shells rather than actual human remains. Such artifacts are found in large numbers in many caves and overhangs throughout the region. However, the vast majority of these sites have not been excavated by archaeologists and very few of the artifacts have been scientifically dated. The first rigorous archaeological survey was carried out in 1965-72 by Chester Gorman who excavated three sites in Pang Ma Pha – Spirit Cave, Banyan Valley Cave and Steep Cliff Cave. Gorman uncovered evidence indicating that these sites had been occupied by human groups from the late Pleistocene to the late Holocene, 11,350 - 1,100 years ago. Gorman found flake stone tools and other artifacts which resembled those associated with the Hoabinhian culture in Northern Vietnam. He also found polished stone tools, animal bones, shells and seeds of domesticated plants, suggesting that during the later stages the people who occupied these caves probably practised some form of agriculture.

Gorman's work has directed the attention of other archaeologists to the region, which has lead to the discovery of many more habitation sites throughout Northern Thailand. Three other sites dating from the Late Pleistocene to Middle Holocene have been excavated so far – Tham Lod and Ban Rai Overhang in Pang Ma Pha as well as Ob Luang Overhang in Hod District. In addition, at least 20 sites in Pang Ma Pha and more than 40 sites from other areas in Northern Thailand have been discovered but not yet excavated. However, the number of sites found in each area has been heavily influenced by the thoroughness of the survey effort – nobody can yet tell the exact number and distribution pattern of habitation sites and there are no clear conclusions regarding who these people were, how they lived and what happened with them. Archaeologists are still trying to put together the whole story from these fragments of data. New discoveries or more advanced techniques of dating and DNA analysis may help to shed light on these questions.

Facing page: Pha Phung Overhang (Mae Hong Son) was used as a habitation site and contains an interesting group of rock paintings.

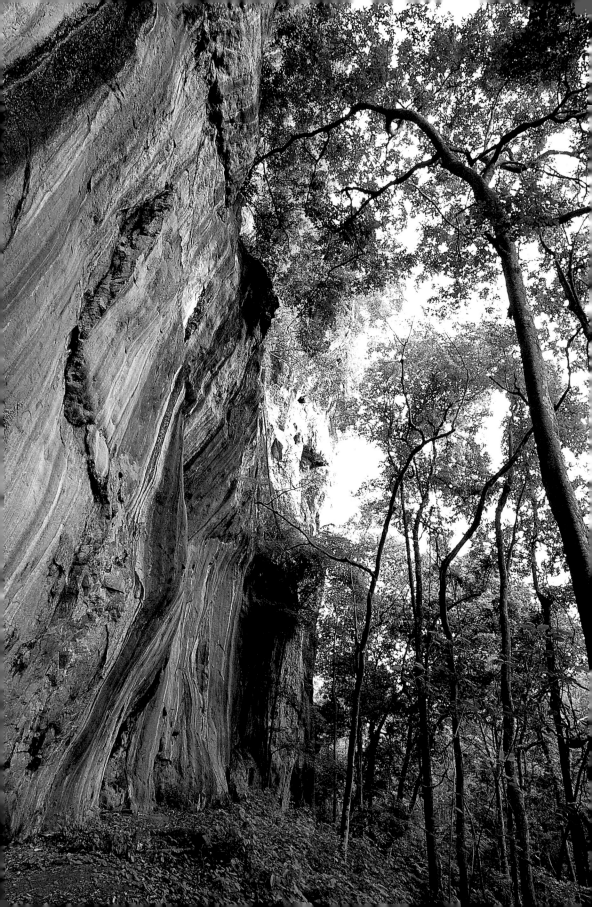

Spirit Cave, Pang Ma Pha

Spirit cave is a good example of a stone age habitation site which was made internationally famous by the excavations of Chester Gorman in the 1960s. The cave is situated on the southern slopes of an isolated limestone mountain in the upper watershed of the Nam Khong River. The area today is thickly forested with a mixture of deciduous species including teak and bamboo with evergreen species in moister areas. The entrance to the cave is 650 m above sea level and is a steep 300 m climb up from the Nam Khong River. The considerable distance between the cave and the river is unusual for a habitation site, although it is possible that a closer water supply used to exist in prehistoric times. The cave itself has a large entrance and three main chambers. The entire cave is currently dry and inactive.

The aim of Gorman's work was to find evidence for the transition to agriculture in hunter-gatherer societies. He focused his attention on the middle chamber which has the deepest floor deposits, where he dug two excavation pits at right angles to each other. He divided the deposits into fire soil layers and two cultural layers which spanned a 4,000 year period between about 12,000 and 7,500 BP. The first cultural layer (11,690-9455 BP) contained flake stone tools and knife blades made from river pebbles which Gorman thought were similar in style to tools of the Hoabinhian culture in Northern Vietnam. A wide variety of animals bones were found in association with the tools, including bats, rats, reptiles, birds, squirrels, monkeys, pigs, fish, turtles and shells. This layer also included sign of fire and the remains of several plant species including betel nut, legumes, Canarium sp. and Terminalia sp. The second cultural layer (8,776-7,622 BP) contained a similar range of plant and animal remains, signs of fire, flake stone tools and knife blades but there were also a number of more advanced tools as well, such as polished stone tools and stone axes. Some of the knife blades were made from slate which is not found in the area. Fragments of 'cord mark' pottery with burn marks on the inside were also found.

Spirit Cave, showing Gorman's excavation site.

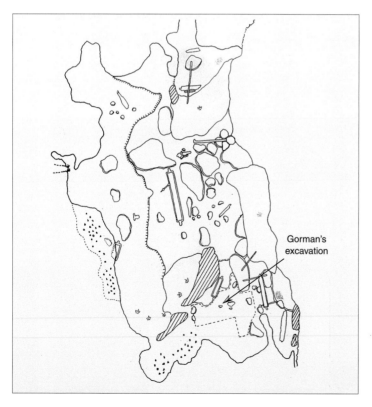

Gorman's
excavation

Subsequent analysis revealed that at least 20 genera of plants were used by the prehistoric inhabitants of Spirit Cave. These plants could have been used for a wide variety of purposes including food, lighting, condiments, stimulants and containers. Several of these species are probably not native to the area and may have been cultivated by these ancient people. If this interpretation is correct, Spirit Cave is one of the oldest records of plant domestication in the world.

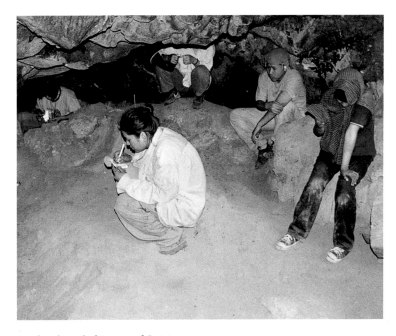

Visit to Gorman's excavation site at Spirit Cave by Shoocondej's team in 2000. (© Highland Archaeology Project)

The animals which were eaten by the inhabitants of Spirit Cave came from a wide range of habitats including streams (fishing cat, otter, fish and shells), cliff faces (serow, goral) and dense forests (wild pig, deer, cattle). Many of the bones belong to small mammals such as squirrels, monkeys, langurs, lorises and gibbons which are only found in the forest canopy and are difficult to catch without the use of sophisticated tools such as bows and arrows or traps. The absence of such tools in the archaeological deposits can be explained by the fact that they were probably made from bamboo or wood rather than stone and have long since rotted away. Surprisingly, most of the bones were not charred despite the existence of camp fire remains in the same layer. This could suggest that they were eaten raw but it is also possible that the meat was boiled in some kind of container such as bamboo, not grilled on an open fire.

Spirit Cave was resurveyed in 2000 by Dr. Shoocondej's team. No new excavations were conducted but the entire cave was mapped and surface artifacts recorded. These included 10 pieces from six different log coffins, indicating that Spirit Cave was reused as a burial site long after the habitation period studied by Gorman. The coffins have not been dated but their estimated age is 2,200-1,200 years by comparison with other sites. The pottery fragments collected by Gorman from Spirit Cave have recently been reanalysed by a student from Bradford University (UK). This new research suggests a date of only 3,000 years BP which brings into question Gorman's conclusions about the antiquity of Spirit Cave.

DATING TECHNIQUES

There are two main methods which archaeologists use to date artifacts – absolute dating and relative dating by comparison with similar artifacts from other areas. Absolute dating is the most accurate method but requires complex and expensive equipment which is not widely available. Archaeology projects are normally working on tight budgets which dictate that only a very few samples can been sent for this kind of dating. The vast majority of archaeological artifacts in Northern Thailand have so far only been dated by comparison with other artifacts. The absolute dating technique which is normally used by archaeologists is the Carbon 14 (C14) method. Although several other techniques are now available, Carbon 14 is still the most generally useful because it covers the time period of most interest to archaeologists

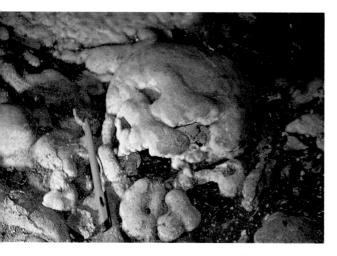

Human skull encrusted with calcite, Pang Kham. (© Pang Ma Pha Cave Project)

and because archaeological artifacts often contain organic carbon in some form. Wood, charcoal, shells, bones or even the latex glue used in prehistoric rock paintings can all be used for dating. All living organisms receive carbon from eating and drinking which has the same proportion of the radioactive C14 isotope as the atmosphere. After they die, the radioactive C14 slowly changes into stable C12 at a constant rate of 50% in 5,730 years. The age of the sample can be determined by comparing the ratio of C12 / C14 isotopes with that found in reference samples of known age. After 37,000 years only 1% of the original carbon 14 still remains, which is usually insufficient for accurate dating. Great care is required when collecting the sample to avoid contamination from other material. A new form of C14 dating technique, Accelerator Mass Spectroscopy (AMS) is more accurate and uses much smaller samples but it is also considerably more expensive.

The comparative technique assumes that the type and style of artifact is related in some way to its age. While this is often a valid assumption for artifacts that use a particular technology, such as bronze or iron making, it is much more difficult to apply to purely stylistic similarities. In Pang Ma Pha, archaeologists initially assumed that the style of coffin heads might be related to their age, with the more complex animal-like carving being more recent than the simpler, abstract designs. However, accurate scientific dating of the coffin wood has clearly shown that there is no such relationship. The style seems to be purely a matter of individual expression or perhaps has some ceremonial or social significance of which we are unaware.

Archaeological Sites in Northern Thailand

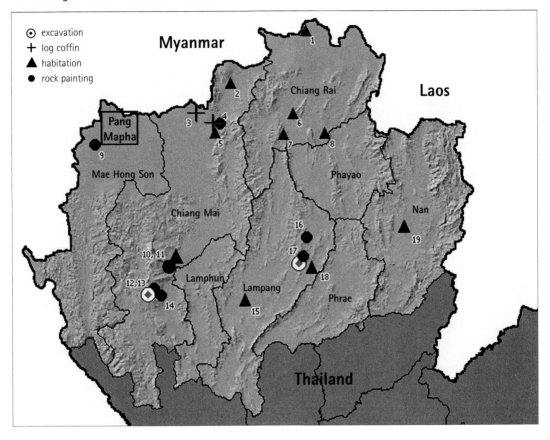

Archaeological Sites in Pang Ma Pha District

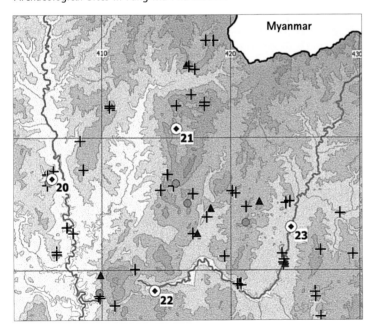

1 Liang Pha
2 Huai Bon
3 Klang Muang
4 Long
5 Ngung Chang
6 Russi
7 Phra Pha Ben Pun
8 Pha Jarui
9 Pha Phung and Tham Wua
10 Pha Kanna
11 Borichinda
12 Ob Luang
13 Pha Chang
14 Pha Mai
15 caves in Amphoe Kor Ka
16 Ban Jampui
17 Ban Tase
18 caves in Mae Mo
19 Amphoe Muang, Nan
Pang Ma Pha District
20 Spirit cave
21 Boong Hoong
22 Ban Rai Overhang
23 Tham Lod

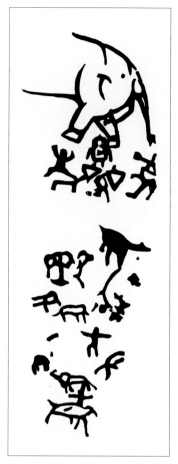

Pha Chang Overhang, Ob Luang, is one of the few rock art sites in Northern Thailand with white paintings. The scene shows human figures with an elephant and other unidentified animals.

Prehistoric Rock Art

Prehistoric people all over the world left evidence of their existence and beliefs in the form of rock paintings in overhangs and caves. This ancient art form incorporates a wide range of styles, from simple hand prints to complex abstract designs and extremely realistic depcitions of animals. In Europe, Africa and Australia, rock paintings have been found over 30,000 years old. These ancient paintings help us to understand not only the people themselves, but also gives us clues as to the natural environment.

A wide variety of techniques were used in prehistoric rock art, including line drawing, printing, spraying and painting. The paint was applied to the rock using fingers, sticks or with brushes made of plant fibres or animal hair. Sometimes these hairs can still be found stuck to the painting and may be used for C14 dating. Most prehistoric rock paintings throughout the world are red, with the colour being derived from ochre or blood. Black, white, browns and yellows are also found, made from charcoal, limonite and other substances. The paints usually contain a binding material such as latex or animal fat. The preponderance of red could be due to the fact that this is the most stable colour and other colours have since faded away, or in some cases thay may have turned to red, as has been observed happening in Bushman paintings a few hundred years old in South Africa.

Prehistoric paintings can help to reveal factual details about extinct animals which are very difficult to obtain from any other source. Although paleontologists have known of the existence of extinct mammoths in Europe for a long time, nobody imagined what these animals really looked like until the discovery of magnificent prehistoric rock paintings from France and Spain which showed that they were covered with long woolly hair. Only much later, when mammoth specimens were found frozen in ice, was this confirmed.

The animals in the rock paintings are often an excellent indication of the large scale environmental changes which have taken place during the past 30,000 years. For example, drawings of hippotamuses on sandstone overhangs have been found in the middle of the Sahara desert at Tassili-n-Ajjer, Algeria. These paintings are only 7,000 years old but they are a clear indictation that this area which is now nothing but sand used to support lush vegetation and lakes in the relatively recent past.

Rock paintings also help us to understand the life style of prehistoric humans. Many paintings seem to depict hunting scenes or ceremonial activites. However, we should be wary to interpret these events as strictly real. Many of them may be scenes from myths or are intended to illustrate experiences under trance. Alternatively, they may be individual artistic expressions with no attempt at realism.

Prehistoric rock paintings in Northern Thailand have only very recently received expert interest. Over the past 20 years at least 25 sites have been discovered scattered across the whole of the region. There are almost certainly many more sites still waiting to be found as many areas are still unsurveyed. The vast majority of the paintings are in monochrome red, although there are a few examples with black or white details. Most of the paintings are handprints, in a wide variety of styles, many of which have been found all over the world. Animals, human figures and apparently abstract designs are also found.

   None of the rock art from Northern Thailand has been directly dated using scientific methods. The most commonly quoted age range is 5,000-3,000 BP, but it is quite possible that at least some are very considerably older than this.

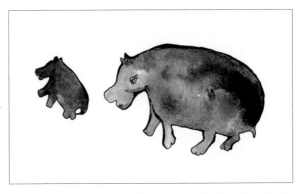

Hippo in desert, Tassili-n-Ajjer, Algeria (7000 BP).

Pang Ma Pha District

At least eight small groups of rock paintings are known from Pang Ma Pha, scattered right across the area. A wide variety of styles are represented, including hand prints, human figures, animals and abstract designs. All of the paintings are in red, except for a single picture of an animal in Tham Lod which is the only known example of a three-coloured (red, white and black) painting in Northern Thailand but is probably of recent origin. The compostion of the paintings from Tham Sri Sophorn are also unique but are likewise almost certainly not prehistoric in

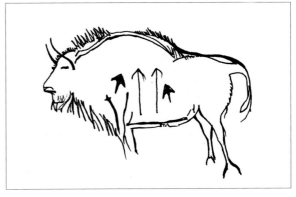

Bison, Niaux cave, France (Late Pleistocene).

origin. Rock paintings are not normally in the same location as log coffins, the exception being Ban Rai Overhang. The only known hand prints are from Jabo Hands Cave and Pha Dtao Daeng which is surprising as this is normally the most common style in other areas. The Highland Archaeology Project surveyed and copied all eight groups of paintings in 1998.

Pha Phung (Mae Hong Son)

A secluded overhang with an isolated group of about a dozen paintings dotted over a 5 m long section of wall. The site may have been used for habitation, although no archaeological excavations have yet been carried out to confirm this. Today, Pha Phung is one of the most important honey collecting cliffs in the area, with an elaborate network of bamboo scaffolding. The paintings depicted are all executed in red and include a number of seemingly abstract designs, one of which may represent the sun. Another could represent a bee, which would indicate that the site has been used for collecting honey since prehistoric times. At least three of the pictures depict an animal with long ears, thick tail and three toes. This creature does not seem to correspond to any known animal, although interestingly it is also depicted at other honey-collecting sites over 100 km away. It has been suggested that it is a kind of bear, since they are fond of honey.

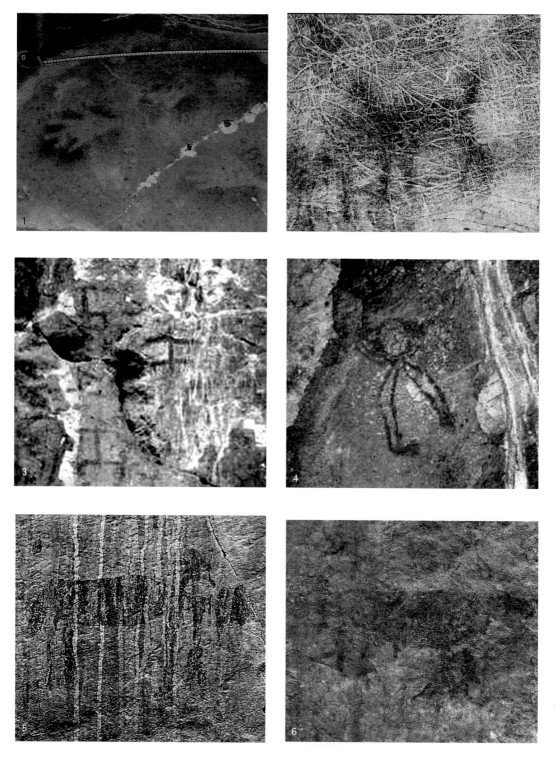

1-4 Pang Ma Pha: 1 Jabo Hands Cave; 2 Tham Lod (probably recent); 3 Dancing Man Cliff; 4 Tham Srisophorn (probably recent). 5-8 Pha Phung, Mae Hong Son. (© 1, 3 and 4 Highland Archaeology Project)

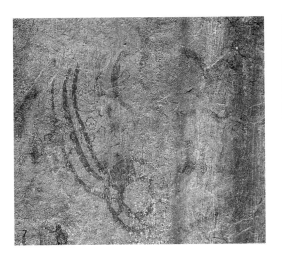

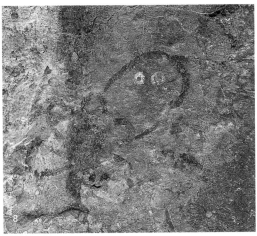

Pha Mai (Hod District)

Another honey-collecting site with an interesting group of red and black paintings. The paintings cover an area 5 by 2.2 m, about 2 m above the ground. Many human figures are shown gathered around triangular objects which could be bees' nests. Several of these objects have irregular gaps in them, such as those made in a bees' nest by honey collectors. The same curious three-toed animal depicted at Pha Phung is also shown.

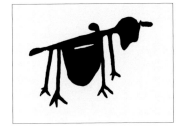

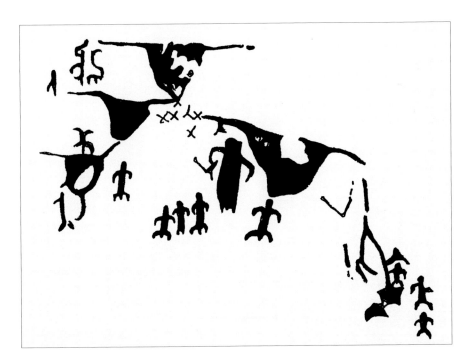

Pha Mai, Hod District.

Phra Tu Pha (Lampang)

Phra Tu Pha archaeological site is situated under an long overhang at the base of limestone cliffs on the west side of Doi Phra Tu Pha. The site is a few hundred metres south of a natural gap in the mountain range forming the border between Mae Mo and Ngao Districts, which has probably been an important travelling route since prehistoric times. Phra Tu Pha harbours the largest collection of prehistoric rock art in Thailand and is one of the few sites where a detailed archaeological survey of the floor deposits underneath the paintings has been conducted. Three survey pits, 2 x 7 m, 2 x 7 m and 3.5 x 3.5 m, were excavated in 1998 by Chiang Mai National Museum. Seven human bodies and associated mortuary objects were found, indicating that the site was used as a burial site at the transition between the Neolithic (New Stone Age) and the Metal Age, 3,000 years ago. The bodies were placed on bamboo mats and wrapped in plant fibres and bamboo matting which remarkably well-preserved. Pottery decorated with a stipple patten, bamboo baskets and a variety of polished stone tools including axes and arrow heads were placed with the bodies. The range of artifacts shows clearly that these people practiced both hunting and agriculture.

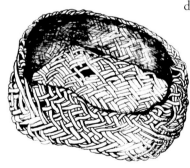

Pottery with stipple pattern.

The rock paintings themselves are located along a 300 m long section of cliff. They seem to be divided into seven distinct groups, but this may be due to the natural layout of the cliff and the suitability of the rock surface rather than to any preconceived design. A total of 1872 individual paintings have been identified, including 930 hand prints as well as human figures, scenes from everyday life, animals and abstract designs. No other known site in Thailand has such an impressive number of paintings or such a unique range of styles.

All of the paintings are in red monochrome – there are no indications that black, white or other colours were used on any of the paintings. The archaeological excavations under the paintings discovered lumps of red ochre with signs of human use, which were almost certainly one of the ingredients in the paint. At least some of the paintings seem to have been executed with a brush, as plant fibres and animal hairs were discovered mixed in with the paint.

Bamboo basket in remarkably good condition, 3,000 BP.

Hand prints are by far the commonest type of painting, completely dominating certain sections of the cliff and in many cases seem to have been overlain on top of older drawings. The majority of the prints were directly printed on to the rock – the person's hand was coated with paint and then pressed against the wall of the cliff. Several of the hands are encircled by a variety of markings such as circles, crosses and dashes which probably have a symbolic significance. In a few cases the hands are drawn onto the rock rather than directly printed. Another technique was to spray paint around a hand pressed against the wall, creating an inverse outline of the hand. This method was commonly used in other sites in the region, although it is very rare at Phra Tu Pha.

Although the paintings have not been directly dated, several pieces of evidence point to the conclusion that at least some of them were made at the same period as the organic remains found in the excavation. The lumps of red ochre found in the same layers as these remains is strong circumstantial evidence. In addition, some of the artifacts found, such as arrow heads, seem to be represented in the drawings. However, there seem to be many layers of painting on top of each other, in a variety of different styles. This might mean that the paintings were done over a long period, and we do not yet know if the dated archaeological remains fall at the beginning, the middle or the end of this range.

Several scenes from the paintings depict groups of human figures who seem to be engaged in everyday activities such as hunting, herding domestic animals and planting crops. Other scenes are more enigmatic and may relate to ceremonial activities or symbolic events. It is also possible that none of these scenes actually represent real life, but may be depictions of spiritual experiences under a state of trance, or that they are episodes from myths and folklore.

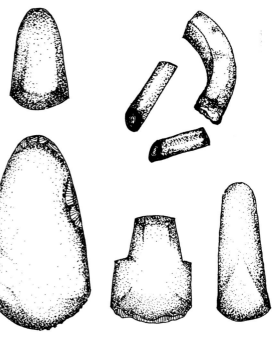

Polished stone tools, hollow stone beads and fragment of stone bracelet.

The rock paintings at Phra Tu Pha give us an invaluable insight into the natural environment of the region 3,000 years ago. Many of the animals depicted are clearly recognisable, such as elephants, serow, wild cattle, peacocks, monkeys, barking deer and squirrels. All of these animals were probably still found in the region until recent times, suggesting that the natural environment may not have changed substantially over a long period. Over the last 100 years much greater changes have taken place than during the previous 3,000 years and over half of these animals are now extinct or critically endangered in this area. A number of the animals represented do not seem to correspond to any living animals, or to any known fossil animals from this region. A long-necked animal with horns is very clearly depicted, in the same style as two elephants shown close by. These animals could be mythical creatures of spiritual significance to prehistoric people, but it is also possible that they are in fact real animals which have since become extinct.

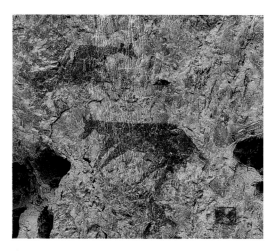

Two serow, or mountain goats. Serow are still found on limestone peaks in Northern Thailand.

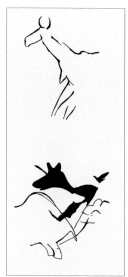

Large line drawing 2m high of an unidentified animal with long neck and horns. The bottom part of the drawing appears to represent two elephants. These original line drawings have been overlaid by hand prints and other paintings, such as the solid unidentified animal in the centre.

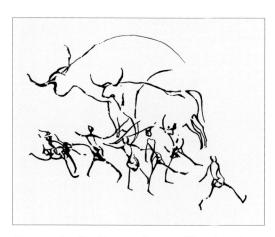

Seven people with sticks surronding at the point of capture.

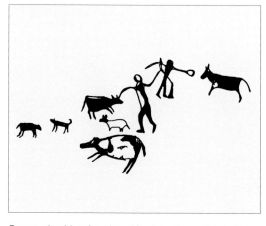

Two people with a ring-shaped implement and sticks in their hands trying to catch some domestic cattle. One of the smaller animals has its tail in the air and is probably a domestic dog.

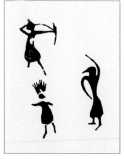

Three dancing figures and two people on a swing? The meaning of these paintings is unclear and may be mythological.

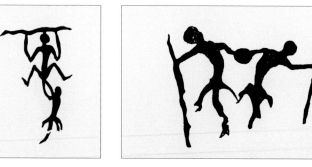

Two people wrapped with cloth holding long sticks and trying to take a round object from each other, possibly representing some kind of ceremony?

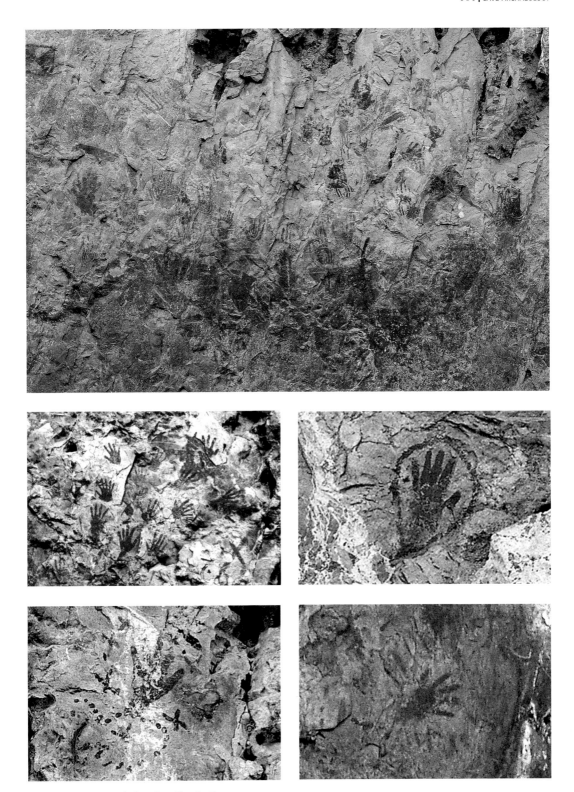

Variety of handprint paintings from Phra Tu Pha.

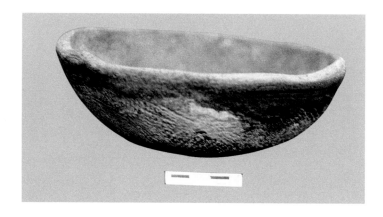

Caves as Burial Sites

Caves and overhangs in Northern Thailand have been used as burial sites for over 10,000 years. Although there are some examples of burial sites inside caves, the majority have been found under rock overhangs. The oldest known human skeleton in the region was found by the Highland Archaeology project at Tham Lod Overhang in 2001. This burial site dates from the Late Pleistocene (13,640 BP) and is the earliest confirmed human use of caves or overhangs in Northern Thailand. Almost all other

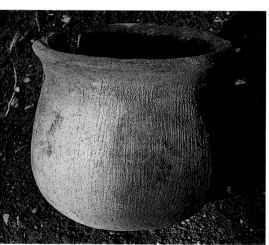

known burial sites are much more recent and date from the Middle to Late Holocene, which corresponds with the transition between the Neolithic (New Stone Age) and the Metal Age in this region. During this period, people started to change from mobile hunter-gatherer groups into more sedentary agricultural communities and many of the human skeletons were buried with pottery, bronze ornaments and iron tools.

The pattern of mortuary practices varies from area to area, even within Northern Thailand. The status and wealth of the individual also undoubtedly played a significant role in the type and quantity of artifacts placed with the body. A burial site dated 5,000-3,500 BP in a cave at Doi Pha Chi (Mae Moe, Lampang) includes many human bones together with 'cord mark' pottery, polished axes and ornaments

Top and Above: 'Cord mark' pottery.
(Top: Highland Archaeology Project)

made from beads, shells and animal teeth. This particular cave is probably a secondary burial site, where the bones were dug up from another site and placed here together with the ornaments. The artifacts found indicate that this was a Stone Age agricultural society that was not yet acquainted with metal.

In contrast, the site above Ob Luang Gorge in Chiang Mai Province was probably a primary burial site where the body was placed directly into the grave. The site is not in a cave but was positioned close to Pha Chang Overhang. The skeleton was buried in the lying position with his head to the west and was decorated with bronze bracelets and shells. Several pottery vessels were placed around the body. The site has been dated as 3,500-2,500 BP.

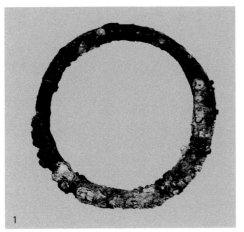

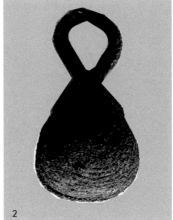

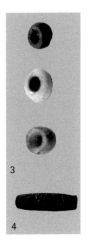

At Phra Tu Pha Overhang in Lampang Province, seven human skeletons were found, including both primary and secondary burials. For primary burials, the body was wrapped with plant fibre and buried with pottery and bamboo wickerware. For secondary burials, the human bones were placed in a wooden coffin after exhumation.

A distinctive type of mortuary practice in caves and overhangs in Northern Thailand is the use of log coffins. It is unclear whether these coffins were used as primary or secondary burial sites. The coffins were placed above the ground and are often found in association with pottery and iron tools. The greatest concentration of log coffins is in Pang Ma Pha, where they have been scientifically dated as 2,200-1,200 BP.

Examples of artifacts found associated with burial sites in Pang Ma Pha:
1 and 2 Bronze ornaments.
3 and 4 Beads.
5 Iron tools.
(© All photos Highland Archaeology Project)

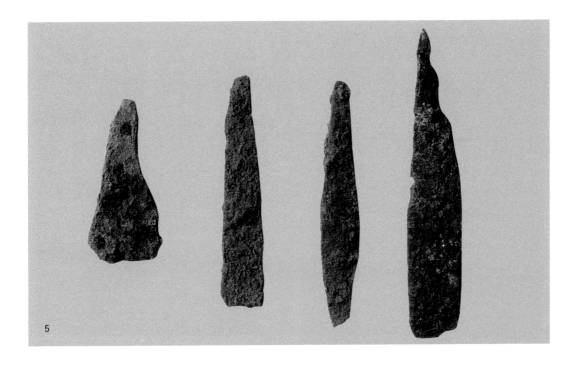

Ban Rai Overhang (Pang Ma Pha)

Ban Rai Overhang is a large horseshoe-shaped overhang perched 200 m above the Nam Lang River at the base of 30 m high limestone cliffs. The site is 900 m above sea level and faces north towards the river. The surrounding forest is disturbed mixed deciduous/evergreen with bamboo and there is still some natural teak (*Tectona grandis*). The amphitheatre enclosed by the overhang is 142 m long by 105 m wide with a slight depression in the centre which supports a small patch of dense evergreen forest. Ban Rai is unique amongst archaeological sites in Pang Ma Pha because it is the only place where a prehistoric habitation site, log coffins and rock paintings have been found together in the same location.

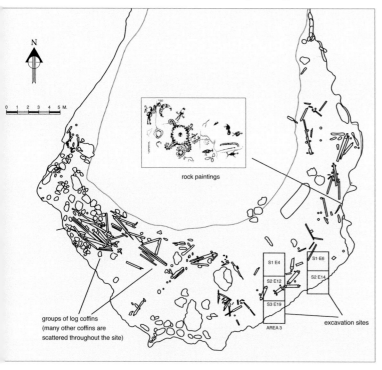

Ban Rai Overhang, showing excavation sites, log coffins and rock paintings. (© Highland Archaeology Project)

The most obvious character of the site is the abundance of log coffins and columns scattered through- out the flat area. A total of 15 coffins and 95 columns can still be seen, which is one of the largest collections known from a single site. However, the overhang does not completely protect the coffins from the rain and many of them are in poor condition. In addition, some coffins have been burnt for firewood or vandalised. Human bones, flake stone tools and fragments of pottery were found in the floor deposits in the vicinity of the coffins. Several recesses in the wall of the overhang have prehistoric paintings in red ochre depicting human forms, animals and symbols.

The Highland Archaeological Project, headed by Dr. Shoocondej from Silpakorn University, excavated the site in 2001-2002. Nine 4 x 4 m pits were dug in three separate locations at various distances from the back wall of the overhang. Analysis of the results revealed that the site had been used by humans between the late Pleistocene and the late Holocene (10,600-1,200 BP). At the beginning of this period, the site was predominately used for habitation, although one human skeleton was found dated 9720 ± 50 BP, indicating occasional ceremonial use of the site. After a long period of no apparent use between 6,000 and 3,000 years ago, the overhang was again used as a ceremonial site during the log coffin period which ended around 1,200 years ago.

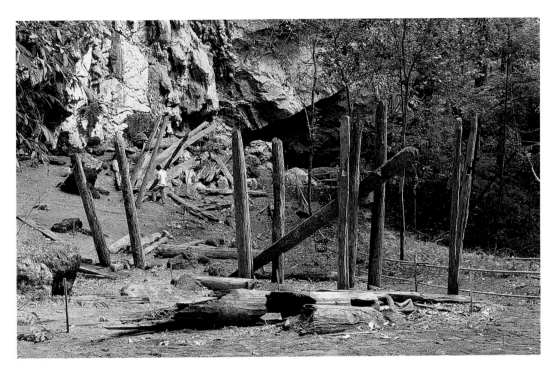

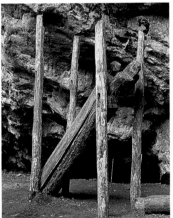

Top and left: *Log coffins in Ban Rai Overhang.*
Below and bottom: *Archaeological pit S2E14 where a 9,700 year old human skeleton was found.* (© *Highland Archaeology Project*)

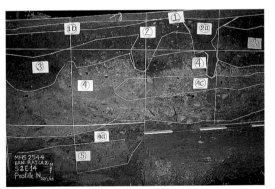

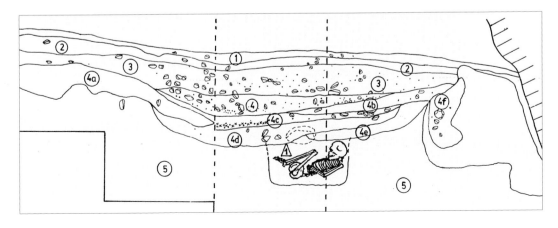

Cultural layers in Ban Rai Overhang

10,600-7,250 BP (Late Pleistocene). During this period the overhang was used as a habitation and tool production site by hunter-gatherer groups. Signs of fire hearths, bone tools and charred animal bones were found as well as large quantities of flake stone tools in an 80 cm thick band. The animals included a wide variety of both large and small mammals (wild cattle, deer, monkeys, squirrels) as well as turtles, fish, river shells and land snails. The abundance of stone flakes (the chips knocked off from the stone core when making flake stone tools) and unfinished stone cores suggests that the overhang was used as a stone tool production site. The vast majority of the stone tools were made from river pebbles, despite the fact that the overhang is a steep 200 m climb up from the river bed. We can only speculate as to why these people carried tons of river stones half way up the mountain rather than process then into tools nearer the stream, as seems to have happened at the Tham Lod Overhang. In area 2 (pit S2E14) a male human skeleton dated 9,720 ± 50 years BP was found in a half-lying position with knee and arm bent inwards towards the body in a circular pit. No ornaments were found associated with the body.

6,000-3,000 BP (Middle Holocene). hardly any archaeological evidence was found relating to this 3,000 year period. Possibly the site was abandoned for unknown reasons or the entire region was depopulated, although this seems unlikely.

2,500-1,200 BP (Late Holocene). The overhang was used by a group of people who used metal and built log coffins. The site was probably only used for burial and other ceremonial purposes as no signs of habitation have been found dating from this period. Human bones, cord pattern pottery, iron tools and bronze ornaments were found as well as the log coffins. The coffins themselves have not been dated but probably fall in the range 2,200-1,200 BP.

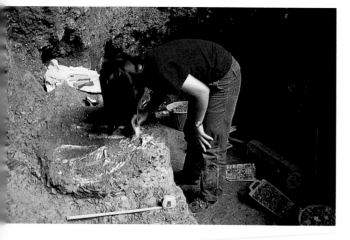

Excavating a human skeleton in Ban Rai Overhang.
(© Highland Archaeology Project)

In addition to the excavation pits, 122 core samples were collected from the log coffins and columns in order to study the paleoenvironment. However, no firm conclusions can be drawn from the data because the tree ring referencing sequence for this region has only been established for the last 300 years whereas the teak wood in the coffins is at least 1,000 years old. The lack of meaningful results has led many people to question the necessity of taking so many samples without any clear idea of what could be done with the results.

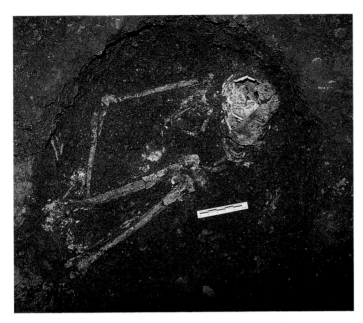

Human skeleton found in Ban Rai Overhang, dated 9720 ± 50 BP. (© Highland Archaeology Project)

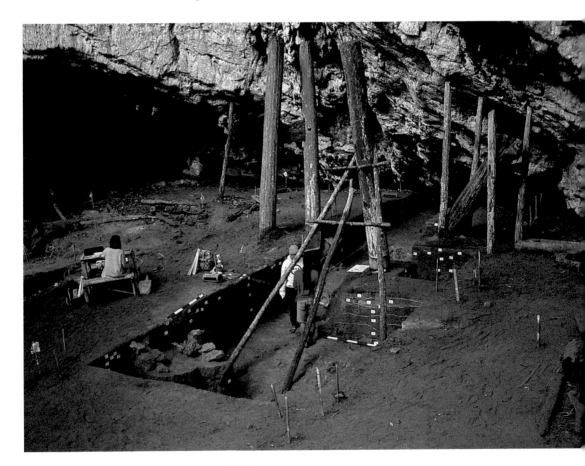

Excavation site in Ban Rai Overhang 2001-2002. (© Highland Archaeology Project)

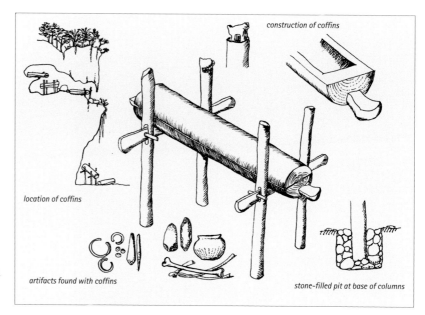

construction of coffins

location of coffins

artifacts found with coffins

stone-filled pit at base of columns

Log coffins in Pang Ma Pha.
(Adapted from Highland
Archaeology Project)

Log Coffin Sites

Log coffins, known as "long piman" in the local dialect, are a distinctive feature of many overhangs and cave entrances in Northern Thailand. The greatest concentration of these remarkable artifacts is in Pang Ma Pha where over 80 sites have so far been found. A much smaller number of log coffin sites are also known from Pai, Chiang Dao and Mae Sariang Districts. However, these areas have not been intensively surveyed so it is possible that many more sites are still waiting to be discovered. Log coffins resembling those in Northern Thailand have been found throughout Southeast Asia from Southern China and Vietnam down through Western Thailand to Borneo and the Philippines.

Samples of nine coffins from Pang Ma Pha have been dated as between 2,200 and 1,200 BP, using Carbon-14 techniques. Most of the coffins were made from teak wood, although there are a few pine coffins and a single known example made from *Eugenia*. The logs vary in size between 1,m and 9,m in length and resemble hollowed out log canoes, for which they were mistaken by early explorers. Each individual coffin consists of two parts, a base and a lid, which were typically constructed using iron tools. A single large log was cut into two lengthways, the inside scraped out and the two pieces placed back together again. In some cases a rebated edge was used to ensure a tighter fit (see photo 5). Most of the coffins in Pang Ma Pha are carved at both ends into distinctive patterns. More than 30 carving styles have been recorded, ranging from simple geometric designs to much more complicated patterns resembling animal heads. There is no apparent relationship between the style and the age of the coffin. In Pang Ma Pha, most of the coffins are raised up from the floor using a variety of column and beam techniques. Many of these support structures are extremely well made with peg joints and beams carved to match the curvature of the base of the coffins. The foundations for

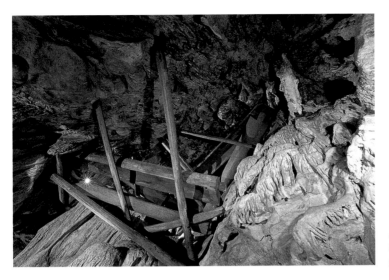

Log coffin on suspended cross beams resembling scaffolding (Tham Pha Daeng Scaffolds).

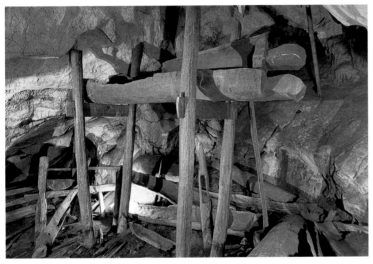

Log coffin on free standing columns and beams (Tham Bor Krai).

some of the columns were prepared by digging holes 1 m deep and filling in with large stones.

Pottery, beads and iron tools have also been found in log coffin sites and may have been burial goods associated with the coffins. In Nam Lang Cliff Cave, over 1,700 fragments of pottery were recovered in the vicinity of a single coffin. In a few sites these artifacts are still inside the coffins but in most cases they are scattered over the floor of the cave and their association with the coffins is not certain. Fragments of human bones and teeth have also been found with the artifacts, both inside coffins and on the cave floor. In some case the remains of several different people have been found in the same coffin. Researchers from the Highland Archaeology Project were able to conclude that the bones and teeth resembled those found in Southeast Asian people today, based on 500 samples from 17 sites in Pang Ma Pha. Further studies will try to extract DNA material from the samples in order to analyse the genetic background of these unknown people in more detail.

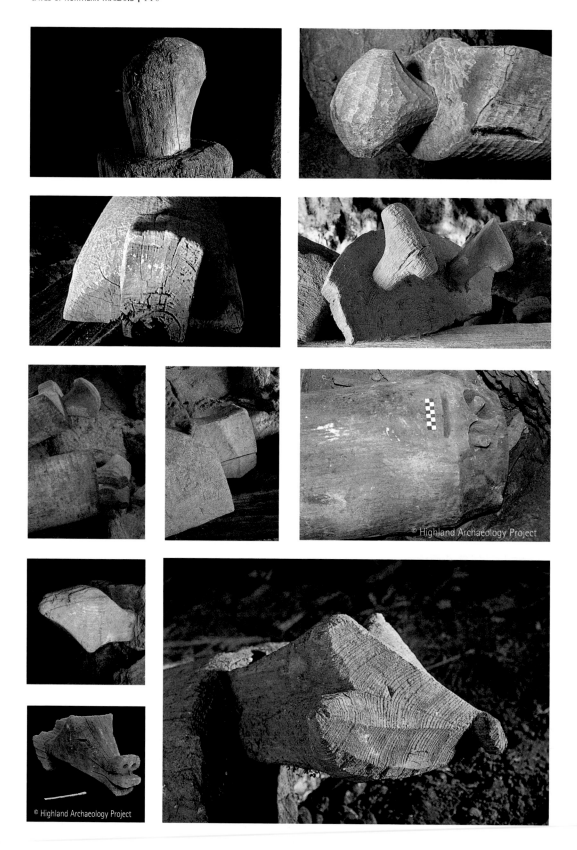

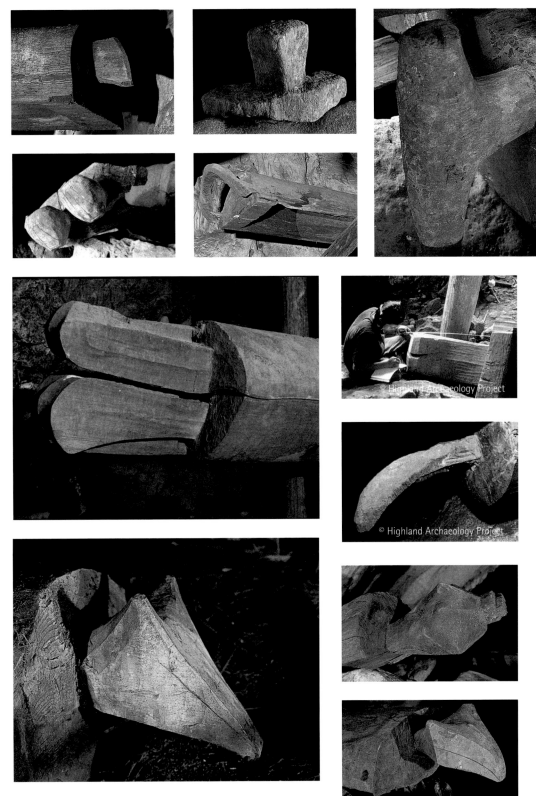

Variety of coffin head styles from Pang Ma Pha and Chiang Dao Districts.

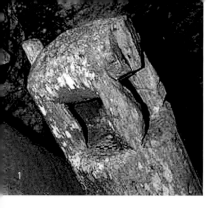

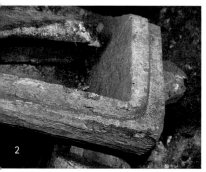

1 Carving at the top of a column
(Tham Long Yak).
2 Rebated edge on a coffin to ensure
tighter fitting of the lid
(© Surapong Lertthusnee).

The log coffin sites have been studied by both Thai and foreign researchers over the past 10 years but there is still no general consensus on their interpretation. Who made these coffins and what happened to them? Did they live in mobile hunter-gatherer groups or in settled agricultural communities? What beliefs did they hold and what kind of burial rituals were associated with the coffins? None of these questions can be conclusively answered without a great deal more archaeological survey work. Although most authorities agree that the log coffins were probably associated with burial rites of some kind, considerable doubt remains as to whether the caves were primary or secondary burial sites. In primary burial sites, the human corpse is placed directly in the coffin, whereas in a secondary burial the corpses are first allowed to decompose by being buried underground or by other means. Only once all the flesh has decomposed are the bones are transferred to the coffins, which perform the function of gigantic mortuary urns.

The Highland Archeology Project team have speculated that the communities who made these coffins could have been the same group of people as the hunter-gatherers who occupied the caves in the late Pleistocene. About 5,000-3,000 years ago these people may have moved to the flat plains, coinciding with the development of agriculture. The team is now exploring these flatter areas but have not yet managed to find any evidence of cultivation or habitation dating from that period. It will probably be many years before we can piece together the full story of this fascinating group of people.

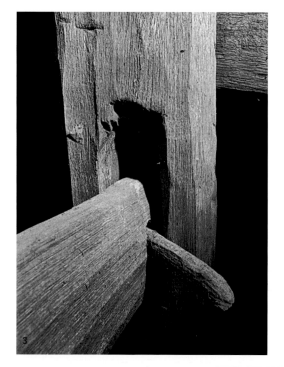

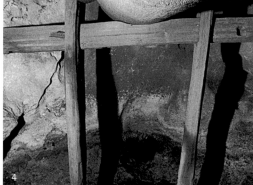

3 and 4 Peg joints used to connect columns and beam
(Tham Pha Daeng Outflow).

Damage and Threats to Log Coffins

Log coffins which have lasted for thousands of years relatively undisturbed by natural or human interference have recently come under increasing threat. Over the last ten years a significant proportion of this unique cultural heritage has been irretrievably damaged. The photographs on this page illustrate both the variety and severity of the destruction. In addition to the coffins themselves many associated prehistoric artifacts of incomparable archaeological value have been damaged or removed by local people, unwitting tourists and unscrupulous traders from outside the area. Pottery, metal artifacts and beads have been indiscriminately smashed or stolen for private collections. If we do not conserve these sites, there will be no evidence left for future generations to study this local history which is part of the priceless cultural heritage of Thailand.

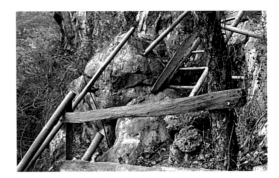

Tham Long (Chiang Dao). Part of the steps on the nature trail built to take tourists to see the coffins was constructed from the coffins themselves!

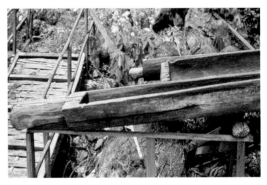

Pang Kham Piman (Pang Ma Pha). Several coffins have been taken out of the caves and put on display in the open air, where they will quickly deteriorate due to exposure to the elements. (© Highland Archaeology Project)

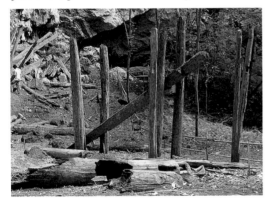

Ban Rai Overhang (Pang Ma Pha), showing 1,200 year old teak columns which have been burnt for fire wood. Many columns and coffins in other sites have suffered the same fate.

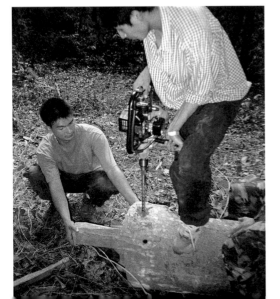

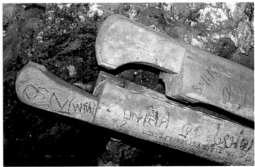

Tham Long Yak (Pang Ma Pha). Graffiti is a common form of vandalism which has defaced many coffins. (© Highland Archaeology Project)

A large number of coffins in Ban Rai Overhang and Tham Bor Krai were drilled by researchers in 2002. Bor Krai has the best coffins in the region with an unsurpassed range of styles. (© Highland Archaeology Project)

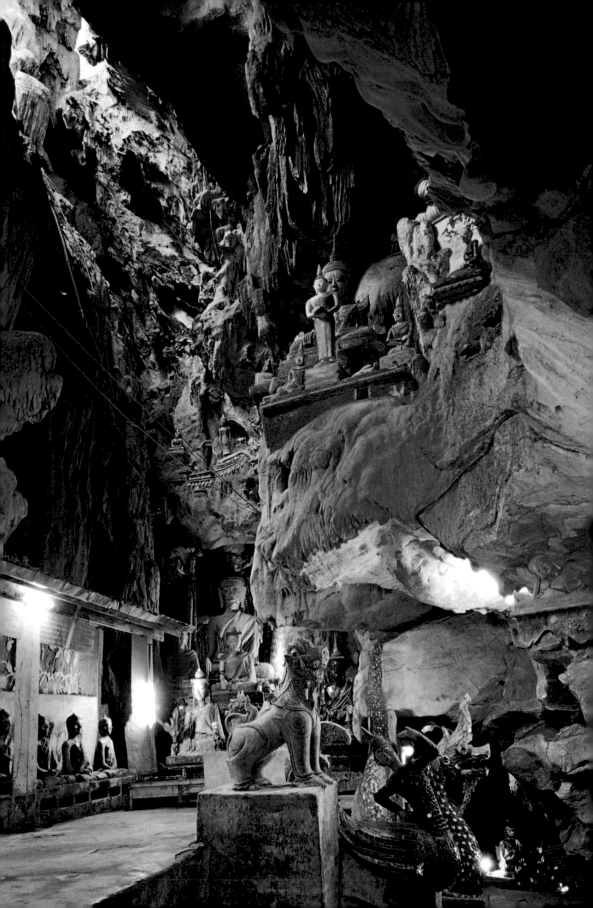

TEMPLES AND FOLKLORE

Caves are an integral part of religious practices and beliefs in Northern Thailand. A large number of caves in the region have some kind of religious significance or play a part in local folklore. Buddhist temple caves such as Chiang Dao and Tab Tao have probably been important centres of worship for many centuries. Over thirty percent of the caves in this book are temple caves or monasteries. Some of these temples are well established centres of learning with a large number of monks in permanent residence. Others are much more modest establishments maintained by a single monk. Numerous caves serve as temporary resting places for wandering monks or are still associated in local folklore with hermits who are said to have lived there long ago. Some caves have no apparent connection with monks at all, but still house some kind of religious imagery, such as a solitary statue of the Lord Buddha in the corner or a altar with offerings to appease the local spirits.

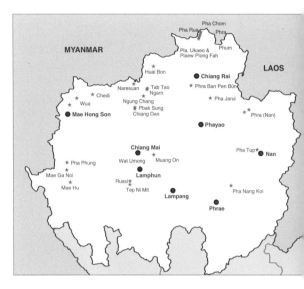

The Pre-Buddhist Era

The dominant religion throughout Northern Thailand today is Buddhism. However, many of the beliefs and superstitions surrounding caves probably date back to the pre-Buddhist era and have been absorbed into popular religious practice. Caves in the region, have been used as burial sites for over 10,000 years reaching a peak 1,200-2,200 years ago during the period of the log coffins (see chapter 4). Many caves are still believed to be inhabited by spirits, known locally as '*pi*', which may be related to this former use of caves as burial sites. In Pang Ma Pha, where the log coffins are still in evidence, this belief is particularly strong and cave spirits have a special name – 'piman'. The *piman* are said to much larger than normal people, perhaps because many of the coffins are 5 m or more in length. Local people believe that *piman* bring bad luck or even death to anyone who attempts to disturb their graves. The decline of the log coffin culture about 1,000 years ago coincided with the spread of Buddhism into Northern Thailand. It seems likely that the Buddhist preference for cremation as a means of disposing of the dead lead to the abandonment of caves as burial sites.

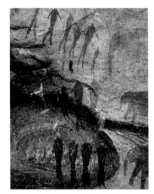

Rock painting of shamans in long cloaks, South Africa.

Several caves and overhangs in Northern Thailand contain rock paintings which may be over 3,000 years old. Such paintings are further evidence that these sites had a ceremonial or religious significance for prehistoric people in the area. In South Africa and Australia, the spiritual beliefs associated with rock paintings have been well documented through dialogue with the aboriginal

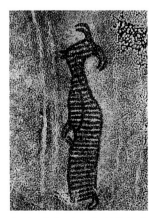

Rock painting from Krabi, S. Thailand.

peoples who still inhabit these areas, or did so until recent times. In South Africa, much of the rock art of the San people is associated with shamanistic practices designed to produce rain. Shamans are often depicted wearing long cloaks and in close proximity to animals, such as elephants, which are associated with rain. Figures in long cloaks are also found in prehistoric rock art from Krabi (South Thailand) and may represent shamans or other figures associated with the spiritual world. Unfortunately, folk memory of the meaning of these rock paintings has long been lost, so it is no longer possible for us to understand the symbolic meaning and wider spiritual or social significance of this art form.

Caves in Non-Buddhist Religions

Hinduism, or Brahmanism, was well established in Thailand by the 4th century AD and was probably the dominant religion over much of the region during the early Khmer period, 800-1,100 years ago. Many aspects of Hinduism were absorbed into Thai Buddhism and are still an important element in several royal ceremonies even today. Despite this influence, there are very few examples of Hindu caves in Thailand. One of the most interesting

Stalagmite wrapped with a 'Ji Won' cloth, Tham Ki Nok.

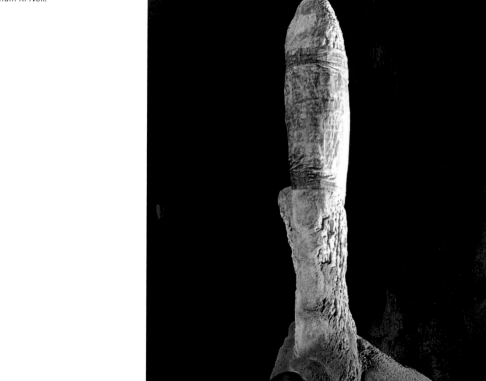

is the rock-hewn cave at Khao Khuha in Southern Thailand, dating from the 6-7th century AD.

In Northern Thailand, Hinduism does not appear to have been a dominant force and there are no known examples of this art form. It has been suggested that the reverence of certain cave stalagmites may be related to the worship of the phallic lingam of the Hindu god Shiva. These stalagmites may form the focal point of a cave and are draped with orange 'Ji Won' cloth to emphasize their sanctity. The enormous perfectly-shaped stalagmite which forms the central attraction of Tham Muang On is one of the best known examples, but there are numerous others, such as in Tham Ki Nok above Tham Chiang Dao. However, it does not seem necessary to attribute this phenomenon to Hindu influence, since the desire for fertility is a central concern in all societies and phallic-shaped stalagmites are a natural fertility symbol. An alternative view is that these stalagmites are revered for their resemblance to a Buddhist *chedi* or stupa. A curious revival of the association of caves and Hindu deities can be seen at Tham Tep Ni Mit, Lamphun, where the sumptuously decorated main chamber is dominated by a large statue of the Hindu god Shiva. His son, the elephant-headed god Ganesha, stands guard at the cave entrance next to a smooth black lingam. Christian and Buddhist iconography are also evident in this eclectic cave.

Statue of Shiva holding a trident, Tham Tep Ni Mit.

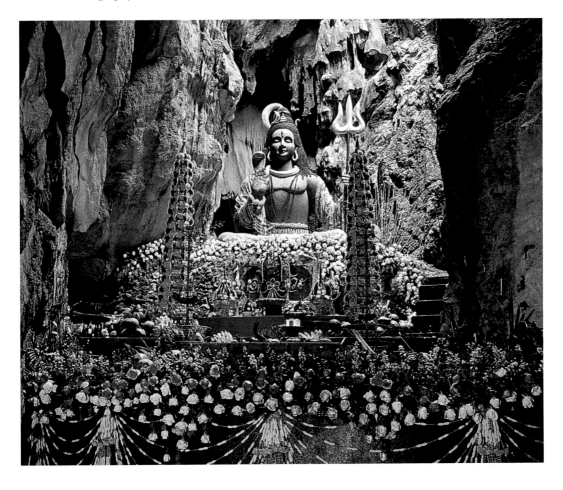

(Ajan Mun) "made up his mind to go to the cave and face whatever would happen there: instead of being overwhelmed by fear, he looked on it as an opportunity to develop his mindfulness.....after several nights, the demon appeared before him. He threateningly approached the venerable Ajan and said: 'if you don't leave this place, I will crush you to death with my huge club, which can crush an elephant to the ground with one blow.' The venerable Ajan communicated with him telepathically, and asked: 'Why do you want to kill me? I appear to have done no wrong. Why should I deserve such capital punishment, when I have not harmed anyone up here?' 'I have long been authorized to safeguard this mountain', answered the demon, 'and will not tolerate anyone who dares to challenge, or attempt to overpower me'. 'But I challenge no-one, nor did I attempt to overcome anyone', the venerable Ajan replied. 'It is just to challenge and overpower the defilements which rule human minds, that I have come here. If you are really someone who possesses skilful powers, as you have boasted about, do you have power beyond kamma and Dhamma, which are the great laws ruling the mases of beings in the three realms?' 'No' the Demon answered. 'The Buddha possesed the power of eradicating from his own mind the desire to dominate and harm others, do you have such power?' 'No!' confessed the Demon" *(from "The Venerable Phra Acharn Mun Bhuridatta Thera" pp. 177-178, quoted in Munier 1998)*

Caves and Buddhism

The association of caves with Buddhism goes back to the very beginnings of the religion, over 2,500 years ago. According to Buddhist tradition, the first council of 500 followers which took place after the Master's death at the start of the Buddhist era in 543 BC was held in Saptaparna Guha Cave near Raja Guha in present day Nepal.

The separation of caves from the outside world makes them particularly well suited for meditation and spiritual detachment. The Buddha himself almost certainly used caves during his long wanderings as an aesthetic in search of enlightenment and caves continue to play a important role as spitiual places for Buddhist monks in Thailand today. Many cave chambers possess excellent acoustic properties which help to amplify the chanting rhythms used by many monks during meditation.

Caves are often believed by lay persons to contain powerful and dangerous spirits which will certainly attack intruders or cause them to get lost and never return. Monks use caves to eradicate their personal fears and to refine their powers over the Self, which is an essential prerequisite to the ultimate goal of achieving enlightenment. The sanctity of individual monks is often believed to have been proved for lay people by their ability to overcome these obstacles. The encounter in Sarika Cave between Ajan Mun, one of the most famous monks in Thai history, and a demon well illustrates both the monk's sanctity and the importance of self-control.

During 'Kao Pan Sa', the agricultural season between August and October, monks are obliged by Buddhist Law to remain in one place. Many wandering monks choose to take up temporary residence in a cave, sometimes returning to the same cave year

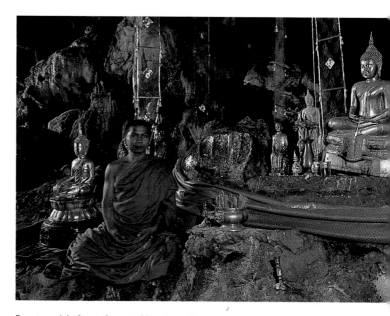

Forest monk in front of a sacred handprint, Tham Jak Dtor.

after year. During the rest of the year, monks are allowed to wander at will, which is seen as a means to cut connection with worldly possessions and desires. Caves are also used during this period as convenient resting places for shelter or an overnight stop.

History of Temple Caves

The first historical records of the use of caves as Buddhist temples is from the time of the Emperor Ashoka, 268-239 BC, the first Indian monarch who adopted Buddhism. During the next thousand years, Buddhism reached its peak in the Indian Subcontinent and there was a great increase in temple caves. The finest surviving examples are the Ellora and Ajanta caves, which were built over an 800 year period between 200 BC and 700 AD. In common with many temple caves in India, this group of over 1,000 individual chambers were hewn from solid sandstone cliffs and are completely man-made. Following the decline of Buddhism in India, the main centre of cave temple construction shifted to China, where many magificent examples survive in the northern interior along the former silk trade routes. The wealth of the trade supported large monasteries which lasted until the ascendancy of Islam in the region.

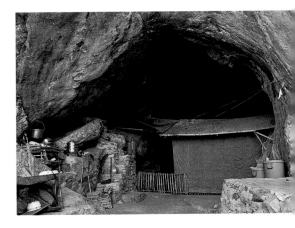

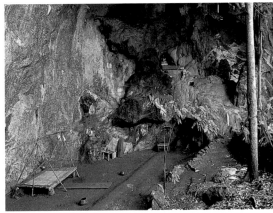

Tham Hu (above) and Tham Huai Bon (below) are inhabited by forest monks.

In Thailand, Buddhism was probably first introduced during the 3rd-5th centuries AD when Southern Thailand was under the control of the Buddhist Kingdom of Sri Vijaya which had its capital at Palembang in present-day Sumatra. The oldest surviving evidence of the use of Thai caves as Buddhist sanctuaries dates from the 6-8th centuries AD. A 6th century inscription in Tham Narai, Saraburi, relates links with the kingdom of Anuradhapura in Sri Lanka. Mon (Dvaravati) bas-reliefs of this same period can be found in Tham Phothisat (Tham Phra Ngam), Saraburi, as well as in four caves in the Khao Ngu range, Ratchaburi. These early artist expressions are from the Mahayana tradition, in contrast to the Theravada (Hinayana) tradition which is the dominant form of Buddhism in Thailand today. They are of immense cultural importance because very few examples of the early art of Thailand survived the Ayuthaya period.

In Northern Thailand, there are no surviving examples of rock carving or any other indications of the use of caves as temples from such an early period. This may be partly explained by the later date at which Buddhism reached the area but perhaps also by the lack of a stone-carving tradition in the region. The first concrete evidence of cave temples comes from the Chronicles of Doi Tung, which details donations made to the temple of Tham Pla-Tham Phum near Mae Sai. In 1,377 AD, King Phaya Ku Na of Chiang Mai gave fields, ten families of slaves, a gold embossed certificate, Buddha images and stupas to worship. In 1,525 AD, King Phaya

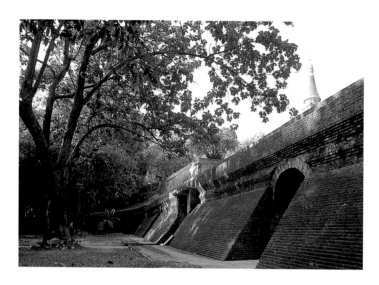

Wat Umong (Chiang Mai).

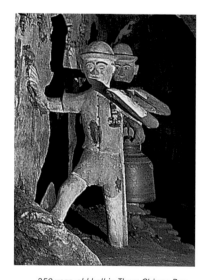

350 year-old bell in Tham Chiang Dao.

Ket of Chiang Mai gave a Buddha image, fields and three families to commemorate his son's birth. The size of the donations and the importance of the occasion suggests that Tham Pla-Tham Phum was already a well-revered site of political significance. The practice of giving fields and families of slaves to work them helped to ensure the long term viability of these remote temples which would have found it difficult to support a community of monks on daily donations of food alone.

The remarkable subterranean network in Wat Umong, Chiang Mai, was constructed between 1,380 and 1,450 AD. These passages represent the only example in Northern Thailand of the man-made caves which are so widespread in India. The inside of the passages are decorated with wonderful paintings of flowers and birds which are still discernable today. Many Buddhist writings refer to a mythical heavenly forest, 'Pa Hinmaphan' where celestial beings live surrounded by beautiful birds, flowers and caves. It seems likely that the paintings and even the cave itself is a stylistic representation of this magical place. According to temple lore, the caves were built by King Meng Rai because he wished to keep track of Phra Jan, a much revered monk whose opinion the king valued. This important site has recently been sensitively restored by The National Archaeological Department.

The oldest religious object from any cave in Northern Thailand which can be authenticated is a 350 year old bronze bell in Tham Chiang Dao. This 200 kg bell bears an inscription naming the benefactor who donated the money for its construction and the monk who consecrated it. No other temple caves in the region can be traced back further than 150 years through surviving artifacts, although there is little doubt that important temple caves such as Chiang Dao and Tab Tao have been revered for many centuries. Even in Tham Chiang Dao, there are no other surviving references to the cave itself until the 1880s when Phraya Intaphiban initiated an ambitious period of construction which included blasting the new entrance which is still used today. Prior to this period, the

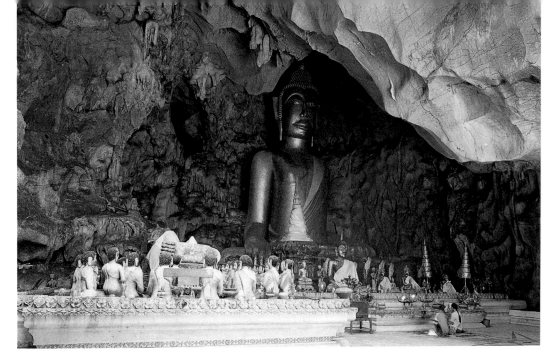

only entrance was through the cave window known as Plong Sawan (chimney to heaven), which involved a risky 30 m vertical descent on bamboo ladders. In 1913 the hermit Aukantha from Burma built the fish pond, the chedi near the entrance and the lying Buddha figure at the end of Tham Phra Non. Since that time, the temple has gone through so many periods of reconstruction and development that most of the alterations known to have taken place in the 1880s have since been obliterated or incorporated into subsequent designs. It is hardly surprising that so little evidence of older designs have survived this frentic activity.

The first written record of Tham Tab Tao comes from the journal of an early European Traveller, Carl Bock, who visited the temple in 1883. The vivid description in his journal makes it clear that Tab Tao was already an important and venerated site by that date. Another European traveller writing a decade later refers to the statutes deep inside the cave which can still be seen today. The large Buddha at the entrance is said to be 1,000 years old although there is no concrete evidence to substantiate this claim.

Outer chamber of Tham Tab Tao, described by Bock in 1884.

To the left of the entrance was a narrow niche or recess in which was a broken figure of Buddha, at whose feet the pious pilgrims had laid a collection of the characteristic clay decanters or water-jugs, pots, and jars of the country. The cavern itself was of interesting formation, the roof being a fantastic array of stalactites of various forms. High up in the centre of the roof, which was probably sixty feet in height, was a natural skylight, through which a dusky beam of light fell on the head of a gigantic recumbent figure of Buddha, in, as the legend runs, his 'dying attitude', lying on an elevated platform in the centre of the cave, but this light from above only made the rest of the great cave appear darker by comparison........ The great Buddha is made of brick-work, covered with a coating of thick varnish, and once heavily gilt, but now in a dilapidated condition........ All around the central god was an assemblage of figures, half life-size, of the same material as the great Buddha, representing disciples of the great Master, sitting and praying or listening, with uplifted folded hands. Every one of these smaller figures was covered with a great number of yellow rags and skull-caps on their heads, left by the priests as tokens of their devotion. At the feet of the gigantic idol lay another heap of clay pots and jars, with rice-trays containing rice, small wooden and stone figures of Buddha, brought all the way from the Ngiou States, and deposited there by worshippers; while heaped upon an adjacent altar was an immense collection of representations of Buddha, together with a curious assortment of priests' clothing, water-jars, streamers, tufts of hair, spittoons, and other odds and ends."

(Bock 1884, pp 288-291, quoted in Dunkley 1995)

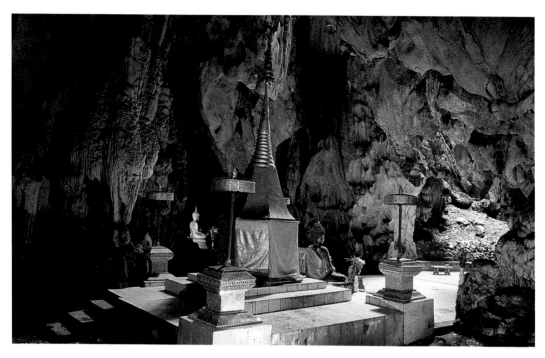

Chedi lit by both natural and artifical lighting, Tham Plaew Plong Fah.

Temples and Monasteries Today

Temple caves can be found all over Thailand and often form the dominant man-made feature in limestone landscape. Many temple caves are perched high above the plain at the end of a long flight of steps, which ensures that all visitors will gain merit through their exertions in reaching the cave. This elevated position provides a spectacular setting with fine views over the surrounding countryside and further emphasizes the distinction between the caves and the everyday world below. The majority of temple caves are dry caves rather than stream caves, although some of them have sacred pools or other water sources within the caves. Temple caves are often associated with a permanent spring, such as at Tham Pla near Mae Sai and Tham Chiang Dao, although only the dry higher levels of the cave can be visited.

Caves are often linked in the popular imagination with wilderness and vast tracts of forests. There are certainly plenty of natural wild caves in such places, but most temple caves are not situated in extremely inaccessible locations. This is not surprising given the reliance of monks on the lay population for food and the need to attract donations from visitors for construction and maintenance. It is no coincidence that the great temple caves of Chiang Dao and Tab Tao are both located close to one of the main trade routes through the region. Tham Pha Nang Koi is another important site just off the main route between Phrae and Nan, whereas Tham Muang On is in the most accessible limestone outcrop to Chiang Mai Town. As has already been mentioned in China temple caves seem to particularly flourish along trade routes. In the days when travel was still fraught with risk, the strong association between temple caves and the natural environment may have been thought to provide the special protection needed by travellers on their journeys through the wild countryside.

Each temple cave is unique both in terms of the physical environment of the cave and the design of the man-made component. Religious objects are typically located in the first sections of the cave where there is at least some daylight, although the most sacred objects are sometimes hidden much further into the cave in the complete darkness. The placing of statues within a chamber is directed as much by the physical properties of the chamber as by religious precepts. The majority of chambers have a single Buddha statue or a chedi placed in the most prominent position, usually on the far side of the chamber and facing the main entrance or line of approach. This focal point is typically raised on a platform with a flat terrace in front for worshippers to prostrate themselves. Other Buddha figures, hermits, chedis or mythical creatures may be located throughout the rest of the chamber, either in rows around the perimeter or in natural alcoves. Advantage is usually taken of natural lighting, such as cave windows, to provide a dramatic atmosphere for the scene. The deep shadows in the further recesses of the caves help to further enhance the air of mystery and other-worldliness. Tham Plaew Plong Fah near Mai Sai is a particularly fine example where rays of light from a narrow slit high in the ceiling illuminate the rich golden colour of the main Buddha which is further enhanced by the deep shadows behind. The advent of electricity has greatly increased the range of lighting possibilities for cave temples, although this has rarely been used to good effect.

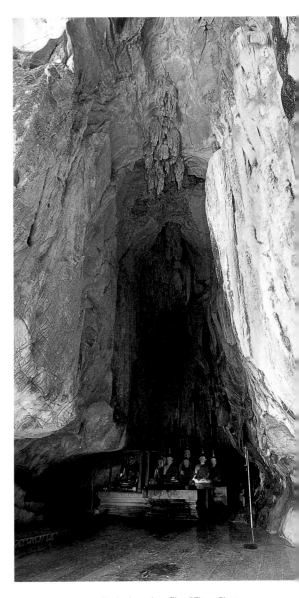

Flask-shaped profile of Tham Phum.

Religious Symbolism

A wide variety of objects are venerated in caves for their shape, beauty or supposedly miraculous powers. What is seen in temples today is a compromise between the austerity of the Dhamma, with its emphasis on discipline and meditation, and the popular desire for magic and mystery.

The natural form of the cave itself can be regarded as symbolic of inwardness, the eradication of Self and detachment from the external world, which are key concepts in Buddhism. The effectiveness of the cave as a receptacle of cosmic energy is enhanced by symmetry and resonance of form. The graceful flask-shaped cross-section of Tham Phum, Chiang Rai is a particularly fine example. The flask motif also re-occurs in stories surrounding the legendary visit of the Buddha to the site.

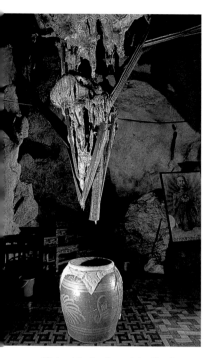

Water is an important element in Buddhist caves and plays a major role in channelling the natural energy of a cave, often forming a focal point for religious devotions. In Wat Tham Pla-Tham Phum the legendary visit of the Buddha to the site culminated in the release of fish into a sacred pool emerging from the cave. In Tham Bor Nam Tip near Nan, the water from a small gour pool hidden in the inner recesses of the cave is said to be endowed with miraculous properties. It is predicted that the pool itself will never run dry, despite the large volume of water drunk by countless visitors eager to partake of its secrets. Water dripping from stalactites is also considered to have magical properties and is collected in many temple caves for distribution amongst the faithful.

Many natural features such as stalagmites, columns or curiously shaped rocks are venerated due to their resemblance to Buddhist symbols, mythical creature or real animals. The respect paid to phallic or chedi-shaped stalagmites has already been mentioned. The long list of venerated objects in Tham Chiang Dao includes natural features thought to resemble an elephant, a lion, a rhino, an owl, a lotus, a cloud and many others.

Water dripping from stalactites is believed to possess magical properties, Tham Pbak Sung.

"When Kakusandha, the first Buddha, came, there was no water to drink. After his alms meal was finished, the four gods of the four directions gave him four jars of water to drink. To thank them he gave them a palace which is said to be on the top of a hill. He also created a pond and freed fish in the pond as a witness."
Another version of the story states that the jars which were given to Kakusandha were wrongly thought to contain alcohol and were stolen, but were later miraculously brought back.
(quoted in Munier 1998)

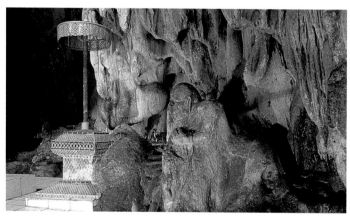

Stalagmite decorated to resemble Pu-tai (Metteya).

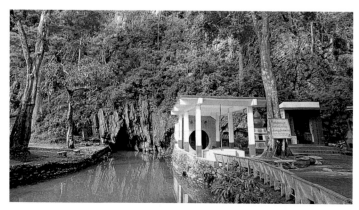

Sacred pool at outflow of Tham Pla (Chiang Rai).

Manifestations and Relics of the Lord Buddha

The most sacred objects are those associated directly with the Lord Buddha himself. The most sacred site in the whole of Thailand is the footprint of the Lord Buddha at Phra Phuttha Bat between Saraburi and Lopburi in Central Thailand. This highly venerated impression in the rock is believed to be an authentic footprint of the Lord Buddha himself, known as a "Paribhoga" footprint. Only five such footprints are thought to exist in the world – the other four are in Afghanistan, India and Sri Lanka. Most other footprints are of the "Udessika" type, which are objects of worship made in memory of the Buddha. Another footprint ascribed to the Buddha at Sa Morakot, Prachinburi, has been a revered site since the 8th century AD. An inscription 3 km from it dated 761 AD is written in Pallava script, establishing a link with Sri Lankan Theravadin communities.

In Northern Thailand, Wat Phra Phuttha Bat Tak Pha, Tham Muang On and Wat Tham Pla-Tham Phum are all believed to have been visited by the Buddha. Any objects which relate to this event are endowed with great sanctity. The temple at Phra Phuttha Bat Tak Pha near Lamphun takes it's name from some coloured marks on the rock believed to have been made by the Buddha's

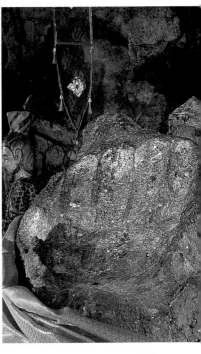

Above and Left: *Handprint, Tham Jak Dtor.*

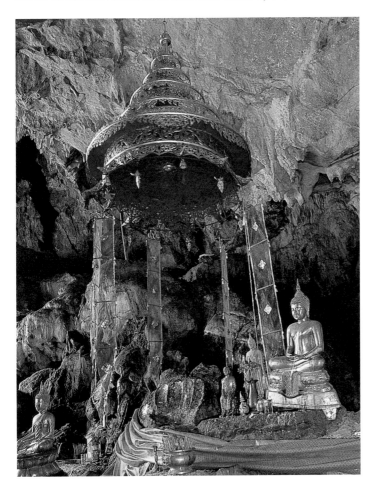

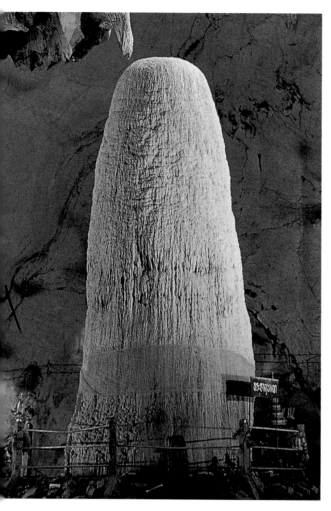

*Stalagmite said to contain a hair
of the Buddha, Tham Muang On.*

robes when they were spread out to dry. At
the same site, a 2.5 m long impression on
the rock is believed to be a footprint of the
Buddha himself, whereas a similar 1.8 m
impression close by is attributed to his
disciple, Ananda. Several other footprints
of lesser personages are also objects of
worship at this site, including those of a
child saint, the hermit Wa Suthep and Phra
Kru Ba, a famous former abbot of the
temple. Several other caves contain
markings on the rock which resemble
footprints or handprints and have become
objects of reverence. Tham Jak Dtor in
Mae Hong Son has a remarkably realistic
handprint beneath a gilded canopy, as
well as a gour pool shaped like a double
footprint.

Tham Pla-Tham Phum is said to have
been visited not only by the Buddha
Gotama but also by several previous
Buddhas and it is predicted that the site
will be visited by Ariyametta (Metteyya),
the Buddha of the Future. All four Buddhas
drank the water from the spring, which is a
symbol of continuity and a sign that the
Dhamma is kept alive in this place: The
only objects which are thought to be
directly related to these visits are the pool
at the mouth of the cave and the fish, which
are naturally considered sacred. The
Chronicle of Phra That Doi Tung states that
the Buddha said that Tham Phum was to be
guarded by monkeys. Consequently, one hundred years after the
death of the Buddha, a legendary king released two monkeys there.
After time, the population of monkeys was said to have reached
500, the number of arahats that accompanied the Buddha when he
went to the cave. Statues of monkeys have been carved along the
stairs leading to Tham Plaew Plong Fah and large troops of
monkeys reside in the temple grounds, giving the cave it's
alternative name – Tham Ling, or Monkey Cave. At nearby Doi
Tung, the Buddha's alms bowl is thought to be embedded in the
rock and the shadow of the mountain at sunset is said to resemble
his robes.

At Tham Muang On, the Lord Buddha is said to have visited
the cave during his lifetime with his disciple Ananda. A local *naga*
turned into a person and gave them alms and they stayed for three
days and three nights. Upon hearing this, many other *nagas* in the
area came to visit the Buddha and he gave them one of his hairs as
proof of his visit. The king of the *nagas* encased the hair inside a
magnificent stalagmite which is the principal object of worship in
the cave today.

Statues of the Lord Buddha

The most prominent and important feature of most temple caves is a statue of the Lord Buddha which often forms the focal point for the entire cave. Many caves have several large Buddhas positioned in different chambers or in alcoves within the same chamber, as well as many smaller statues donated by individual benefactors.

The dominant statues of the Buddha in caves tend to be depicted in a slightly different range of poses than those found in buildings. The poses most frequently seen in caves are the "Meditation" position (with both hands placed one on top of the other in the lap), the "Calling the Earth to Witness" position (with the right hand pointed towards the ground) and the "reclining" position. Although both of these poses are also fairly common in temple buildings, they represent over 90% of all Buddha statues depicted in temple caves. The smaller Buddha statues are often depicted in a much wider range of poses. Many benefactors choose to donate money for a statue associated with the day of the week on which they were born, which accounts for the relative abundance of these eight poses (there are two poses for Wednesday, one for those born in the morning, another for those born in the evening)

The preference for the '**Meditation**' position might be explained by the association between caves and aestheticism, inwardness or the inward quest, of which meditation is the outward form. Caves are seen as particularly suitable for meditation and were probably used by the Lord Buddha for that purpose during his lifetime, so it is not surprising that Buddha statues in caves should be shown in this pose.

The '**Calling the Earth to Witness**' position relates to a crucial event in the Buddha's life just before he achieved enlightenment

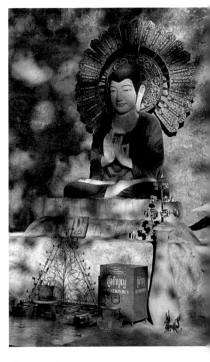

Dhammacakra
(first sermon) gesture, Tham Wua.

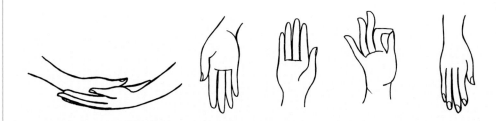

The Buddha's hands may be shown in any of six gestures (mudra):
1. **dhyana** *(meditation) both hands in the lap, one on top of the other*
2. **varada** *(charity) the right hand hanging downwards with the palm facing outwards*
3. **abhaya** *(freedom from fear) the right hand held upright with the hand facing outwards*
4. **vitarka** *(reasoning) the right hand held upright with the thumb and forefinger joined to form a circle representing the "Wheel of Law"*
5. **dhammacakra** *(first sermon) both hands held as in the "reasoning" gesture*
6. **bhumisparsa** *(calling the earth to witness) the right hand held downwards with the palm facing backwards and the fingers extended*

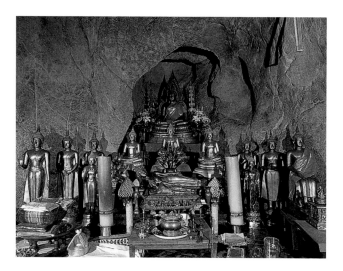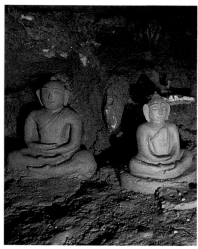

*Left: Variety of Buddha figures
associated with different days of the
week Tham Pbak Sung.
Right: Mud figurines in meditation
posture, Tham Ukaew.*

Facing page:
*The Buddha can be depicted in six
different postures (iryapatha):*
*1-4 virasana - sitting cross-iegged
    with one leg resting on the other
    (the 'heroic' posture) (not ill.)
    vajrasana - sitting cross-legged
    with soles of both feet facing
    upwards ('diamond' posture);
    10 pralambanasana- sitting with
    both feet hanging down
    ('western' posture)*
5-8 standing
9   walking
*11-12 reclining
    Every accepted combination of
    posture and gesture has a
    specific name and eight of them
    are associated with particular
    days of the week.*
1 meditation (Thursday)
2 calling the earth to witness
3 teaching the disciples
4 protected by Mucalinda (Saturday)
*5 stopping the relatives from
    fighting (Monday)*
6 looking at the Bodhi tree (Sunday)
*7 begging for alms (Wednesday
    morning)*
8 in reflection (Friday)
9 walking
10 sitting on a boat
11-12 reclining (Tuesday).

when Mara, the god of illusion, sent his armies to try to tempt the Buddha. In response, the Buddha called the earth-goddess, Mae Thorani, to witness the vast store of merit which he had accumulated over many life times. The goddess came to the Buddha's rescue by sending a vast flood which devastated Mara's armies. Caves are frequently associated with water and the forces of the natural world, which may explain the preference for this particular position.

The '**Teaching**' or 'Reasoning' position is more frequently depicted in buildings than in caves. In this position, the right hand is shown in the *vitarka* gesture, with the thumb and index finger joined to form a circle which signifies the "Wheel of Law", a key concept in Buddhist teaching. Caves are not particularly suitable places to depict this event, as the Buddha conducted most of his teachings in the open air, not in caves. The 'Teaching' position is usually shown with the Buddha in the seated position with the legs crossed and one hand resting in the lap with the palm upwards. Occasionally, the Buddha is shown standing rather than seated. In Tham Wua, a temple in Mae Hong Son with strong Burmese influence, the 'Wheel of Law' is performed with both hands, (the *Dhammacakra* or 'First Sermon' gesture) but this is the only case we have come across in Northern Thailand.

The most important event in the Buddha's life is the moment when he achieved enlightenment under a Bodhi tree. Depictions of this scene are common in temple buildings but very rare in caves. Since the Buddha did not achieve enlightenment in a cave but under a tree, caves do not seem to be suitable places to depict this scene.

The '**Reclining**' pose depicts the moment of the Buddha's death, when he is about to achieve *parinibbana,* ultimate nirvana. This well-known scene is frequently the focal point of a cave chamber and is also found in many open air sites, although it is much less common inside buildings. A particularly fine example is in the outer cavern in Tab Tao, where a giant golden Buddha is shown surrounded by a crowd of life-size disciples. Reclining

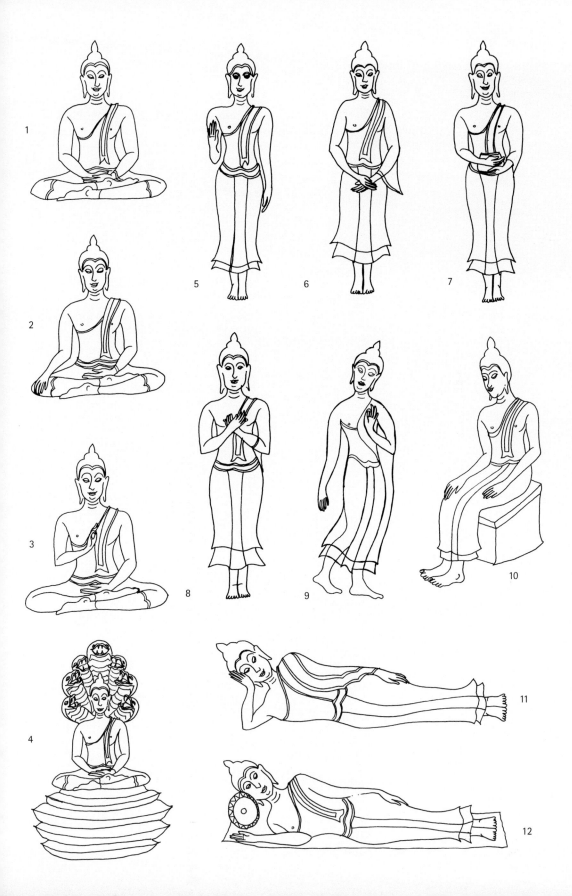

1

2

3

4

5

6

7

8

9

10

11

12

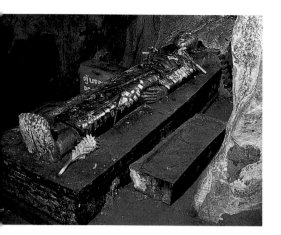

*Phra Phuttha Chao Kassapa,
Tham Chiang Dao.*

Buddhas are also found in less dominant positions in natural recesses in the wall, such as at Tham Phum. The natural proportions of the lying figure conflict with the vertical lines in a building, whereas many caves have suitably shaped alcoves which are ideal for displaying a figure in this position. The Buddha is usually shown lying on his right side, with the head resting on the right hand. At the end of Tham Phra Non in Tham Chiang Dao there is a unique statue called Phra Phuttha Chao Kassapa, where the figure is shown lying on his back.

The combination of the four postures and six gestures form the basis for any accepted representation of the Buddha. The royal degree of King Rama III specified that 40 positions, relating to different episodes in the Buddha life, were legitimate representations. His uncle, the supreme patriarch, described and represented these poses in the 'Pathama Sambodhikatha', an illustrated manuscript or treatise. These 40 poses were later cast in bronze and placed in Wat Phra Kaew in Bangkok. Only a small number of these positions are used in common practice, and local variations from the standard are popular in many areas.

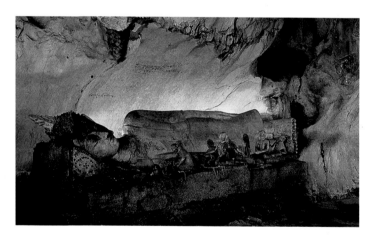

*Reclining Buddha in natural alcove,
Tham Phum.*

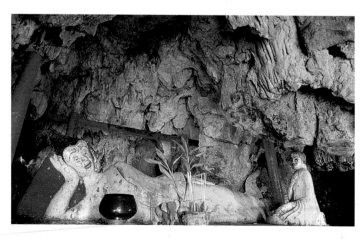

*Reclining Buddha with red pillars,
Tham Ukaew.*

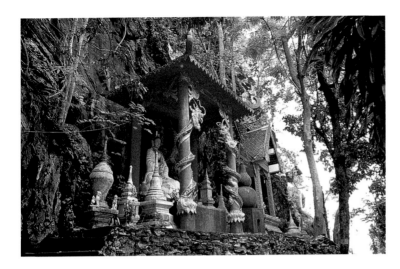

Guan Yin in Chinese pavillion protected by dragons, Tham Pha Chom.

Guan Yin and Pu-tai (Metteya)

Visitors to caves are likely to come across two depictions which have a strong Chinese influence. Pu-tai (also known as Metteya or Ariyametta) is the smiling 'fat' Buddha of the Future who has not yet come and is associated with good fortune. Another popular representation is Guan Yin, the feminine form of *Avalokiteshvara,* the *Bodhisattva* of compassion and mercy (not strictly speaking a true Buddha). Pu-tai and Guan Yin often attract more devotees than the other statues in a cave, even though they are rarely in a dominant position.

Hermits

Statues of hermits are very common in caves throughout Northern Thailand. Hermits are always depicted in a position of meditation and are typically shown wearing animal skins. They can be distinguished from Buddhas by their lack of any of the characteristic marks of the Buddha, such as a protrusion on top of the skull (*'usnisa'*), the 'wheel of the law' mark on the sole of the feet, elongated fingers and ear lobes. Hermits are often portrayed seated at a lower level than the main Buddha statue on his left hand side. They may also be placed on their own in the dark recesses of the cave. Many caves are called Tham Russi, or Hermit's Cave, which may refer to a particular hermit or to a succession of hermits that used to live in the cave. In some cases the statue is supposed to represent a particular person, such as the turbaned yogi at Tham Pbak Sung near Chiang Dao.

Shrine to Guan Yin, Tham Phak Sung.

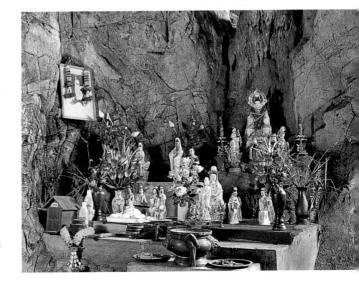

The association of caves with hermits is probably extremely ancient and almost certainly pre-dates the introduction of Buddhism. The portrayal of hermits in temple caves could be seen as symbolic of the absorption of pre-Buddhist beliefs into mainstream religion. This is well illustrated in the legend surrounding the founding of Lamphun and the hermit Rishi Va Sudeva, better known as Wa Suthep. The hermit protected and advised the future queen before she emerged from the forest to rule Lamphun as a Buddhist monarch. Doi Suthep near Chiang Mai Town is named in his honour. The cave at Wat Phra Phuttha Bat Tak Pha near Lamphun is said to have been used by Wa Suthep and an impression on the rock in the temple grounds is believed to be his footprint.

1. Hermit, Tham Tab Tao.
2. Hermit at entrance to Tham Pla (Mae Hong Son).
3. Hermit Wa Suthep in Wat Phra Phuttha Bat Tak Pha.
4. Buddhas and hermit, Tham Mae Hu.
5. Yogi in Tham Pbak Sung.

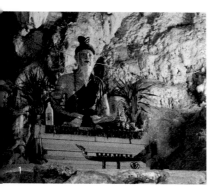
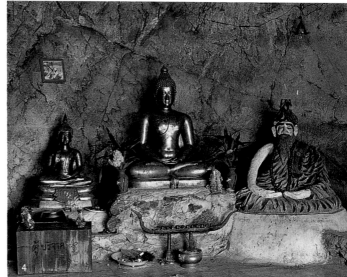
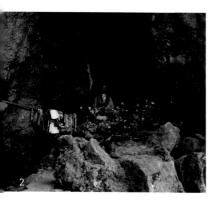
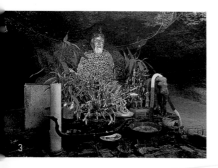
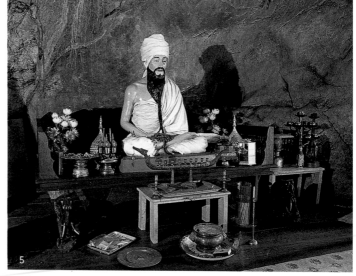

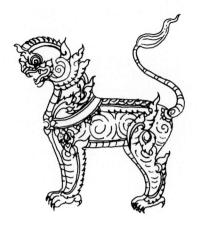

Singha

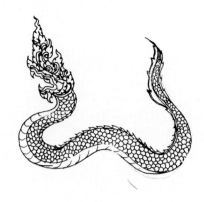

Naga

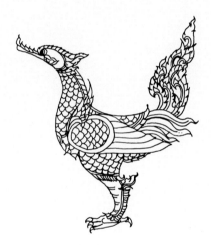

Kinnari

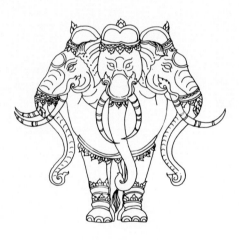

Erawan, a three-headed elephant

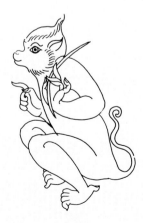

Monkey

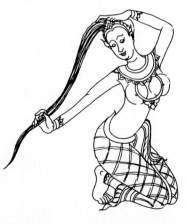

Mae Thorani wringing the water out her hair

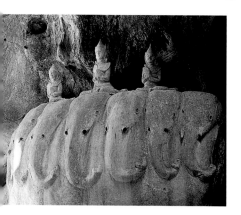

Elephants in Tham Tab Tao.

Mythical Creatures

Buddhism has adopted a number of mythical creatures which serve to illustrate key points of doctrine and help to enliven popular folklore. Nagas are serpent-like creatures associated with rain and fertility which are thought to live in caves and other underground places. The steps leading up to temple caves are usually flanked with the sinuous bodies of a pair of *nagas,* who serve to protect the cave and it s contents from evil spirits. In one of the episodes of the Buddha's life he was sheltered from a storm by a benevolent seven-headed *naga* (Mucalinda). The natural form of alcoves enclosing two of the statues in Tham Phum have been accentuated to emphasize their resemblance to *nagas,* in reference to this event.

Yaks are giants with terrible expressions and large swords held with both hands. Despite their terrifying appearance, Yaks are benevolent creatures who also protect the cave against evil spirits and will not attack anyone with a righteous mind. Evil demons can also take the form of giants and are sometimes represented in caves. A sculpted rock in Tham Chiang Dao is said to represent the demon Phanthurat who was turned to stone by a hermit whose elephant he had eaten.

Phra Ya Yom looks exactly like Christian depictions of the devil, with a trident in one hand and a human skull in the other. He is the guardian at the entrance to hell but righteous people need have no fear of him. Kinnari (half woman and half bird) and Singhas (mythical lion-like creatures) are other types of guardian commonly depicted at the entrances to caves or at the base of platforms leading up to Buddha statues.

Mae Thorani is the earth goddess who drowned Mara's armies by wringing the water out her hair (see page 129). A prominent statue of her can be seen next to the pool near the entrance to Tham Chiang Dao. Erawan is a three-headed elephant who is the vehicle of the Hindu god Sakka (Indra). Several caves contain rocks which are thought to resemble him.

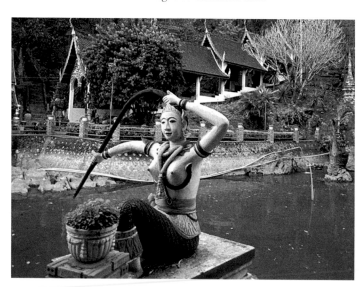

Mae Thorani in front of entrance to Tham Chiang Dao. (© Dean Smart)

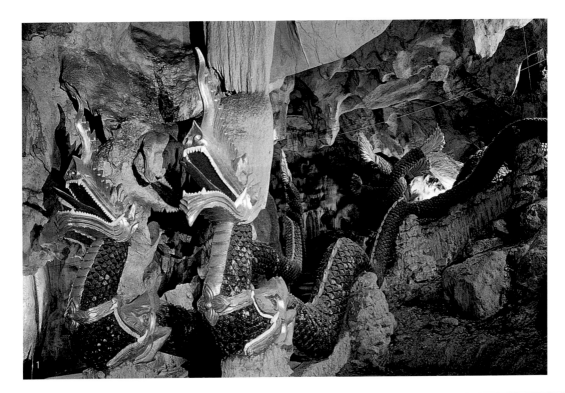

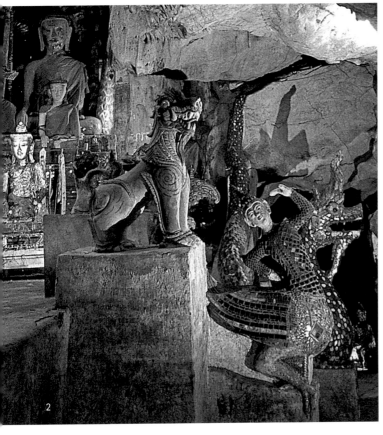

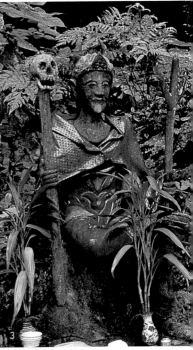

1. Nagas flanking staircase, Tham Tab Tao.
2. Singha and Kinnari, Tham Chiang Dao.
3. Phra Ya Yom, Tham Phra (Chiang Rai).

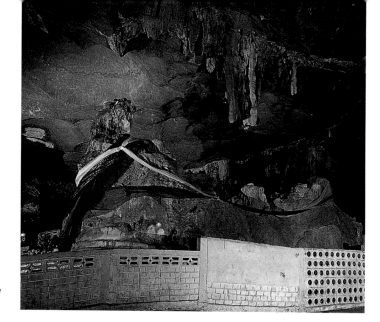

The Princess Aranyani, a natural rock formation resembling a woman holding a baby, Tham Pha Nang Koi.

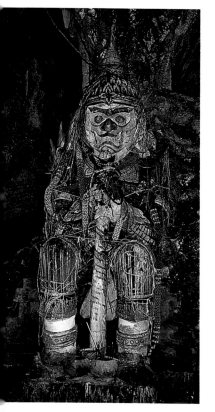

Yak protecting the entrance to Tham Tab Tao.

Animals and People

Many real animals also appear in Buddhist stories and have a religious significance. Elephants, lions, monkeys, fish, turtles and crocodiles are well known examples. Although these animals are not usually depicted in statues, many curiously shaped rocks or cave formations are said to resemble them. In some cases the natural likeness is enhanced to increase the resemblance or the rock itself is incorporated into a man-made statue. At the entrance to Tham Sao Hin, an elephant-shaped rock serves as the base for a statue of the Buddha.

Statues of deer and other forest animals are sometimes seen in caves, which may be an illusion to the heavenly forest, 'Pa Himaphan', where gods & other celestial beings reside. Tham Chiang Dao is particularly associated with a white deer that figures in one of the popular legends surrounding the cave. It is said that on certain nights, balls of fire can be seen emerging from Plong Sawan, the old entrance to the cave, which are believed to be the spirits of the beautiful Inlao and her lover Lord Khamhaeng.

In Tham Pha Nang Koi, the name of the cave comes from a rock shaped like a beautiful woman with a baby in her arms. According to local legend, the lady is Princess Aranyani, daughter of the King of San Wi (see page 362).

"Once upon a time the Lord Khamhaeng went forth to hunt in the jungles of the Doi Angsalaeng mountains, where he came upon a beautiful deer. He pursued the deer for a whole day but was unable to run it down. Eventually the deer came to the mouth of the Chiang Dao Cave. Lord Khamhaeng dismounted from his horse and followed the deer into the cave. Suddenly the deer changed to a beautiful maiden, who told the Lord that her name was Inlao. Lord Khamhaeng was smitten by her charm and immediately fell in love with her. The two spent the rest of their lives together in the cave. People believe that the spirits of the Lord Khamhaeng and of the beautiful Inlao protect the cave"

(translated from a guide book in Thai ca. 1980, quoted in Dunkley 1995)

Other Objects

Certain objects are typically associated with Buddhism and are
frequently placed in temple caves. In other cases, natural
phenomena in the cave are said to resemble these objects. The
commonest of these objects is the bell, which visitors to the cave
ring to gain merit. The ancient bronze bell in Tham Chiang Dao is
the oldest known religious artifact from any cave in northern
Thailand. Tham Phra near Nan has two fine bronze bells mounted
on scaffolding near the entrance. Perhaps the most impressive
example is the bronze bell buried deep inside Tham Jak Dtor at
the top of a steep slippery slope, which must have taken a major
operation involving many people to get it to such an inaccesible.
location.

Chedis or stupas are also key Buddhists symbols which are
often built in caves or form part of the imagined natural
environment. The group of stupas in the inner chamber of Tham
Tab Tao are a particularly fine example. Although these particular
examples are man-made, they have been eroded by the ravages of
time to such an extent that they now appear almost to be part of
their natural surroundings. The area in front of some cave
entrances is filled with small white stupas that contain the ashes of
deceased people. In Tham Jarui, Phan District, a large group of
such stupas are clustered near the cave entrance, whilst a skeleton
hangs next to a platform with many Buddha statues in the inner
chamber. The association of caves with death and mortuary
practices still continues.

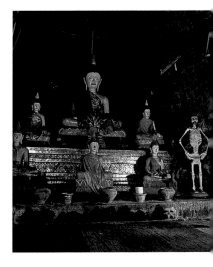

Altar with skeleton, Tham Jarui.

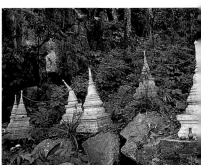

White funeral chedis, Tham Jarui.

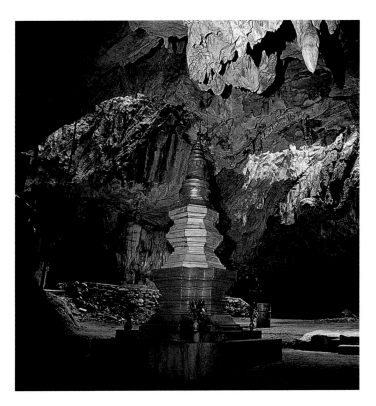

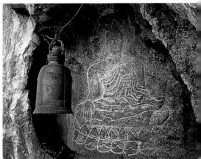

Bronze bell at Tham Pbak Sung.

Golden chedi in Tham Pha Rua.

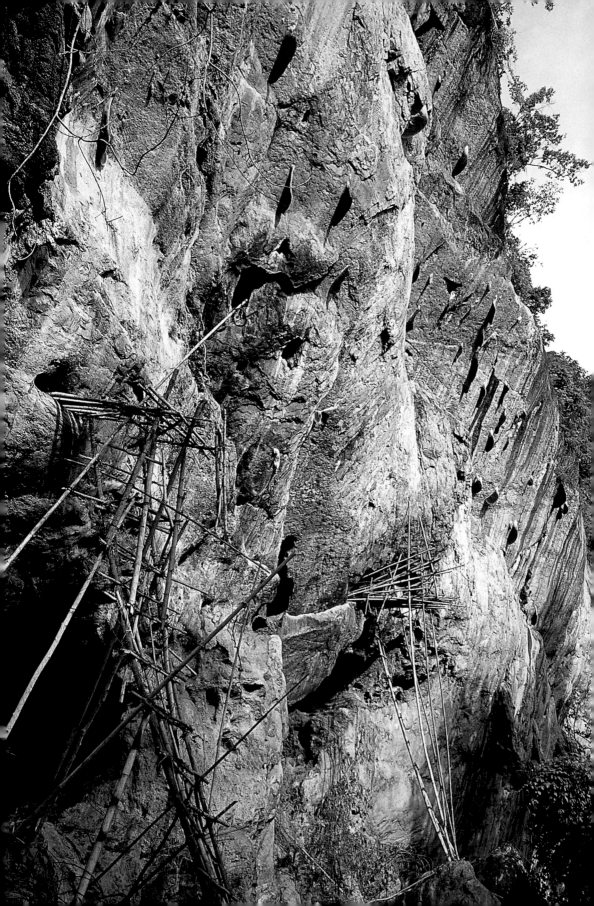

USES AND CONSERVATION

This chapter outlines the potential value of caves, the uses to which caves have been put both in the past and in the present, the impact that these uses and other human activities have on caves and the ways in which these impacts can be minimized.

Caves are extremely fragile environments which are easily damaged even by well-intentioned visitors as well as by changes to the surface environment outside the cave. Most caves and their contents are non-renewable in human timescales and the concept of sustainable use is not applicable. Each and every visit to a cave has an impact and it is the responsibility of any visitor to a cave to make sure that his impact is kept to a minimum.

Value of Caves

Uses and values are often difficult to separate because different people often ascribe different values to the same cave, which is usually an expression of the use which they perceive for that cave. Visitors may consider the beauty of a particular cave as its primary value for themselves, while at the same time valuing its educational potential for their children. A local villager is more likely to see the same cave in terms of its economic value as a tourist attraction or as a source of bat guano. A scientist may consider the primary use of the same cave as a key to our understanding of landscape formation processes whilst a person who has never visited the cave at all may still consider the cave to be a part of their national heritage which they have the right and duty to protect. The cave also has a value as a habitat for other organisms which may have no human use but nevertheless have the right to exist in their own terms. The challenge of cave management is to maximize the use value of the cave for all these varying groups while at the same time not endangering the long term survival of the cave and its ecosystem.

Caves represent an excellent opportunity to study how the world works and our place in it. In few other environments are the ecological chains of cause and effect, and the environmental determinants on human society, so clearly evident. The processes of landscape evolution, erosion rates and climate change over long time periods are well preserved in caves and are particularly relevant to present day human society. Cave sediments often preserve evidence of past environments long after they have disappeared from the surface. The temperature in caves is so stable that they are often an accurate reflection of average annual temperatures. A fluctuation throughout the year of 30°C at the surface is reduced to only 1°C at a depth of 11 m. At greater depths the effect of a marked rise or fall in average surface temperatures may not become apparent for hundreds or even thousands of years.

Summary of cave values (IUCN 1999)

as habitat for endangered species of flora and fauna

as sites containing rare minerals or unique landforms

as important sites for the study of geology, geomorphology, hydrology, palaeontology, archaeology and other disciplines

as spiritual, religious or culturally important sites

as sources of groundwater for tourism and its associated economic benefits

Opposite: Bamboo ladder used to collect honey at Pha Phung.

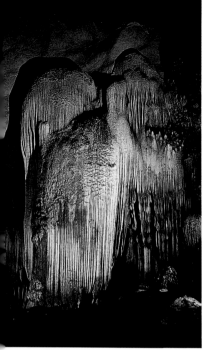

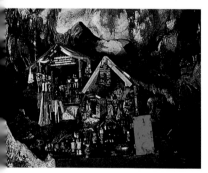

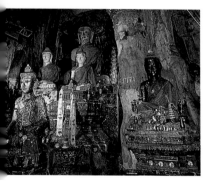

Tham Chiang Dao is a good example of a multiple-use cave. The cave has been an important religious site for centuries, is Northern Thailand's most popular tourist cave attracting over 100,000 visitors per year, is an important water source, contains outstanding speleothems and supports a wide variety of endemic troglobitic species.

USES OF CAVES

Caves have been used by humans and our ancestors for many millions of years and continue to serve many functions today. The earliest known human use of a cave is from Zhoukoudian near Beijing where evidence of *Homo erectus* occupation including stone tools, bones and fire hearths has been dated as 700,000-1,500,000 years old.

Shelter

Throughout most of human history the main use of caves has been as shelter and a great many caves in Northern Thailand contain stone tools and other evidence of habitation during prehistoric times (see chapter 4). Plant remains dated 10,000 years BP have been excavated from Spirit Cave (Pang Ma Pha) which indicate that this site was used by early humans and may be one of the oldest records of plant domestication in the world.

During the Thai-Burmese wars of the Ayuthaya period caves were used to shelter whole armies, such as at Tham Naresuan near Muang Na which may be the last place that King Naresuan stayed on Thai soil before he died in 1605.

In the Second World War, caves were an important refuge for both Japanese and resistance forces and evidence of their occupation can still be found. Weapons have been uncovered in Tham Borichinda, whilst a platform and drawings in Tham Srisophon appear to date from the same era. During the 1970s insurgent communist groups also made use of caves to escape detection, the most famous example of which is Tham Pha Daeng in Nan which was used as a field hospital with separate chambers serving as male and female dormitories. Even today, caves in Northern Thailand are still used as shelter by wandering monks and local hunters as well as illegal immigrants and smugglers in border areas. Many caves are called Tham Russi after the hermits which live there or used to do so in the past and Tham Jhon (Lampang) is named after an infamous group of bandits who used the cave as a hideaway.

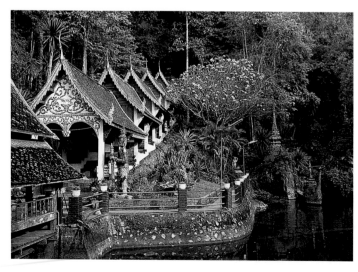

Spiritual and Ritual Uses

Caves are considered highly significant as spiritual sites in many cultures and have been used for ritual purposes in Northern Thailand for at least 10,000 years. The abundance of prehistoric rock paintings at sites such as Phra Tu Pha can only be explained as some form of ritual, and the highly distinctive 'log coffins' of Pang Ma Pha are almost certainly associated with burial rites. The oldest dated cave burial in Northern Thailand is at Tham Lod Overhang where a 13,000 year old human skeleton has been excavated, together with associated burial artifacts.

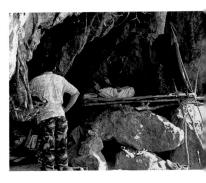

Temporary shelter at Tham Srisophon.
(© Pang Ma Pha Caves Project)

Many of these ritualistic uses of caves have been superseded by or absorbed into Buddhism, which has a great number of associations with caves (see previous chapter). Many of the spiritual values currently ascribed to caves seem to relate more to local superstitions than Buddhist religion, such as the use of scratches on cave walls to predict lottery numbers or the magical properties ascribed to sacred pools and water dripping from stalactites. Certain types of mud from caves are believed to be endowed with special properties and are sometimes used to manufacture clay amulets imprinted with an image of the Buddha or a famous monk. A large number of caves in Thailand are currently used by temples and monks, to such extent that some areas have been aptly termed "monastic karst". There are probably well over 100 temple or hermit caves in Northern Thailand alone and almost a third of the caves in this book fall into this category. Religious uses range from a single statue placed in the corner of the cave, such as in Tham Mae La Ga or Tham Pha Thai, to large permanent monasteries such as at Tham Chiang Dao and Tham Tab Tao. The primary religious reason for lay people to visit a cave is to gain merit, although many caves may be a spiritual experience in their own right. Monks themselves use caves as shelter during their wandering periods, as tranquil places for meditation and as tests of their control over physical and mental fear. Many monastic caves are still close to their natural state but some have been so extensively modified and sumptuously decorated that the impression is more akin to a drawing room than a real cave.

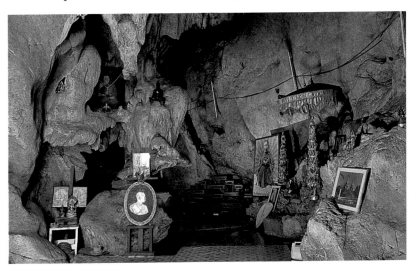

A variety of religious and secular imagery in Tham Pbak Sung.

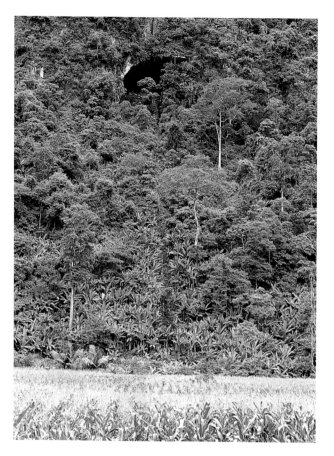

Tham Sakoen, an important site for collecting bat guano. The straight trail used to ferry bags of guano down the mountain is clearly visible.

Collecting

Caves have been exploited as a source of valuable commodities since prehistoric times and continue to be used for this purpose today. At Khao Tab Kwai and Non Pa Wai near Lopburi 3,500-4,500 year old pits excavated with stone tools were used as a source of copper-ore at the beginning of the Metal Age in Thailand. American Indians have probably been using cave minerals as food flavouring or as body paint for many thousand years. A 4,000 year old skeleton of an Indian man who was killed by a rock fall while collecting minerals was recently found 4 km deep inside Mammoth Cave (USA). During the American war of Independence, caves were a vital source of saltpetre used for gunpowder and evidence of these workings can still be seen today.

Guano from bats and birds can be found in vast quantities in many caves in Northern Thailand and has long been collected by local people as an excellent agricultural fertilizer. Guano is particularly rich in phosphates and nitrates which are essential nutrients for plant growth. The practice is so prevalent in certain areas that the local authorities have had to restrict access to prevent over-exploitation. In Tham Sakoen, the National Park authorities auction collecting permits to the highest bidder only once or twice a year.

In some caves the bats themselves are also collected for food. Each individual bat represents a rather meagre morsel but by setting up traps at a constricted cave entrance hunters can quickly capture a reasonable meal, although they run the serious risk of disturbing the whole colony. Such practices are not common because local people are generally afraid of caves, which they believe to be inhabited by powerful spirits who also protect the bats, fish and other creatures. A sign at the entrance of Tham Lod (Pang Ma Pha) posted by the local abbot informs that anyone who catches fish in the cave will be cursed for seven generations – a heavy penalty indeed! The large numbers of enormously fat fish at such sites as Tham Lod and Tham Pla are testimony to the effectiveness of these beliefs. Many of these animals only use the cave at certain times of the year and can be hunted with impunity as soon as they emerge from the shelter of the cave. The cave serves as an important refuge and ensures the long term sustainability of the wider ecosystem.

Overhangs and cave entrances are one of the best sources of wild honey, which is collected by intrepid local villagers using precarious bamboo scaffolding. Prehistoric rock paintings which seem to represent bee hives suggest that some of these honey collecting sites have been in use for several thousand years. In Southern Thailand similar contraptions are used to collect the extremely valuable birds' nests used to make the famous soup. The business is so profitable that the collecting rights to individual caves are auctioned for several million baht. Regrettably for the local economy, the swifts in Northern Thailand do not produce edible nests.

Caves contain minerals and rock forms not found on the surface which are valued for their rarity or special character. No large scale use of cave minerals has been reported from caves in Thailand, although there is a small black market in cave crystals and speleothems for decorative purposes. This form of collecting is illegal and perpetrators can face heavy fines.

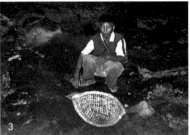

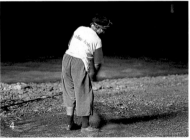

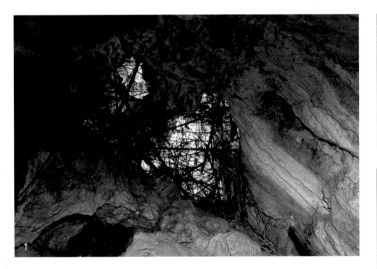

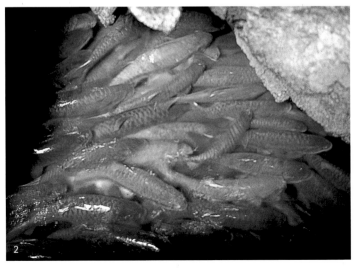

1 Trap used to catch bats at the entrance to Tham Huai Bon Nam.
2 Extremely well-fed carp in Tham Pla.
3 and 4 Collecting guano in Tham Lod.

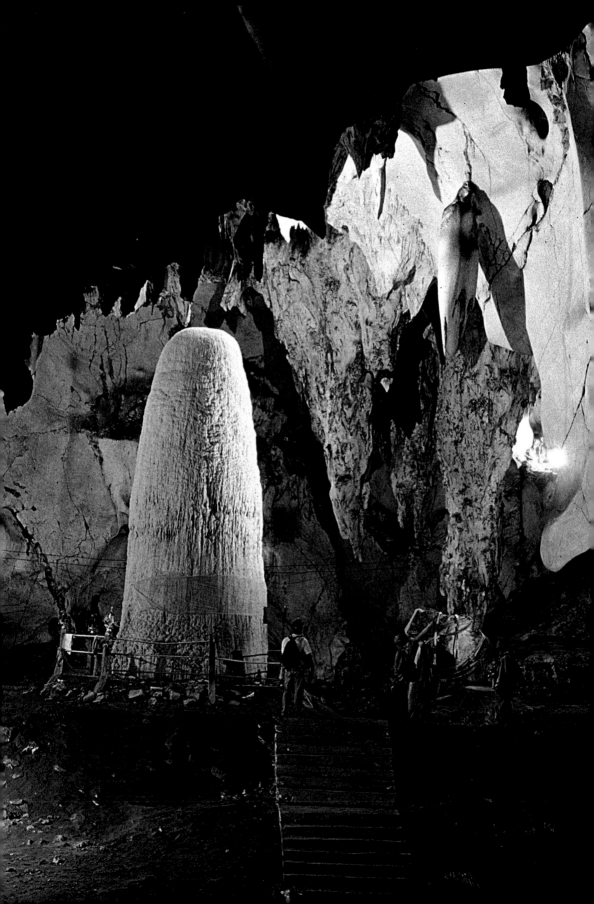

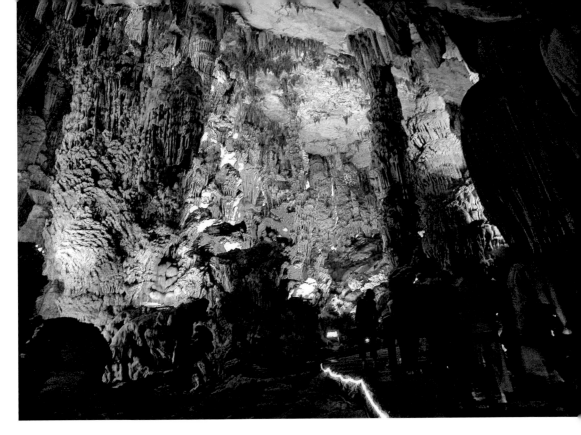

Tourism

Caves are mysterious and fascinating places outside the daily run of human existence so it is not surprising that tourism has become one of the most important use of caves worldwide. An estimated 650 caves throughout the world have electric lighting systems and are visited by over 20 million people every year. The most popular cave in the world is Mammoth cave (USA) which receives 2 million visitors a year. As many more countries such as Thailand start to exploit the tourist potential of caves visitor numbers are likely to increase very substantially in the near future. The most popular caves in Northern Thailand are Tham Chiang Dao and Tham Lod which both receive over 100,000 visitors annually. In other parts of Thailand, caves such as Tham Khao Luang (Phetburi) are also extremely popular and it seems likely that several million people visit caves in Thailand every year.

Tourists visit caves for a whole variety of reasons ranging from aesthetic appreciation of beautiful speleothems to the physical challenge of adventure. In Thailand many visitors combine recreation and enjoyment with making merit so well-known temple caves are extremely popular with local holiday-makers. For local communities and tour operators, caves represent the potential to earn hard cash, often in remote localities where there are few other money-making opportunities. Tham Chiang Dao and Tham Lod are the mainstay of the local economy both directly from incomes earned by guides but also through services such as restaurants, souvenir shops and hotels.

Spectacular lighting in Reed Flute Cave, Guilin, China.
(© Surapong Lerdthusnee)

Opposite: *Sacred stalagmite in Tham Muang On, believed to house a hair of the Lord Buddha.*

Pearl Returning Cave, Guilin, China.
(© Surapong Lerdthusnee)

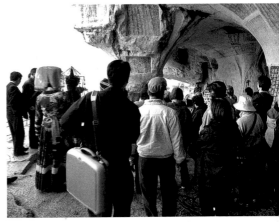

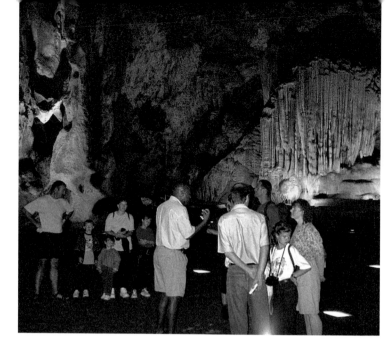

Cango Caves, one of the top tourist attractions in South Africa.

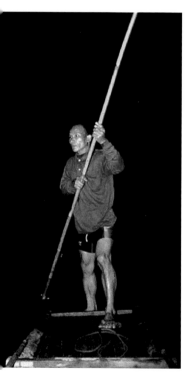

Rafts used by tourists in Tham Lod, Pang Ma Pha.
(© Pong Ma Pha Caves prjcet)

Water

Springs issuing from underground caves have been important water sources since time immemorial and continue to be of vital use today. The IUCN estimate that 25% of the world's population gain their water supplies from karst, either from discrete springs or from karst ground water. Chinese karst springs were important documented water sources in the Shang dynasty (1,600-1,100 BC) and the Jinci spring was being used for irrigation in 450 BC. Many Chinese springs are now multiple-use sites exploited for hydro-electricity, irrigation and as tourist caves.

Famous brands of bottled mineral water are often derived from karst springs. Kentucky bourbon from the Jack Daniels distillery still relies partly on cave spring water. The flow of cave water is usually more consistent than surface supplies and is often cleaner because filtration through limestone bedrock removes most of the sediment load and adds important minerals such as calcium. Contrary to popular belief, this natural filtration process is surprisingly ineffective at removing bacteria and other pathogens so spring water may not be safe to drink despite its clarity.

Many villages in Northern Thailand rely heavily on water issuing from caves, such as Tham Mae Hu (Pang Ma Pha) which is the principal water source for Ban Mae Hu. Deforestation and the use of agricultural fertilisers or pesticides in the water catchment above the cave is likely to have adverse effects on the quality of these water sources.

Storage

Caves are also used for storing a wide range of commodities, being particularly suitable for items requiring cool dry conditions. Tham Klaeb (Chiang Mai) was used until quite recently to store rice husks, which can still be seen all over the floor. Caves are also useful for hiding items to escape detection as they often have many complex passages that are very difficult to find. In Burma caves were used to hide family heirlooms to escape looting by the Thai army during the

Ayuthaya Period, and it seems extremely likely that the same thing happened in Thailand when the Burmese armies invaded in their turn.

Caves are particularly useful for hiding contraband and other illegal activities where there is a need to avoid detection. During the recent drugs push in 2003 the largest single haul in the whole of Thailand was from a cave near Ban Mae Suriya (Mae Hong Son) where 1,500,000 tablets of amphetamine and 80 kg of heroin were uncovered from a single cave. In October of the same year 200 soldiers from the Thai Army finally managed to oust a group of militants from the Wa State Army who were using a cave in Chiang Dao District as their base for forays across the border into Burma. Large quantities of weapons, ammunition and satellite phones were discovered in the cave which was said to be big enough to house 100 people.

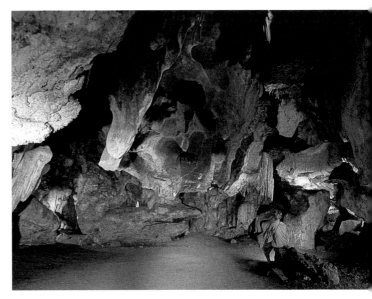

Tham Klaeb derives its name from rice husks that used to be stored here.

A widespread rumour prevalent throughout the whole of Southeast Asia relates that the Japanese forces hid large quantities of gold in caves at the end of World War II. Many caves in Northern Thailand have been extensively dug over by hopeful treasure-seekers but the whereabouts of the gold remains elusive.

Other Uses of Caves

Caves are also important for the role they play in the wider environment, which may have a direct economic value, or an indirect but no less important one. The significance of caves as refuges for animals has already been mentioned. In Malaysia, the vital importance of caves for the local economy was only discovered after massive failure of the durian crop. Exhaustive investigations revealed that the bats which used to pollinate the durian flowers roosted in nearby caves which had been heavily disturbed to such an extent that the bats were no longer able to live in the area.

Many of the values ascribed to caves are difficult to place in an economic context as they relate to spiritual, cultural and other riches rather than simply economic ones. Even very direct benefits of caves may not have a clearly defined economic value. The use of geological, biological and cultural diversity can be expressed in terms of economic value such as the potential for medicinal uses or tourism but for many people is better expressed in spiritual rather than economic terms. The world is an incomparably richer place with this diversity than without it and the purpose of our lives should be to make the world a richer place, not a poorer one. Any natural ecosystem has an intrinsic value in its own terms and has the right to exist irrespective of any human value or scientific objective.

Main effects of visitors on caves

- *alteration of the physical structure of a cave*
- *alteration of the water chemistry*
- *alteration of cave hydrology*
- *alteration of air movement and microclimate*
- *introduction of artificial light*
- *compaction or liquification of floors*
- *erosion or disturbance of cave sediments and their contents*
- *destruction of speleothems*
- *destruction or disturbance of fauna*
- *introduction of pollutants, nutrients, animal species, algae and fungi*

HUMAN IMPACTS ON CAVES

Caves are extremely fragile environments which are easily damaged by well-intentioned visitors or management schemes as well as by vandalism or overcollecting. Human impacts on caves are very various and often far-reaching. Some forms of impact may be serious in the short term, such as heavy trampling of a stream bed, but in active stream caves these impacts are quickly erased by annual flooding. Other impacts may appear small but have extremely far-reaching consequences. The accidental breaking of a single calcite straw will take many thousand years to heal because geological formations grow so slowly. Other forms of damage, such as the burning of a log coffin, are completely irreversible on any time scale. Not only the object itself but also the insight it could have provided into our own prehistory is lost forever.

The most significant form of human impact varies from cave to cave. Many types of impact are accidental or may even be due to well-intentioned but misguided management. For many caves, the damage caused by visitors is undoubtedly the single most important factor. Most incidents of visitor damage are avoidable and caused by ignorance or clumsiness. The majority of visitors go to caves because they like them, appreciate their beauty and would be horrified if they knew the full extent of the damage they cause.

Collecting

All forms of collecting have an impact on caves but many of these activities are much less detrimental to the cave environment than accidental damage by well-intentioned visitors and management schemes. Local people have often been collecting guano and other products in caves for many centuries and are very well aware of the sustainability of their actions. No scientific studies have been carried out on the potential impacts of collecting activities in Northern Thailand. Although a certain level of collection is probably sustainable, guano is an extremely important part of the food chain in many caves and the ecosystem would collapse without it. In the bird-nest business in Southern Thailand there are strict rules as to how much can be collected from each site. Each pair of birds only has enough energy to build two nests, so the collectors will only take the first nest, leaving the second nest for the birds to use for its natural purpose. Only after the birds have finished using this second nest will it be collected, although these used nests fetch a much lower market price.

The most destructive form of collecting is the deliberate removal of crystals, speleothems and archeological artifacts. All such forms of collecting can be highly damaging and are illegal. The responsibility

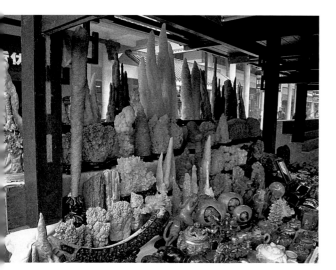

Collecting speleothems and crystals is particularly damaging as these natural wonders take thousands of years to regrow. (© Surapong Lerdthusnee)

for such destruction lies as much with people who buy these objects from markets like Chatuchak to decorate their living rooms as with the local people who use this activity to supplement their meagre incomes.

Vandalism

Graffiti is a particularly obvious form of vandalism and can spoil the aesthetic quality of a cave which is often its primary value. In ecological terms, such vandalism may have minimal impact, although some spray paints contain highly toxic ingredients which could have an adverse effect not only on the cave ecosystem but also on human visitors in constricted passages with little air movement.

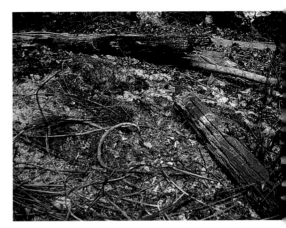

A very serious form of deliberate damage to caves is the burning of log coffins and the wanton destruction of archaeological artifacts, often without any apparent purpose. Such vandalism is one of the most serious threats to archeological sites, in particular the log coffin sites in Pang Ma Pha. Only through an intensive education programme focusing on the value of this irreplaceable local heritage can such tragic waste be halted. The promotion of archeological sites as tourist attractions may be the most effective means to curb this activity as local people perceive that these sites have greater potential to attract visitors when they are in an undamaged state.

Log coffins are an irreplaceable part of our national heritage.
(© Highland Archaeology Project)

Visitor impacts

In many caves the most serious damage to cave systems is done accidentally by visitors through carelessness or ignorance of the consequences of their actions. It is the responsibility of every person visiting a cave to ensure that the impact of their visit is kept to an absolute minimum (see next chapter for codes of conduct). Both living organisms and geological formations are naturally protected deep in a cave and would not survive normal conditions on the surface. Fragile crystals and speleothems can only exist in caves because there are no large animals stomping about or green plants breaking them up.

Individual caves vary enormously in terms of the visitor levels which they can withstand. Caves with active streams or seasonal flooding are subject to continual natural changes which quickly eliminate the evidence of other lesser changes, such as those caused by entry of visitors. Other caves or even sections of the same cave may have extremely low energy throughputs and the impact of human entry, no matter how carefully managed, may be considerable. The destruction of delicate speleothems can undo 10,000 years of growth in seconds. Archeological sites are also extremely fragile and contain artifacts that are absolutely irreplaceable.

"In all caves where there are delicate features present such as speleothems, calcite floors, crystal pools, sedimentary deposits etc. the passage of each and every caver does some damage, however small. This damage of course varies from one individual to the next depending on such factors as clumsiness, physical fitness, awareness, personal lighting system and concern for the environment. These factors are also influenced by the environment itself – a more strenuous trip will lead to more fatigue and therefore less consideration for the immediate surroundings. Nearness to the cave entrance can be an important factor. A highly decorated passage which is entered directly from the daylight is particularly vulnerable to accidental damage because the caver's eyes have not fully adjusted to the darkness and many of the finer features are simply not seen".
Judson et al. 1995, p 185.

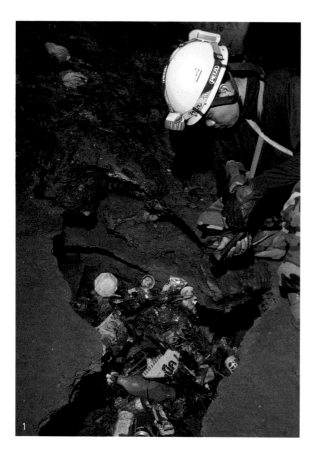

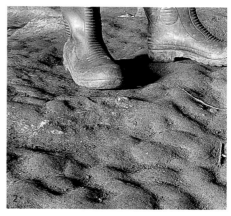

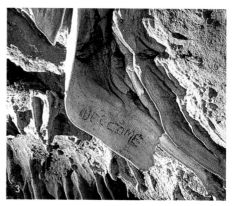

1. *Garbage can seriously upset the delicate natural balance of cave ecosystem. (© Dean Smart)*
2. *Compaction of cave floors by visitors.*
3. *Graffiti on cave drapery.*

External Impacts

Most cave ecosystems rely almost completely on the surface environment for water and food supply and are extremely susceptible to changes throughout their catchment areas. Without water caves would not exist at all. Water is vital to all forms of cave life as well as to the process of cave formation and the growth of speleothems. The acidity of the water is critical to all geological processes in caves because limestone is far more soluble under acidic conditions. Water which has filtered through soil rich in organic matter contains high levels of carbonic acid which has the capacity to dissolve calcium carbonate. This process is an essential precursor to the development of all speleothems and other forms of cave decoration. Activities such as forest clearance and burning which remove vegetation cover will drastically reduce the organic content of soils. These practices are prevalent throughout almost all of Northern Thailand and may have very far reaching consequences on caves. Over 80% of the original forest cover has been cleared from the top of Doi Chiang Dao and annual fires continue to erode the few remaining patches. Scientific studies on the effect of these changes have yet to be carried out but they are likely to be profound.

Forest clearance and agricultural activities often lead to changes in siltation levels and other hydrological patterns which may be vital for cave functioning. The area above Tham Mae Lana has extensive patches of wet rice cultivation, which diverts wet season flow away from the cave. Annual flooding may have an important role in 'flushing out' cave passages and restoring the system to its natural state. Pesticides can have a devastating impact on cave ecosystems which are ill-adapted to withstand such shocks. The population of most troglobitic species is often so small that they can easily be wiped out by a single incident. The greatest threat to the internationally important population of *Cryptotora thamicola,* the 'waterfall-climbing fish' in Tham Susa is probably pesticide poisoning from agricultural activities above the cave. Even fertilisers may have devastating effects on cave ecosystems, which have evolved to function under conditions of low food supply. Sewage, oil leakage and other forms of pollution have caused enormous damage to fragile cave ecosystems throughout the world, although no cases have been reported in Northen Thailand.

The most destructive form of human activity is mining and quarrying which physically removes caves and their contents forever. The impact on caves in Northern Thailand has been limited so far, although many caves in Central Thailand have been lost in this way. In some cases mining has been sucessfully diverted from caves which contain particularly valuable features, such as the caves in the Khao Ngu range (Ratchaburi) which contain some of the earliest Buddha images in Thailand. Mining can also cause serious structural damage to caves as the blasting sends shock waves through the bedrock. All caves in the vicinity of mining sites should be closed to the public for safety reasons until their structural integrity can be assured. Mining activities can also have unexpected benefits as some caves were only discovered by mining. This is particularly the case with caves that have no natural opening to the surface, such as hydrothermal caves formed by hot water percolating up through the bedrock below the cave. Tham Kaew Komol (Mae Hong Son) was discovered in this way and has become an important tourist attraction.

Below left: This very rare hydrothermal cave was recent exposed by mining activities. Unless immediate action is taken, it will soon be lost forever.

Surface construction such as road-building and car parks changes water drainage patterns which can have a profound effect on geological processes such as speleothem growth. The surfacing of the land with concrete or bitumen renders it nearly impermeable, in contrast to the high natural permeability of karst.

Below: Forest clearance can have far-reaching consequences because forest soils are rich in carbonic acid which is essential for many cave processes.

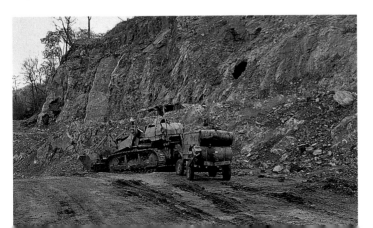

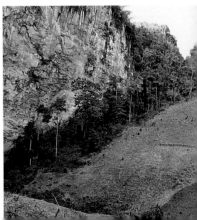

CAVE MANAGEMENT

The primary purpose of natural resource management is to ensure the long-term sustainability of human use and to minimize the impacts of such use as far as possible. The regeneration of many cave features occurs over such long time periods that some forms of use are inherently non-sustainable in human timescales.

Cave management is frequently carried out in ignorance of the functioning of cave systems and in many caves the most serious form of human disturbance has been well-intentioned but inappropriate management practices. During the nineteenth century, many tourist caves in Australia had steel and chicken wire structures installed to keep visitors from touching the speleothems. Over time, these structures have corroded to produce zinc and iron salts which contaminate cave pools and leave unsightly stains over the very speleothems which they were designed to protect.

The best form of management for many caves is probably to leave them as 'wild' caves with no active management at all. This is particularly the case with active stream caves where the effects of visitors are reduced by annual by flooding. With over 2,000 caves and limited personnel and financial resources available to natural resource managers there is currently no other practical option for the vast majority of caves in Northern Thailand. However, some caves are so sensitive to visitor impacts that it is imperative to restrict access, at least to the most vulnerable sections, either through gating or a permit system. Less sensitive caves can be promoted as tourist attractions and appropriate management can play a vital role in reducing the adverse effects of visitor pressures.

Since the threats to caves are from both internal and external sources, proper management needs to take into account a wide variety of factors. Installation of visitor infrastructure is usually the most visible form of management, but education of local communities and reforestation of watersheds can be equally important. Karst systems are effectively delineated by the total watershed area, which may be quite different from surface topography. Activities anywhere in the watershed may have an impact on caves and need to be carefully monitored. For any form of management to be effective it must involve local communities, visitors and other user groups in the decision-making process.

Cave Infrastructure

Many of the adverse effects of visitors can be minimized by the installation of environmentally sensitive infrastructure. On the other hand, such installations can also have a major impact on the cave system. In Carlsbad Cavern (USA) there are more than 1,400 lights, 65 km of electric wiring and 6 km of walking paths. Surface impacts resulting from the construction of car parking areas, walking tracks, kiosks, hotels and interpretive centres may be added to the direct underground impact. The level of visitor enjoyment and appreciation of a cave is often intimately

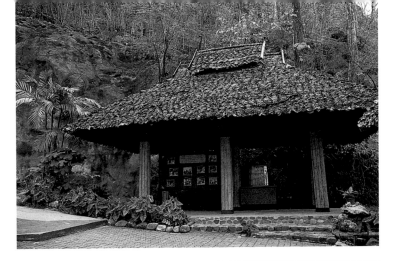

Signboards are an important part of the visitor experience.

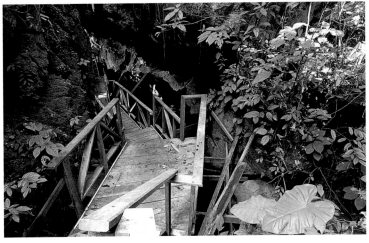

Wooden pathways can upset the ecological balance of caves.

connected to the facilities provided. Signboards, pamphlets and well-informed guides are an important aspect of education and long-term cave conservation through the raising of public awareness of the unique values of caves.

Well-defined pathways are vital to reduce the impact of visitors on easily damaged cave floors and to keep people from straying off into particularly sensitive areas. Paths should be permeable so they do not interrupt drainage patterns and must be removable as far as possible to allow for replacement and design flexibility in the future. Plastic sheets placed underneath concrete paths can help to facilitate their removal at a later date with less damage to the cave floor. Many caves have been irreparably damaged by the installation of concrete paths, platforms and other constructions directly on top of cave formations. A better alternative is to raise paths above the floor. However, many other materials can also be damaging – iron or zinc coated railings quickly corrode in moist cave conditions, whereas wooden steps become a food source for fungi and other organisms which might upset the delicate ecological balance in caves. The best option is probably recycled plastic boardwalk, ceramic tiles, stone slabs or other inert materials.

Lighting

Several dozen caves in Northern Thailand have lighting systems permanently installed and this number is likely to increase significantly in the future. Lighting systems must take into account the effect on visitor perceptions as well as economic factors such as electricity prices and replacement costs. The cheapest method, which has been installed in many temple caves such as Chiang Dao, Tab Tao and Muang On is fluorescent bulbs. However, these systems can seriously impinge on the aesthetic qualities of a cave and produce a harsh, cold light which is accentuated by the grey colour of limestone. Very few caves in Northern Thailand have aesthetically pleasing lighting systems – Tham Pha Thai (Lampang) is an outstanding example. Placement of lights can be difficult because it depends partly on aesthetic perceptions which are largely a matter of personal taste. The best forms of lighting are non-intrusive and hidden from view as much as possible. Lighting should be designed to highlight important features of a cave, such as a Buddha statue or particularly fine speleothem. Backlighting is also very effective at creating a dramatic or mysterious feeling. Many temple caves have traditionally used natural lighting from cave windows high in the ceiling to great effect. The sense of drama is heightened if some areas of the cave are left unlit, this also gives visitors more of a feeling of the natural conditions in a cave and helps to preserve the cave ecosystem. An alternative approach is to abandon any attempt to portray the natural cave environment and create a fantasy world with coloured lights and other effects.

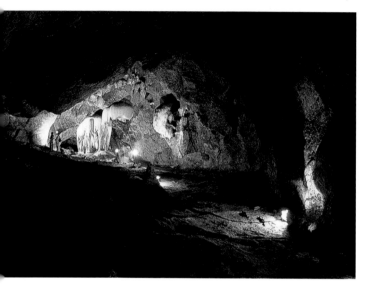

Attractive spot lighting in Tham Pha Thai.

Artificial lighting can have detrimental effects on cave systems. Prolonged artificial lighting leads to the growth of algae and other green plants ('Lampenflora') which causes unsightly green staining of cave formations, corrodes speleothems and can upset the natural ecological balance. This effect can be minimised by using low-wattage bulbs and only switching the lights on while visitors are actually present. Lights should be installed on smooth surfaces wherever possible as this makes plant colonization more difficult. As with any form of installation, all lighting systems should be easily removable as they will eventually degrade and have to be replaced.

The use of hurricane lamps such as at Tham Chiang Dao or Tham Lod (Pang Ma Pha) has some advantages because it avoids the need for any kind of permanent installation. In addition, it ensures that almost all visitors are accompanied by a guide and provides considerable extra revenue for local villagers. Hurricane lamps are only suitable in large caves with plenty of air flow and even here it is likely that high visitor pressures will eventually lead to blackening of cave ceilings and formations which will be very difficult and costly to clean.

Airflow, Temperature and Humidity

Airflow is an important consideration for cave management since many areas may be impassable or dangerous to visitors without adjustments to the natural conditions. Many of the most spectacular speleothems and crystals occur in places with little air movement so the installation of air circulation systems is likely to arrest the growth of these features. This is particularly the case with caves where there was previously no natural opening to the outside world, such as in Tham Kaew Komol where a system was installed but is only rarely used due to fears that it was accelerating the degradation of the outstanding crystals which are the cave's primary attraction.

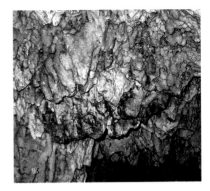

Lampenflora in Cango Caves, South Africa.

Many caves in Northern Thailand have sections with low oxygen levels, which could quickly be depleted by large groups of visitors. Even popular tourist caves such as Tham Pha Thai and temple caves such as Tham Tab Tao suffer from this phenomenon. Visitors also increase temperatures and humidity in confined spaces which can upset the delicate natural balance of cave ecosystems and geological processes. The ideal option in these circumstances is to restrict visitor numbers or fence off potentially dangerous parts of the cave. Any form of barrier in a cave must be carefully designed so as not to restrict air movement, water flow and the passage of cave animals.

Survey and Classification

Scientific surveys are an essential precursor to effective management of all natural resources, including caves. Such surveys need to be holistic in their approach and should be geared towards developing a proper understanding of the functioning of cave systems as well as the human expectations and pressures on those systems and the range of management options suitable for each cave. The recent survey of caves in Pang Ma Pha by a team of scientists from Mahidol and Silpakorn Universities is the first step towards the development of an integrated management system which is crucial to the long-term survival of the caves and the people which rely on them.

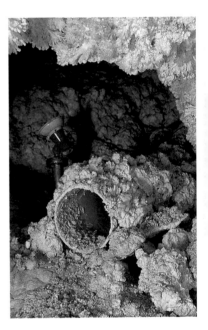

Air circulation system in Tham Kaew Komol.

One of the primary outputs of cave surveys should be the development of a cave classification scheme which is a vital tool for any management regime. Cave classification needs to consider an individual cave in relation to the immediate area that surrounds it, to the rest of the park or reserve in which it is located, and to its national and global context. Geology, hydrology, biology, archeology, culture and geographical location all need to be taken into account.

Several different classification schemes have been adopted around the world to meet the needs of cave management in different areas. All such schemes would need to be adapted to local

*Archaeological survey of
Ban Rai Overhang by a
team from Silpakorn
University. (© Highland
Archaeology Project).*

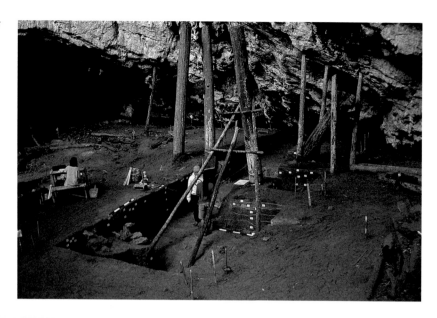

Classification of Caves in Pang Ma Pha

1 Public Access caves	no of caves
1.1 Adventure caves	7
1.2 Tourist caves	4
1.3 Religious sites	4
2 Special purpose caves	
2.1 Reference sites	3
2.2 Sites of special natural and/or cultural value	100
2.3 Dangerous sites	2
2.4 Human industry sites	7
3 Wild and Unclassified caves	
3.1 Wild caves	45
3.2 Unclassified caves	14

(adapted from John Spies)

circumstances before they could be applied in the context
of Northern Thailand.

The Recreational Opportunity Spectrum, developed in
Australia, identifies the range of recreational opportunities at a
site, the basis for identifying these, and the physical resources
needed. This approach requires that caves be classified in terms of
their potential uses, likely impacts and resulting environmental
conditions. These can range from undisturbed natural
environments (wilderness caves) through to highly modified
environments suitable for high density use (tourist caves). A
modified version of this system has been applied by John Spies
to 176 caves in Pang Ma Pha (see inset). Pang Ma Pha is unusual
in the context of Northern Thailand because of the very high
number of archeological sites (class 2.2), and the relatively low
number of religious sites (Class 1.3).

The Limits of Acceptable Change system, developed in
Britain (Stankey et al. 1985), is concerned with defining key
environmental conditions and maintaining them. This system
involves selecting key indicators, setting standards of achievement
for each indicator, monitoring to allow comparison to that
standard, and modifying recreational use and management
strategies in the light of non-conformity to that standard. The
first need in the application of this system is to determine target
features or organisms that can be used as indicators. A physical
indicator might be the soiling of speleothems in a heavily used
passage; a biological indicator might be the maintenance of a
viable population of a particular species. Frequent monitoring of
one key parameter is preferable to the occasional monitoring of
many. It is better to monitor an indicator which is simple and
cheap to measure at many sites than a complex and expensive
one. This system has not yet been applied to caves in Northern
Thailand but could be modified to suit local conditions.

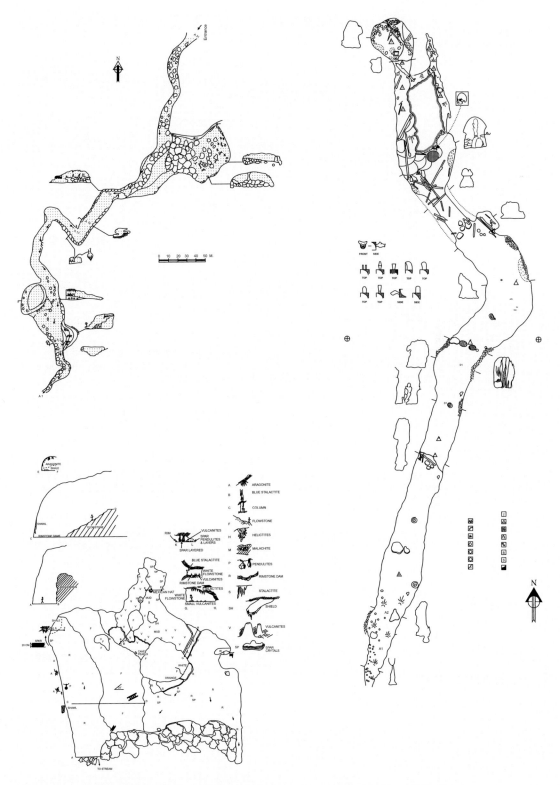

Accurate cave surveys are a vital first step towards drawing up appropriate guidelines for cave management.
Top: *Tham Pang Kham.* Bottom: *Upper level of Tham Pha Mon.* Right: *Tham Pang Kham Pi Man* (© *Pang Ma Pha Cave Project*).

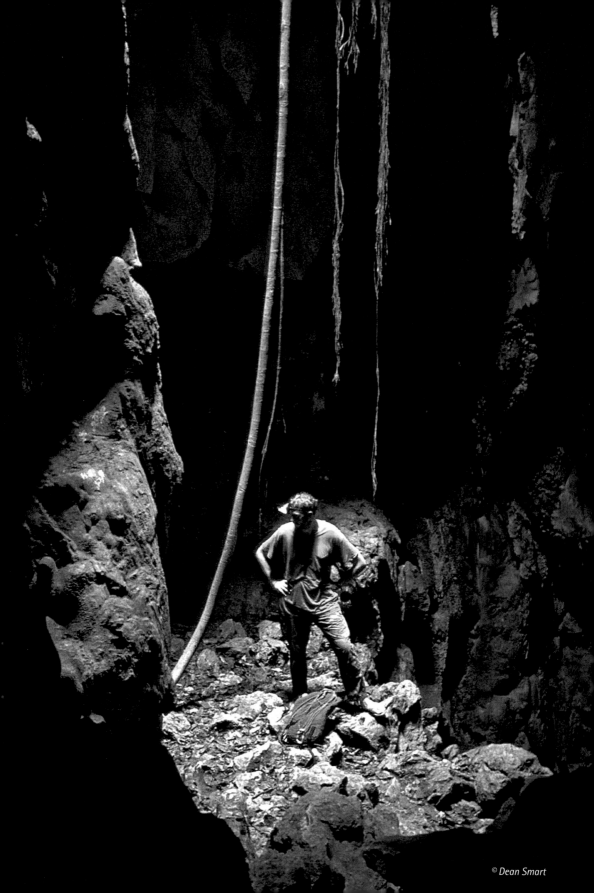

© Dean Smart

CAVE EXPLORATION

Caves are beautiful, exciting and fascinating places. No wonder people like to explore them. Unfortunately, caves can also be dangerous and remarkably easy to damage-aspects of which many explorers seem blissfully unaware.....

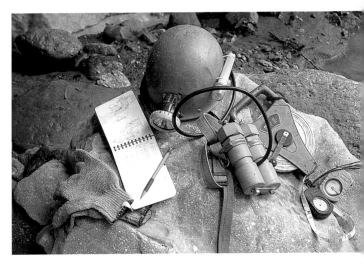

Equipment used for surveying caves.

EQUIPMENT

The equipment required to explore easy caves does not need to be specialised or expensive, but it does need to be of reliable quality. A basic list includes main head torch, backup head torch, helmet, backpack, spare batteries and bulbs, food, spare clothes, boots and appropriate clothing. Other items that are useful include an extra torch, gloves, kneepads, waterproof bag, a cigarette lighter and some empty plastic bags for carrying out litter.

Lighting

The most essential piece of equipment is the main head torch. This provides the principal source of light whilst in the cave and needs to be bright and have good duration. Above all, the main head torch must be reliable. Most head torches consist of two parts – the headpiece where the light is produced and a power source of some kind, usually a battery or a calcium carbide 'generator'. The headpiece is best mounted directly onto the helmet, either with a bracket (better) or elasticated straps. In this way the light will always be shining in the same direction as the head is pointing.

Battery-powered head torches are the best lighting systems for caving. They are simple, clean and reliable. The battery is connected via a cable to a light bulb inside a headpiece and the light is turned on and off using a switch. This design has very little that can break or go wrong and if the light does not work, it usually means the bulb or battery needs replacing – a simple repair. The best battery-powered head torches have a headpiece containing two bulbs, a bright main bulb for normal caving and a smaller bulb for rest periods, such as eating lunch. Switching between the bulbs conserves battery power. Headpieces designed for use by the mining industry are perfect for caving. They are strong, waterproof and can be bought quite cheaply second-hand. Lamp brightness and battery duration are determined by the type and size of battery used and the light output of the bulb. Larger batteries last longer than smaller ones, but are heavier and more cumbersome. Disposable alkaline batteries last longer than rechargeables but are more expensive in the long term and less ecologically friendly. They are however, essential in remote areas where mains electricity is unavailable. Some battery packs attach to the back of the helmet. This conveniently does away with a long cable but the extra weight on the head limits the size of the battery. Wearing the battery on a waist belt allows a larger, longer lasting battery to be used. Waist mounted batteries made up of either a pack containing three or fore alkaline 'D' cells, or a 6-volt rechargeable lead-acid battery provide good compromises between power and bulk. Bulbs must be carefully selected as they need to be bright enough to see the cave by though not so bright that the battery drains too quickly. A main bulb producing about 3 to 6 watts of light is a good choice. The 'lunch' bulb can be less bright – about 1 watt is sufficient. Halogen bulbs are brighter than ordinary bulbs, but it is more important to use good quality, long lasting bulbs rather than cheap, bright ones. The system currently being used by myself is a 4 alkaline 'D' cell battery made from blue PVC pipe with a two-bulb mining headpiece fittedwith a 3 watt halogen main bulb and a 1 watt lunch bulb (see photo page). With a fresh set of batteries, this lamp produces more than 30 hours of light on the main bulb and at least 90 hours on the smaller one. Outdoor stores sell ready-made head torches designed specifically for use in cave exploration. These lamps are generally very good, being reliable and providing up to 10 hours of light from a single set of batteries. They are however quite expensive to buy and are less sturdy than mining lamps. If the task of building a lamp seems a little daunting, buy a manufactured one.

Recent technological advances have enabled the use of white light emitting diodes (LEDs) instead of ordinary filament bulbs. The main advantage of using LEDs is their extremely long life-span, over 10,000 hours compared to 100 hours for a good quality bulb. They can also be dimmed and brightened as required to save battery power. The down side is that they need delicate electronics to work efficiently and this must be protected from water and sudden jolts.

The backup head torch must be bright enough to use as a main caving light in the event of failure in that light. It also needs to last the duration of a caving trip. Head torches sold in outdoor stores make good backups. It is also a good idea to carry a third light, a hand-held torch or similar, just in case the backup fails as well.

A widely used alternative to the electric lamp is the acetylene, or 'carbide' lamp. These lamps produce light by burning acetylene gas, which is created by adding water to calcium carbide (a byproduct of the steel industry). In modern set-ups, the gas is created in a waist mounted 'generator'. This has two compartments – an upper one to hold the water and larger, lower one for the calcium carbide. The water is allowed to slowly drip onto the calcium carbide and the gas produced passes up a tube to a nozzle on the front of the helmet where it is burned. Regulating the water drip controls gas production and this in turn governs the size of the flame and the brightness of the lamp. A single fill of calcium carbide can give up to 10 hours of bright, even light. However, calcium carbide often contains impurities that form poisonous byproducts during the making process. One of these byproducts is hydrogen sulphide gas. This highly toxic substance smells like rotten eggs and is the reason why acetylene lamps stink

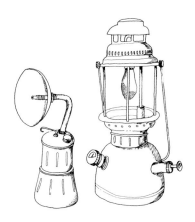

Left: Carbide lamp.
Right: Hurricane lamp.

even though acetylene gas itself is odourless. Careless disposal of spent calcium carbide also poses a pollution risk. For these reasons and the fact that sales outlets for calcium carbide are becoming scarce, electric lighting is considered a better option.

Hurricane lamps are commonly used especially by local people. They are quite difficult to light and require regular maintenance but the fuel – paraffin – is cheap and widely available. Once lit, they are extremely reliable and can be used continuously for at least 8 hours on a single fill. However, they are cumbersome to carry and the mantles are easily broken. Also, burning paraffin creates black smoke and heat that can upset the cave environment, damage speleothems and harm cave life – especially in caves with poor air circulation. As with carbide lamps, electric head torches are preferable.

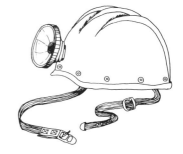

Helmet

Helmets are essential to cave explorers. These seemingly simple devices not only provide protection to the head from falling rocks, the cave roof or during a fall, they also provide a convenient place on which to mount head torches and for keeping things dry. Helmets come in a wide range of qualities and prices. The cheapest are plastic building site helmets. These are low quality, uncomfortable to wear and are not tested for safety, though they are definitely better than nothing. One-time-only explorers in safe, easy caves will find them adequate. Serious cave explorers require something better as they wear a helmet regularly for many hours at a time and go into caves that are more dangerous. The serious caver's helmet needs to have a snug, comfortable fit and not move around on the head. It must feel light and balanced after all the lights are attached and it must have a 'Y' shaped chin strap. Most importantly, the helmet must provide the best protection available. Helmets with good protection qualities carry an international safety standard, such as the European CE standard. Always use one of these helmets.

Backpack and Clothing

Cave exploration backpacks are made from extra strong, PVC coated nylon material and have a minimum of straps and buckles to prevent snags. These are the best, if a little expensive, and will last several years of being dragged through caves. Cheap, locally made backpacks can be used, but carry a sewing kit to make repairs when they fall apart. A caving backpack needs to be large enough to carry spare clothes, food, spare batteries and bulbs and everything else you wish to carry, such as a camera. Everything can be kept dry by lining the backpack with a waterproof, rafting bag.

Clothing should be lightweight and quick to dry. Artificial materials like polypropylene, polyester, rayon and nylon are best. Avoid cotton and wool as these materials stay wet for a long time and as they dry, they draw heat out of the body. Never wear jeans. A good clothing combination for exploring caves in northern Thailand would be a pair of tight fitting tracksuit bottoms, shorts and t-shirt. Wearing a lightweight, one-piece nylon 'oversuit' will

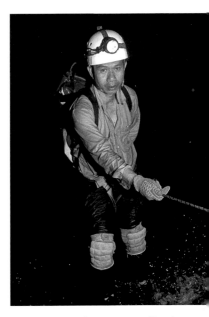

Cave helmet with battery-powered head torch. (© Dean Smart)

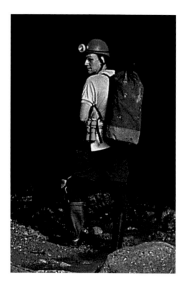

Backpacks need to be strong with very few straps.

help prevent bare arms and legs from being scratched though it can get uncomfortably hot. Boots with heavy treads and ankle support are preferable to training shoes. The most popular footwear are walking boots or Wellington boots. Gloves are useful to protect the hands from cuts and kneepads make crawling a great deal more pleasant. A short sleeved wetsuit top may be required in cold, wet caves or caves with lots of swimming, in which a life-vest may also prove useful. Vertical cave exploration and cave diving require the use of highly specialised equipment and techniques. It is beyond the scope of this book to attempt to describe these activities. Besides, most of the known caves in northern Thailand and all of the caves included in this book are horizontal and air-filled. People interested in taking up these pursuits should consult the books listed in the bibliography and contact a suitable organisation for training.

SAFETY

Safety underground is a very serious consideration for all cave explorers. Caves can be extremely dangerous places and many people have been injured or have died during their exploration. Very few accidents though are due to circumstances beyond our control. By far the greatest majority occur because of a mistake made by the explorer. Inexperience, errors of judgement, poor equipment or not using equipment properly account for many more accidents than the cave roof suddenly collapsing, unseasonal flash floods or being bitten by venomous snakes. By being sensible and taking adequate precautions, cave explorers can reduce the risks to a very low level.

First of all, tell somebody where you are going and what time you are expected to return. Then, if you don't come back, that somebody can organise a rescue effort. Always explore caves as a team – no less than 4 people. If one team member has a problem, another can stay with them while the other two go to get help. Before going underground, make certain that every member of the team is adequately equipped, experienced enough and fit enough to undertake the planned trip. If not, choose an easier cave. If the trip is an original exploration, be aware of each member's limits and do not exceed them. Always move as a team at the pace of the slowest member.

Extremely dangerous section with recent rockfall caused by blasting.

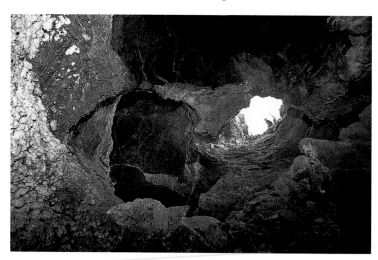

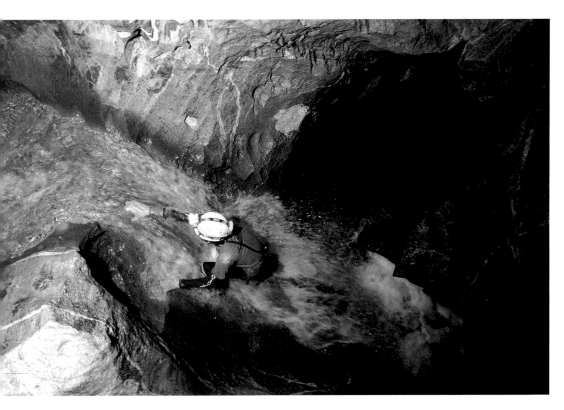

Scaling a waterfall in Tham Mae Lana. (© Dean Smart)

Where there is the possibility of falling down a hole in the floor or a shaft, everybody needs to be securely attached to a firmly anchored rope. Climbing up and down waterfalls, ladders or the cave wall also requires a rope belay. Take special care when crawling through or climbing over boulders. Even very large rocks can be loose and move unexpectedly. Try not to knock debris down onto people below. If a rock does begin to fall, warn everybody by shouting 'ROCK!' Uneven floors and wet, slippery mud can be treacherous to walk on. Pay attention to where you put your feet.

Flooding occurs in many caves. Fortunately, the weather patterns of northern Thailand are reasonably predictable. Most of the rain falls between May and October – a time when stream caves should be avoided. The months with the lowest water levels (and lowest flood risk) are March and April. In the event of a sudden flood, everybody should try to remain calm. Carefully climb up to above the highest flood level (marked by sticks and leaves stuck to the wall) and wait for the water to go back down. If the wait is long one, cold may become a problem and that extra clothing being carried by everyone will become useful. Cold may also be a problem at high altitude and during the cold season of December to February. Again, carry extra clothing and use it if necessary.

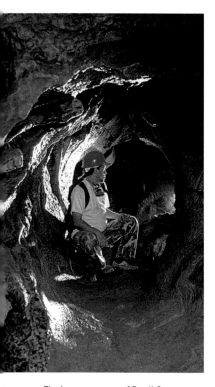

The inner passages of Fossil Cave can have high CO_2 levels.

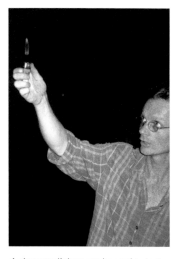

A cigarette lighter can be used to test oxygen levels.

Bad Air

Many caves of northern Thailand have a phenomenon called 'bad air'. This is where the atmosphere of the cave has either an elevated level of carbon dioxide, a reduced level of oxygen or both. Initial signs that high carbon dioxide is present (up to 4%) are shallow, rapid breathing, feeling tired and not being able to concentrate. At higher levels (4 to 8%), very rapid breathing is accompanied by hot flushes. At very high levels (>8%), unconsciousness occurs and may be quickly followed by death. Low oxygen does not initially cause any noticeable physical effects and breathing remains normal down to 16%. A cigarette lighter that has been kept dry may be used to test for low oxygen. It will fail to light at about 17%. At moderate to very low oxygen levels (<16%), similar symptoms to high carbon dioxide will be experienced. If any signs of bad air are met in a cave, the best response is to turn around and SLOWLY walk out. Do not panic and run as the physical exertion will only make the problem worse. Abseiling down a shaft into bad air is particularly dangerous as climbing back up the rope is strenuous and it is very easy to fall unconscious. Advanced rope techniques are required to haul an unconscious explorer up a shaft and novice explorers should not attempt it. Be aware that bad air levels can change rapidly and dramatically depending on the year, season, time of day and location within the cave. Unfortunately, insufficient research has been carried out to enable accurate predictions of where the worst times and places will be. Until that research is completed, take care.

Dangerous animals are occasionally encountered in caves. Generally they do not venture more than 200–300 m in, but great care is needed in this zone to avoid being bitten or becoming a meal. Snakes are often seen. Most are harmless like the Cave Racer and very few are dangerous to people. To be safe though, leave all snakes alone. Snakes always prefer to crawl away and hide if given the opportunity and only cause problems when they feel trapped or threatened. Centipedes, spiders, scorpions, wasps and other large invertebrates can inflict nasty bites or stings. Try to avoid contact with them. In areas where they still survive, bears, large cats and even elephants use caves for shelter. Meeting one of these in a cave would be extremely hazardous. If there is any suspicion that the cave is already occupied (fresh footprints, rub marks, smell, noises, etc.), it is best to go to a different cave. Also, be wary of porcupines in low crawls.

Exploring caves often involves swimming in dirty water, crawling in mud mixed with animal faeces, breathing dust, rubbing against sharp rocks and being bitten by flies, mosquitoes and parasites. Any of these activities introduce the risk of contracting a range of serious medical problems, such as infections, diarrhoea, leptospirosis and malaria to name a few. Cave explorers can reduce this risk by using a little common sense. Wear protective clothing to minimise cuts and scratches. Clean any cuts and scratches that are picked up with antiseptic cream and cover them with waterproof plasters. Also, avoid swallowing unclean water and ward off parasites with mosquito repellent. Should a fever or any other serious medical condition appear during or soon after a trip, it needs to be investigated by a doctor. Explain very clearly what you have been doing, where, when and the kinds of things you encountered.

Vertical cave exploration and cave diving are extremely hazardous without specialised equipment, knowledge and experience. Proper training is essential before carrying out these activities. It is worth noting that most cave diving fatalities are not experienced cave divers carrying out long, deep exploration. Most deaths are actually open water divers who "go in for a look". They assume that diving in a cave is just the same as diving in the open sea. It most certainly is not. The equipment, techniques and philosophies are all different and divers who ignore this fact are placing themselves in great danger.

Another important thing to remember is that there are no specialist cave rescue teams in northern Thailand. The police or army could be called upon for assistance, but it seems more sensible to practise self-reliance. Being sensible and preparing adequately will prevent many emergency situations from happening in the first place. Cave explorers can also make sure that they have the equipment and ability to perform a self-rescue in the event of a situation.

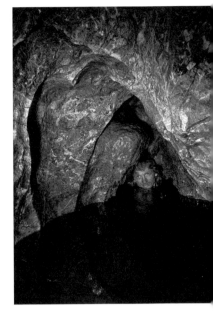

Avoid swallowing dirty water when swimming.

Cave accidents get widely reported in the media (probably because they do not happen very often) and a distorted image of caves as being inherently dangerous has been created. No statistical analysis has ever agreed with this idea. Cave explorers are much, much more likely to die in a road accident whilst driving to the cave than they are during the caving trip itself. Despite this fact, exploring caves, like driving cars, can be dangerous without adequate experience, equipment and forward planning.

BE SENSIBLE – BE SAFE!

Conservation for Cave Explorers

Caves are complex geological and biological systems with an environment that remains stable for a very long period of time. They are also physically confined by walls and a roof and have limited exchange with the surface. All damage caused by explorers is therefore concentrated into a tiny, very fragile space, which has little opportunity to repair itself naturally. This damage begins as soon as the original explorer enters the cave and keeps on accumulating with each subsequent trip. The impacts caused are often unintentional, sometimes deliberate, but once the damage is done it is usually irreversible.

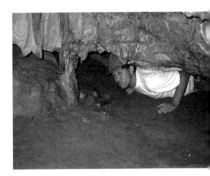

Low crawl in fossil cave.

Modern cave explorers try to minimise their impacts as much as possible by following the guidelines promoted by such organisations as the National Speleological Society (USA) and the Australian Speleological Federation. It is the responsibility of ALL cave explorers to do this, not just the original explorers, but everyone else who follows as well. Only with the cooperation of everyone will the caves of northern Thailand have any hope of remaining in good condition.

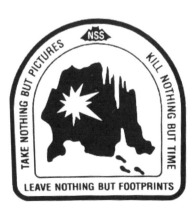

TAKE NOTHING BUT PICTURES
NSS
KILL NOTHING BUT TIME
LEAVE NOTHING BUT FOOTPRINTS

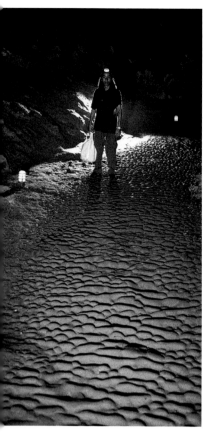

The wrinkled, brain-like surface of a heavily damaged cave floor in Tham Chiang Dao. Sediments floors are extremely important, but are easily damaged by trampling.
(© Dean Smart)

So, what does a responsible cave explorer need to be aware of? How is a cave affected by the actions of visitors? Obviously, deliberate vandalism such as breaking speleothems, writing graffiti, catching and disturbing animals, damaging or stealing archeological remains, lighting fires, etc. is inexcusable and a responsible cave explorer would never do such things. In many caves, such behaviour would be illegal as well and may result in prosecution.

Unfortunately, many of the impacts caused by cave explorers are small and accumulative and it is not always immediately obvious that a problem exists until it is too late. The damage caused to sediment floors by trampling provides a good example of this.

Each time a person walks across a sediment floor, the sediment is compacted a little by the person's body weight. Small pieces of sediment stick to the persons shoes and get carried around as well. Over time, this compaction and sediment tracking build up until the floor is destroyed and its capacity as a habitat is impaired. Sediment floors that have already been destroyed by trampling have an appearance like the surface of a brain.

Other impacts come from smoking in caves and discarding used calcium carbide and batteries. These actions release many toxic substances into the cave ecosystem that can poison cave fauna.

Bacteria and fungi are easily transported between caves in the mud and dirt stuck to boots and clothes. If this mud is not removed it could spread new and invasive species between caves.

Dropping pieces of rice and food in a cave causes a problem called 'enrichment'. This is an artificial increase in the amount of energy available to the animals of the cave, which are adapted to a low, stable supply. Any sudden increase in energy may upset the balance of the ecosystem and cause it to collapse. It may also attract non-cave animals from outside. All degradable material – food, paper, cardboard, wood, candle wax and cloth – will cause enrichment problems.

Human skin produces oils which rub off onto speleothems when they are touched. These oils stop further calcite from being deposited and cause the speleothems to 'die'. Touching also increases the risk of breakage, especially the tapping of speleothems to see if they 'sing'.

Breathing and body heat can alter the chemistry, humidity and temperature of the atmosphere and start the movement of irregular air currents. This not only affects the biology of the cave, but also its preservation qualities and archaeological material may begin to suffer. Speleothems may also dry out.

Loud noises and bright lights easily disturb bats. If the bats move out, their precious guano will go with them and the entire ecosystem will be affected.

Various techniques can be used to aid the conservation-minded explorer:

Use only battery-powered lighting and not calcium carbide, flaming bamboo torches or hurricane lanterns.

Walk in single file and avoid walking on delicate floors (remove muddy footwear where this is not possible). Where the roof is low, remove your helmet to reduce the risk of breaking stalactites – but mind your head! Also watch the heads of your companions and warn them if they get too close.

Always take empty plastic bags for carrying out litter, spent batteries, etc. Also, gather up and carry out the litter dropped by earlier, irresponsible explorers. Rub out their graffiti as well.

Chewing nicotine gum is a good alternative to smoking cigarettes in caves. Lay down a plastic sheet to catch falling food scraps at lunchtime and take it out of the cave.

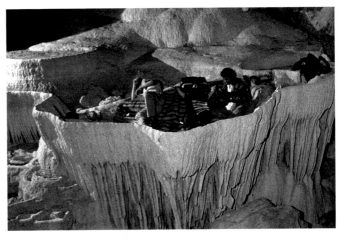

Camping in a cave introduces many potential impacts – unless great care is taken. (© Dean Smart)

Keep the noise down and point torches and flashes away from bats. Try not to remain in any one place for too long and wash muddy caving equipment after every trip.

Try to go to the toilet before entering the cave.

Camping underground dramatically increases the risk of damaging a cave unless special methods are used. For example, ALL waste generated by the camp MUST be removed from the cave and this includes human waste. This thought should put most people off the idea. There is also fact that no cave in northern Thailand is long enough to require a camp, all can be explored in a single day. If necessary, camp outside at the entrance where it is more comfortable and convenient.

Cave explorers are often faced with a dilemma, especially in fragile caves. Knowledge of where the cave goes and what it contains is important, but does that justify the damage that will result from the caves' exploration? Sometimes it does not. If there is any doubt, conservation of the cave takes priority and turning back is the best option.

Minimal Impact Caving Code *(excerpts)*
Australian Speleological Federation 1995

Remember EVERY caving trip has an impact. Where possible, the party leader should have visited the cave previously and should be aware of sensitive features of the cave and reduce the need for unnecessary exploration.

Cave slowly. You will see and enjoy more, and there will be less chance of damage to the cave and to yourself. This especially applies when you are tired and exiting a cave.

Keep your party size small – 4 is a good size.

Constantly watch your head and those of your companions to avoid damage to speleothems.

Ensure all party members stay on marked or obvious paths. If no paths are obvious, define ONE.

Learn to recognise cave deposits or features that may be damaged by walking or crawling through them.

Take care in the placement of hands or feet throughout a cave (oils).

Wash your caving clothes and boots regularly so that the spread of bacteria and fungi are minimised.

If it is necessary to walk on flowstone or rimstone, remove any muddied shoes or clothing first or do not proceed.

Treat the cave biota with respect, watch out for them and avoid damaging them and their traps, webs etc. Also avoid directly lighting cave biota, esp. bats.

If you eat food in a cave ensure that small food fragments are not dropped as this may impact the cave biota.

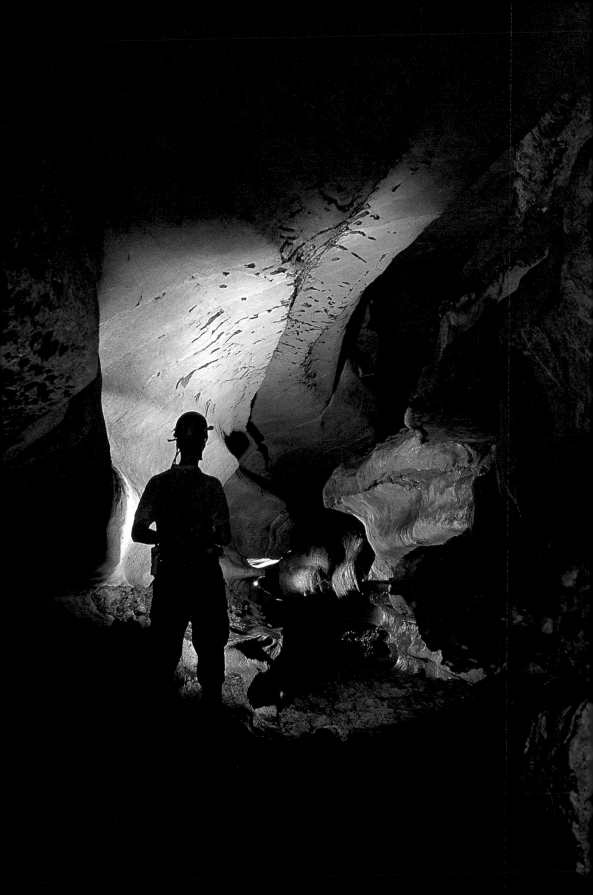

CAVE PHOTOGRAPHY

"The satisfaction and challenge of cave photography lies in the careful and creative use of lighting under difficult conditions. The importance of lighting makes cave photography in some ways similar to studio photography, which can teach us a great deal about lighting technique. The similarity does not go very much further! The studio is a comfortable environment for the photographer and his models, and there are no serious limitations on the weight and bulk of his equipment. In contrast, the cave photographer works in cold, wet and often muddy surroundings, his helpers quickly become cold and bored, and mud and damp get into the equipment. He has to carry and use his camera and lighting in constricted passages, and this leads to compromise on its size and weight. Overcoming these difficulties gives immense satisfaction and can produce photographs completely different from anything achieved on the surface."   Rowland (1995)

Equipment

All photographic equipment used in caves must first and foremost be able to withstand the rigorous physical conditions to which it will be exposed. Caves are typically very dusty or muddy places and often extremely damp – much of the time you will be working in conditions of 100% humidity. You must crawl with your equipment through narrow passages and wade or even swim along waist-deep streams. In addition, you have to operate in situations where the floor is very uneven and you cannot see what you are doing clearly – the risk of equipment crashing to the ground at some point is very high.

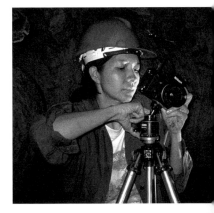

Many of these risks can be avoided by having a good protective casing for when the equipment is not actually in use. Tough, purpose-made plastic boxes with foam linings do the best job. Ammunition boxes with a home-made lining are equally protective and much cheaper, although rather heavy. Photographic dry bags are another cheap option, but do not protect against knocks. If you are only going to be using your equipment in dry, open caves, any normal camera bag wrapped in a thick cloth and plastic bag gives some degree of extra protection.

Camera Given the arduous conditions of caves you will probably find that cameras with complex electronics are more of a liability than an advantage. The last thing that you need after an exhausting several hours crawl into an awesomely beautiful cavern is to find that your camera has stopped working. Mechanical shutter over-ride in case of electrical failure is a godsend in such circumstances. The only really indispensable camera feature for cave photography is a 'B' shutter setting for very long exposures. Remember that most of your work will be in the dark so the less complicated dials and settings the better. Most of the photos in this book were taken with Nikon F3 and FM2 cameras using manual exposure.

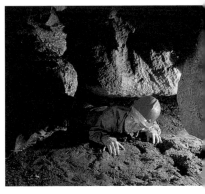

Lenses If you are going to be doing a lot of cave photography, you should definitely consider purchasing a wide-angle lens. In the constricted space of a cave you often cannot get far enough away from your subject to get the whole picture with a standard 50 mm lens. A 24 mm lens is probably the best single wide-angle lens, a combination of 20 mm & 35 mm is more flexible. A standard

Equipment used for this book
Nikon F3 and FM2 bodies
20 mm Nikkor
28 mm Nikkor
55 mm Micro Nikkor
105 mm Micro Nikkor
1.4X teleconverter
Nikon SB26 flash (X2, guide no 36)
Metz flash (X2, guide no 32)
TTL flash sync. cord
Cokin filter 1B (pale blue)
Manfrotto C190 tripod
Fuji Provia 100 slide film

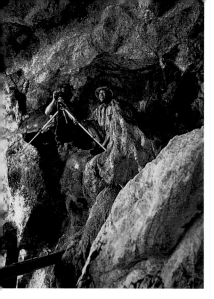

Tripods are indispensable for cave photography. This is a Manfrotto C190 with legs that can be set at almost any angle.

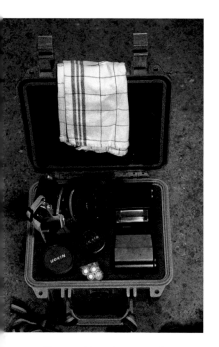

Photographic equipment needs to be well-protected in a sturdy waterproof container.

50 mm lens is useful for speleothems and medium-sized subjects. Long telephoto lenses are of limited use in caves because the air is often full of moisture and dust which accumulates over long distances causing the shot to appear blurred. In addition, the limitations of artificial lighting mean it is very difficult to shoot anything over 15 m away. However, a short telephoto lens, such as 100 mm, is very useful to avoid getting too close to jittery cave life or fragile speleothems and to photograph inaccessible upper chambers. A short telephoto lens with macro facility is particularly useful for close-ups as the extra lens-to-subject distance makes flash much easier.

Tripods are essential equipment for most cave photography, an invaluable aid to composing your picture when you cannot see the whole scene at once. Set up the shot roughly then use a torch to pick out the corners of frame and make minor adjustments. Channel-section legs are better than enclosed-section legs – it is difficult to clean out the latter which can lead to early failure of the locking devices on the legs. Tripods whose legs can be angled independently of each other are useful on uneven floors and tricky situations.

Lighting systems

Except right in the cave entrance, you will need some kind of artificial lighting system. Deep inside the cave where it is pitch black you rely completely on your own lighting. In many ways this presents exciting opportunities – in no other natural situation is it so totally and utterly dark as in a cave, not even on a moonless night outside or even in outer space. If you left the shutter open for a year, there would still be no image on your film at all. This means that the lighting is completely under your control. Only what you choose to illuminate will be in the photograph – everything else will be black.

When you are actually in a cave you only see a small part of the overall scene with your torch at any one time, whereas in the photograph you have the opportunity to combine the whole scene into a single picture. One of the most exhilarating aspects of cave photography is that the image you capture in a photograph has the potential to be considerably more beautiful than what you actually experienced in the cave.

The most difficult zone for cave photography is the twilight zone close to the entrance. In this zone your lighting system has to compete with the natural light from outside. Short exposures will normally be needed to make sure that the area lit by natural light is not overexposed, but this does not give you time to manually set off the flashes or to use lamps. You will usually have to attach the flash directly to the camera body or use a series of sync. cables. Slave units are another possibility, but may not work in bright conditions.

There are four main ways of providing artificial lighting in a cave:
1. Electronic flash is by far the most flexible lighting system for
cave photography and is balanced to natural daylight so colours
will appear natural. Compact lightweight models with a minimum
of complex programmes are the best – sturdiness and reliability
are again of paramount importance. Most automatic exposure or
through-the-lens (TTL) programmes are of limited use in caves
because of the highly variable reflectivity of cave surfaces and the
large flash-to-subject distances involved. A notable exception is in
close up work – for crystals and cave invertebrates TTL exposure
systems make life a lot easier.

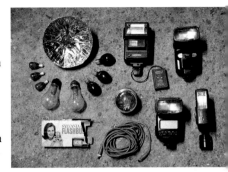

*Electroric flash, sync. cords, slave units
and flash bulbs used for photographing
caves.*

Any flash for use in cave photography must have an 'open
flash' or 'test' button which fires the flash when pressed – do not
consider any flash that does not have this feature. Another useful
feature is the ability to manually control the flash output to 1/2,
1/4 or less of full output. This helps you to balance each flash
according to the reflectivity of the subject or the flash-to-subject
distance and is especially useful for close-ups. You should buy the
most powerful flash that you can afford – even so, there will be
many times when you will wish you had more! More power means
smaller apertures and better depth of field, which can be critical if
the light is so dim that you cannot focus properly. The power of a
flash is conventionally given as a guide number – the higher the
number, the more powerful the flash. You can use the guide
number to calculate the maximum distance that the flash
can illuminate using the formula:

*Lighting this enormous chamber in
Tham Tab Tao required four people each
firing an electronic flash manually up
to four times.*

Guide Number/Aperture = flash-to-subject distance
(in metres, using ISO 100 film and 50 mm lens)

For example, using a flash with a guide number of 40
and aperture of f4, you can illuminate objects 10 m away,
whereas a flash with a guide number of 20 will illuminate
only up to 5 m at the same aperture. If you want to
photograph even moderately sized cave passages you will
need a flash with a guide number of at least 36. There are
very few portable flashes with a guide number greater than
40 and they are all very expensive and hefty to lug around.
The most powerful flash commercially available in Thailand
is the Metz 70 Mz-5 with a guide number of 70 but even this
flash can only reach 20 m at f4 under ideal conditions.
At over 35,000 baht this mega-flash is well beyond most
personal budgets and is in any case too hefty to lug
around a cave.

The only practical solution for photographing larger
caves is to use several powerful flashes simultaneously. The
use of two or more flashes can add tremendously to the
atmosphere and feeling of depth produced in a shot. You
may also need to fire each flash several times for each shot.
You can do this by using a tripod and simply keeping the
shutter wide open for as long as it takes to fire all your
flashes, using the 'B' setting-provided that the cave is in
pitch darkness and your subject is absolutely stationary.

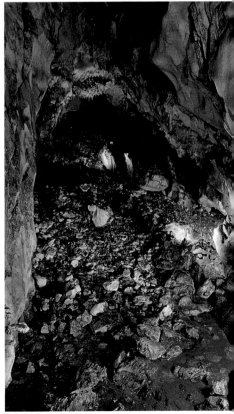

This technique considerably increases your lighting possibilities with the same equipment. The working distance of each flash increases with the square root of the number of times you fire the flash. To reach 20 m with a flash which reaches 10 m on a single firing you need to fire the flash four times, to reach 30 m nine times. In practice you will frequently have to add some extra illumination because over larger distances the dust and moisture reduce the reach of the flash and because many objects in caves have light-absorbent surfaces – the standard guide number is calculated in a normal room with flat, pale walls and perfectly clear air.

Any flash fired nine times in succession will get pretty hot – be careful not to burn yourself or overheat the unit. You will also need to use fairly fresh batteries otherwise you will be waiting around a long time for each flash to recharge nine times. If you are doing a lot of cave photography, rechargeable batteries are well worth the initial outlay. You will still need to take spare normal batteries as rechargeable ones do not last as long and you may not have chance to recharge them in a remote area.

2. Flash Bulbs are hard to get nowadays but are still very useful for certain situations in cave photography. The power of the largest bulbs is far greater than even the most powerful flash. Guide numbers of up to 180 are possible, which theoretically enables you to illuminate a subject 45 m away at f4 and is five times more powerful than the best electronic flash on the market. The other main use of flash bulbs is for photographing moving water. Electronic flashes have a duration of less than 1/1000 th sec which 'freezes' the water and destroys the sense of motion, whereas the duration of flash bulbs is around 1/30 th which gives a pleasing blurred effect. Each bulb can only be used once and has to be transported with care since it is easily broken. Flash bulbs are most suitable for the occasional special shot of large caverns or waterfalls rather than for regular use on a day-to-day basis. A further advantage of bulbs is that they fire at only a few volts, so that they can be used successfully and safely in very wet conditions, such as beneath waterfalls, and even under water. Electronic flashguns typically generate a few hundred volts, so for safety and reliability they should be kept as dry as possible.

3. Lamps, Gas Lanterns & Torches can all be used to provide extra light for a photograph and are especially good for black & white photography because the shadows are less harsh than with electronic flash. All of these light sources will produce an orange glow when used with normal film which is colour-balanced for daylight conditions. However, the effect can be quite pleasing if you are not aiming for a naturalistic shot. The exposure times are likely to be several seconds for even moderately-sized chambers, which can create problems if your human scale is not absolutely still. If you are using only lantern light,

4. Existing lighting systems An increasing number of caves have lighting systems installed to illuminate path-ways and features of interest. Chiang Dao, Pha Thai, Muang On, Tab Tao, Phum and Kaew Kamol caves all have extensive lighting systems whilst many other caves have lighting in certain sections. These systems are a great aid to casual visitors but cause difficulties for photography. Many systems make no attempt to hide the light bulbs, making it almost impossible to avoid burnt patches given the long exposures which are usually need to provide enough light for the rest of the shot. This problem can be partly overcome by using flash on short exposure so that the effect of the lighting is minimized.

Most of these systems consist of halogen bulbs or fluorescent tubes which appear white to the human eye but play havoc with the colour balance of normal photographic film adjusted to daylight. These colour-balance problems are not detectable with the human eye but show up clearly on photographic film. Fluorescent light imparts a blue-green tone to the whole scene, giving it an underwater feel and making people look extremely ill. To correct this distortion you can use a purple filter but any non-fluorescent light, such as flash or daylight, will then appear purple. Halogen lights give the whole scene a orange-red glow which can be rather attractive if you don't mind your photo looking like the inside of a volcano. Use a blue or pale blue filter to overcome this, although again any non-halogen lights will appear blue.

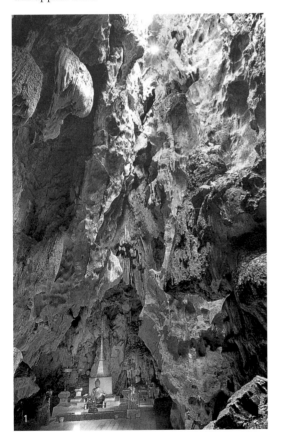

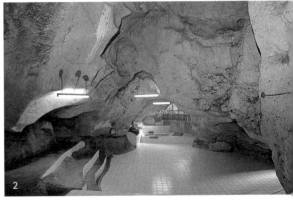

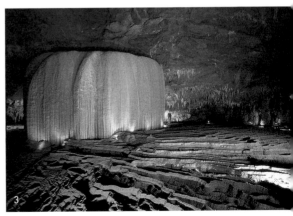

1 Difficult lighting conditions in Tham Plaew Plong Fha. Electronic flash and lanterns were used to complement natural light from a cave window.
2 Florescent lighting in Tham Phra Bam Pen Bun, imparting a bluish tint.
3 Halogen lighting in Tham Pha Mon, casting a dramatic reddish glow.

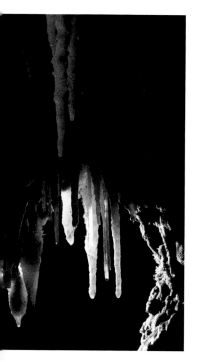

Back-lighting brings out the beauty and rich colours of these speleothems from Tham Pha Mon.

Positioning lighting

Many shots will need one or more flashes positioned away from the camera to provide dramatic back-lighting or to illuminate features at the farther end of the cave. Back-lighting is particularly effective in bringing out the magic of semi-transparent objects such as crystals. It can also help to highlight the subject, and produces attractive reflections from wet or highly textured surfaces such as streams or dripping spelothems. A further benefit of back-lighting is that it can give the impression that the cave goes on beyond the section being photographed, and so adds interest and depth. A flash placed directly on the camera not only produces flat lighting with no shadows, but will also reflect light from atmospheric dust and mist directly back into the camera. Indirect lighting is aesthetically more pleasing and may be the only way to illuminate large areas without burning out the foreground. There are several ways to position flashes off the camera:

Flash cables are very convenient if the flash is fairly close to the camera but rapidly become unfeasible at camera-to-flash distances greater than 3 m. Much longer cords are commercially available but are potentially extremely hazardous under cave conditions – someone is highly likely to trip over them in the dark sending the whole apparatus crashing to the ground.

Slave units are small units which attach to the bottom of a flash (or are inbuilt in some advanced flashes) and will fire the flash with a pulse from another flash, which can be near the camera. There are two types – normal slave units (including all inbuilt ones) are triggered by visible light but are usually only effective within 6 m of the source flash and must be in direct line of vision, which seriously limits their usefulness. 'Cats eyes' are similar cheaper and more readily available. They are quite useful for lighting up small spaces such as rock crevices but are equally difficult to trigger. You will need to use several in conjunction to create any serious lighting effect. The second type of slave units are triggered by infra-red light and are far more useful in cave conditions since they work up to about 500 m from the source flash and do not need direct line of vision. There is

Four separate flash units were needed to create a sense of space in this complex multiple layer passage in Tham Kaeng Khao.

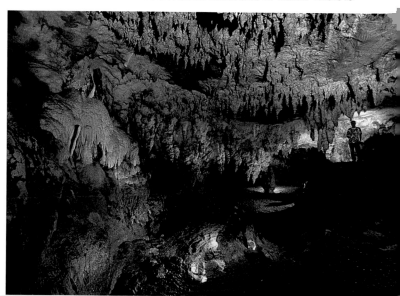

currently no supplier for the infra red type in Thailand, although they are available from abroad.

People who can manually trigger each flash are undoubtedly the best method in many circumstances, provided you have enough willing helpers prepared to put up with hiding in corners. There are difficulties if you want to light up a constricted corner and communication is a problem over large distances, especially with running water. At least one person should be in the shot to act as a scale. This person can also have a flash which they hold away from the camera, which can create an effective silhouette effect. To get a true silhouette don't flash the front of the person.

Techniques and other considerations

Conservation Photographers often seem to assume that they are somehow exempt from the usual rules of cave conservation because they are recording the beauty of the cave for posterity. This is definitely not the case – photographers can be a major source of damage especially in photogenic pristine environments because they often try to get off the beaten track to 'get that special angle'. Don't do it under any circumstances!

Scale With no normal objects such as trees to give a sense of scale, it can be difficult to convey an impression of size in caves. Geological features look remarkably similar at widely different scales – a 1 m long stalactite is very close in form to a 10 m one and a rockfall could be composed of pebbles or massive boulders the size of houses. The most convenient way to convey the true scale of a cave scene is to have a person in the shot. The person should be large enough to be visible but not so large as to dominate the shot. Brightly coloured clothing is an immense benefit especially in large chambers and can make the difference between a good shot and an indifferent one. Another alternative is for the person acting as a scale to hold a flash on their side away from the camera, creating a silhouette effect much favoured by professional cave photographers.

Reflectivity of the subject causes considerable difficulties and is often very hard to judge under poor lighting conditions. The surfaces in caves vary widely in their reflectivity and need correspondingly different levels of lighting to achieve correct exposure. Active speleothems with bright white crystalline surfaces will need many times less light than muddy walls and would appear completely burnt out if illuminated at the same intensity as the rest of the cave.

Over-illumination is when all parts of the photograph are evenly lit, creating a rather flat effect in which the sense of scale and distance is easily lost in a complex 3-dimensional space such as a cave. Caves are naturally dark places, so a good photograph should try to preserve this quality by having at least some areas dark. The balance between shadows and highlights is of course a subjective matter but a shot with no dark areas at all will rarely communicate the special character of caves.

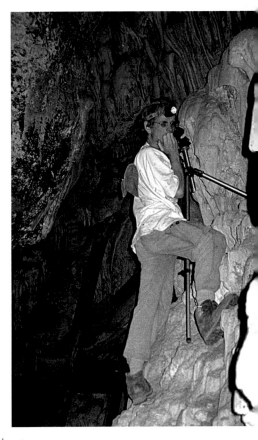

Photographers can cause serious damage in their quest for an unusual angle.

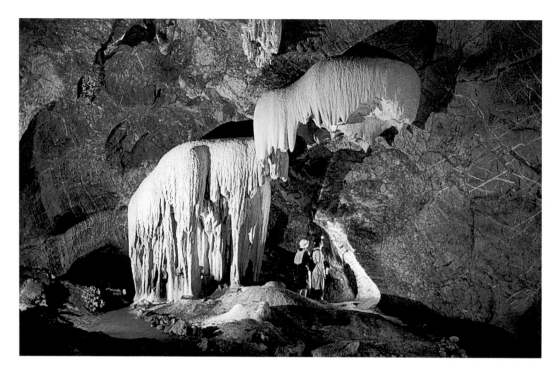

Eroded flowstone in Tham Pha Thai. Without the two people to act as a scale it would be very difficult to judge the true size of this feature. The orange tone is caused by halogen lighting.

Misting up of lenses can be a real headache in very damp conditions such as are prevalent in many caves. The problem is particularly acute when you pass quickly from dry, cooler air outside a cave to the moist, warmer areas within. Under such conditions, the best thing you can do when you arrive at a potential photo opportunity is to immediately take all the lens caps off any lenses you are likely to use. You can then spend time setting up the tripod and lighting and hopefully after 10 minutes or so the mist will have cleared. If your patience runs out and you have a heat source, such as a lantern, to hand you could consider using it to heat up the lens very slightly – this will rapidly clear the mist but great care is needed to avoid damaging the delicate optical elements. NEVER attempt to wipe clean a misty or dirty lens in a cave. You cannot see clearly what you are doing and dust or tiny rock fragments are likely to have mixed in with the moisture so you seriously risk permanently damaging your lens by scratching it.

Colour balance Many situations in caves result in light quality which is not colour-balanced for daylight film. As well as fluorescent lights which give a green tone and halogen lamps or lanterns which give a orange-red tone, there are some other situations where even natural light will not appear 'natural' on the film. This is particularly the case when most of the available daylight is reflected from the limestone surface, or when the air is full of dust. Both these situations produce a distinctly blue tone on film which is quite different from what you experience in reality. To correct this effect, you need to use a pale orange warming filter. If you will be using the same type of light source for the whole film, the best option would be to use film balanced for artificial light, but such film is usually expensive and only available from specialist suppliers.

Digital adjustment of photographs

Computer technology presents an alternative solution to the problem of colour balance without the need to resort to filters.

If you are using a digital camera or you scan your slides or prints onto a computer you can use image processing software such as Adobe Photoshop to correct many of these colour balance problems, although it is difficult and time-consuming to achieve a naturalistic effect. Software can also be used to convert colour photographs into black and white, which is useful if you do not normally use black and white film but would like to use it for specific circumstances, such as archeological sites, where black and white is more dramatic.

Computer software can also be used to 'push' an underexposed shot, which can be very useful if your photograph is only slightly (less than one stop) underexposed. It is worth remembering that it is much easier to push an underexposed shot than to tone down an overexposed one because in an overexposed shot the bright details have been 'burnt out' and cannot be retrieved, whereas in a slightly underexposed shot the details are still there in the shadows and can be brought out. If in doubt, it is better to slightly underexpose a shot than to risk overexposing it. In practice, even with digital adjustments, you will need to take each shot at a range of exposure levels to be certain of getting an acceptable result.

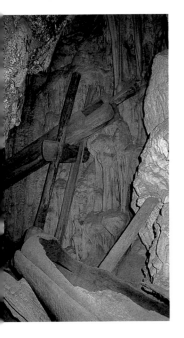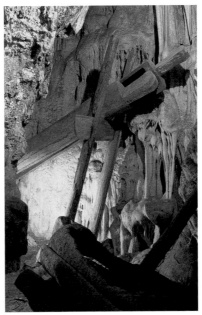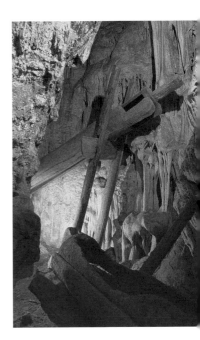

Log coffins in Tham Jabo.
The shot on the left was taken with electronic flash and is a reasonable representation of the true colours and very dusty conditions.
The middle shot was illuminated using hurricane lamps and brings out the carving on the coffin better, although the colours are less realistic.
The black and white shot was created digitally from the lantern shot and preserves the detail of the scene without giving a false impression of colour.

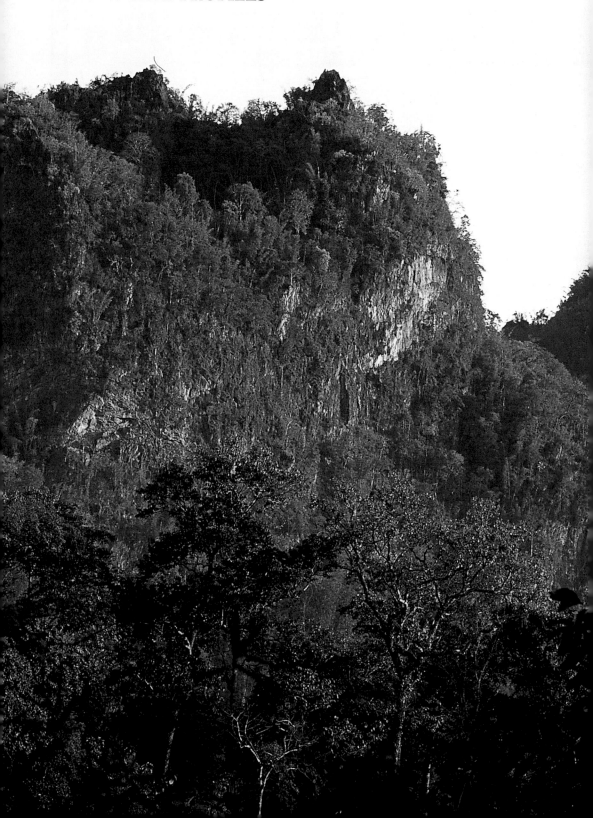

PART 2: CAVE PROFILES

PANG MA PHA

Tham Pha Daeng   ถ้ำผาแดง

A sporting through cave formed by a permanent stream which emerges at the base of a towering red cliff high above the Nam Khong River. The outflow entrance is occasionally used by wandering monks and has a charming bamboo shrine. A small coffin cave is situated in the cliff face 30 m to the south (see p. 184).

The main route through the cave is 1,200 m long and mostly easy going along a gravel stream bed with three minor bypasses at slightly higher levels and good airflow throughout. The passage is 5 m wide and 10 m high near the outflow but gradually becomes more constricted upstream with a couple of low crawls and a neck-deep wade which would certainly be impassable in the wet season.

There is some fine decoration in places, particularly in slightly higher side sections which are not subject to annual flooding. The main stream passage has a good collection of active stalactites and flowstone.

The upstream entrance is usually blocked by logs and other flood debris, but a cave window 30 m downstream provides an exit to a magnificent natural amphitheatre hemmed in by massive overhanging cliffs. From here you can visit several interesting coffin caves (see page 184-187) before returning back through the stream cave or taking a much longer path round the mountain.

Tham Pha Daeng supports a good variety of cave life, although no troglobitic species have been found. The inaccessible cavern visible in the cliff above the outflow is home to five species of bats and is one of the largest colonies in the region. The guano from this cave may be a vital food source for the animals in Tham Pha Daeng.

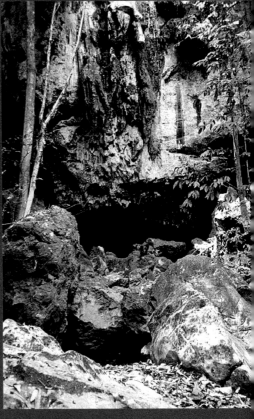

© Pang Ma Pha Caves Project

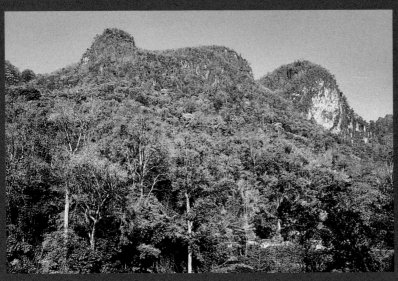

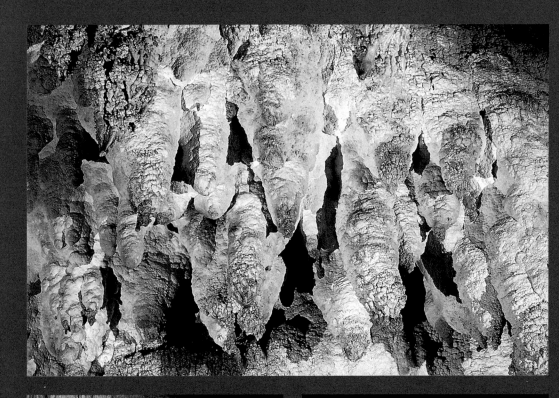

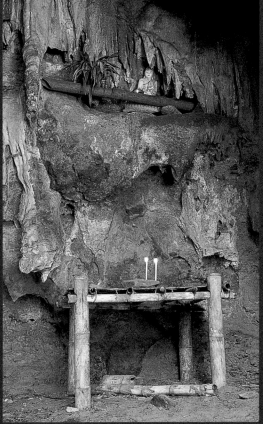

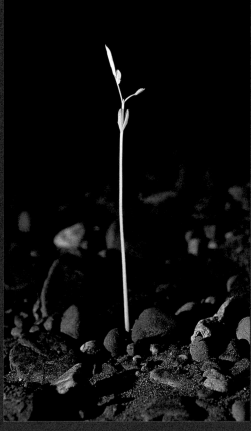

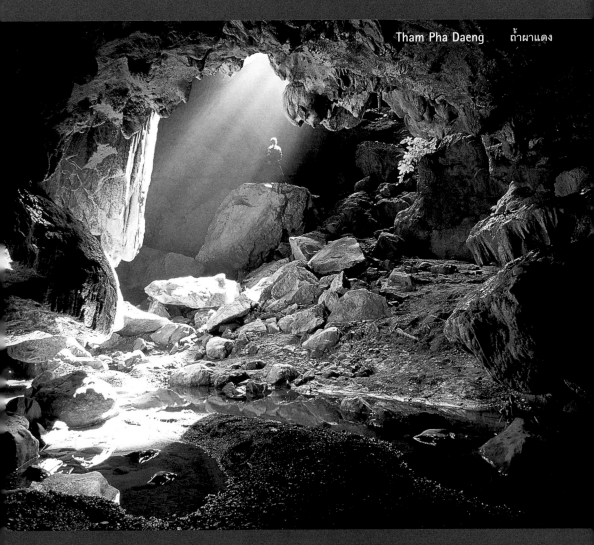

Tham Pha Daeng   ถ้ำผาแดง

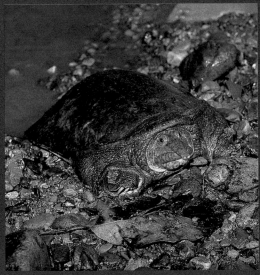

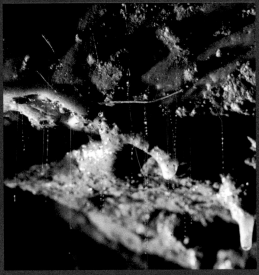

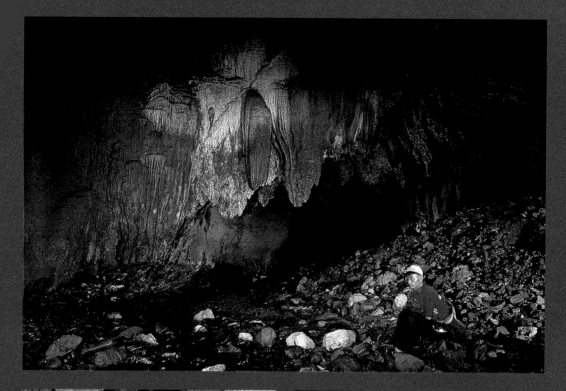

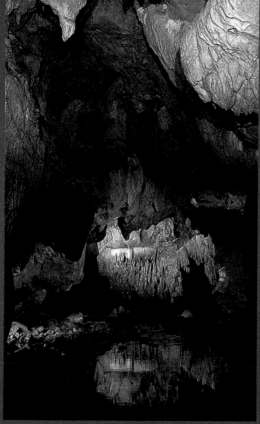

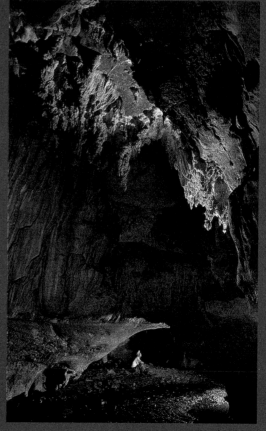

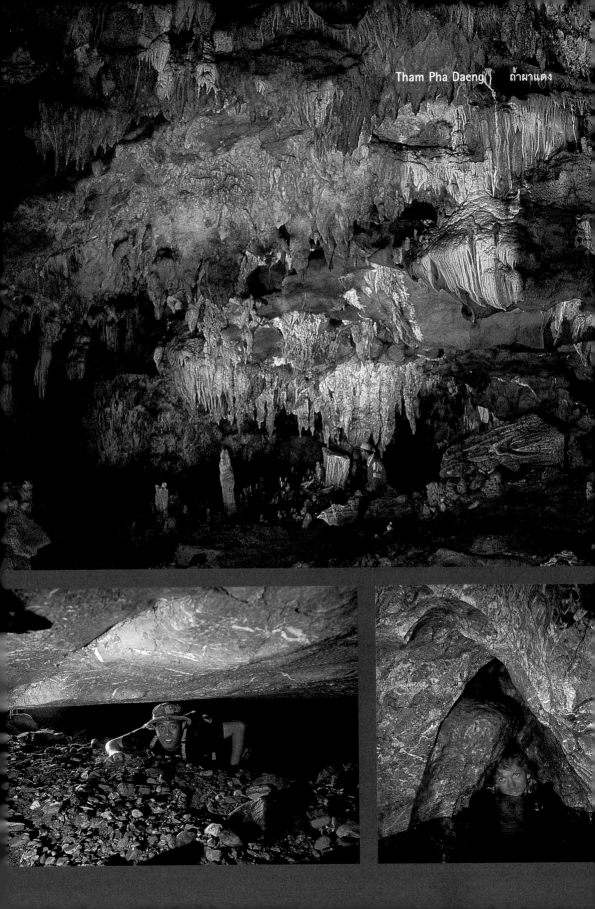

Tham Pha Daeng   ถ้ำผาแดง

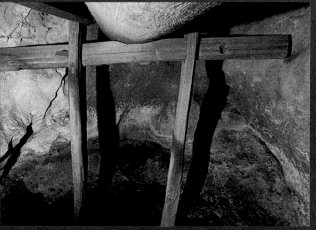

The cliffs around Tham Pha Daeng house a large number of small dry caves, several of which contain log coffins and signs of Stone Age occupation. Many of these caves are remote and inaccessible but there are two interesting examples in the immediate vicinity of Tham Pha Daeng.

Tham Pha Daeng Resurgence
ถ้ำผาแดงผีแมน is a small cave 30 m south of the stream outflow. The single narrow chamber is only 6m long but houses two coffins in excellent condition, both complete with their original lids. The joints on the column and beam supports are particularly fine.

Tham Pha Daeng Scaffolds ถ้ำเก้าชิ้น is half-way up a steep slope to the north of the stream inflow. Two small entrances open into the same high chamber which has an impressive array of well-preserved coffins stacked up to the roof on a remarkable set of supports. There are nine coffin pieces with six different styles of head carving, including a unique style not found in any other cave. One coffin has been carbon-14 dated as 2,000 years old. Stone tools, pottery and human bones have also been found at this site.

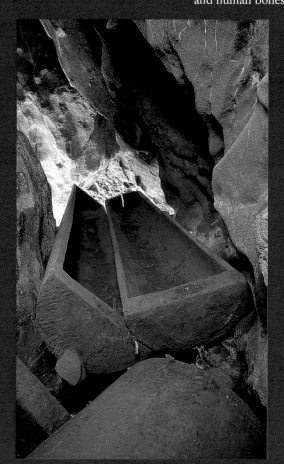

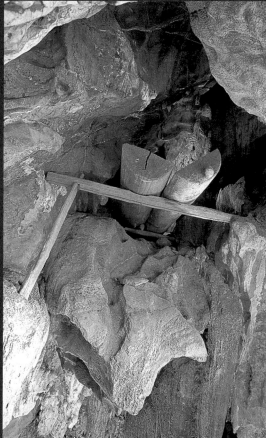

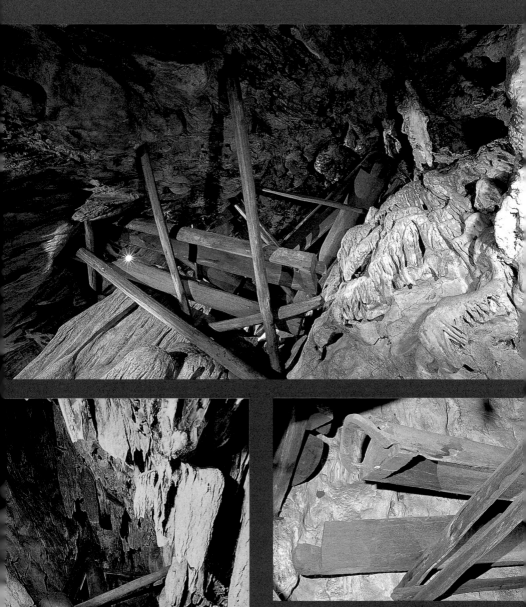

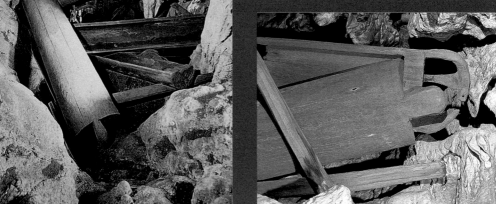

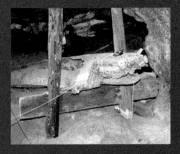

Tham Chan Long   ถ้ำชันโลง

Open dry cavern 12 m wide and 10 m deep, 100 m south of Tham Srisophon. There are three teak coffins in rather poor condition, only one of which is still on its supporting column and beam. Another column has been defaced by graffiti. All three coffins have the same style of carved head. Fragments of pottery, iron artifacts, stone tools and human bones have also been found on the site.

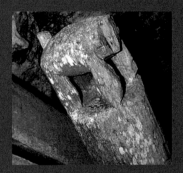

Tham Long Yak   ถ้ำโลงยักษ์

An important archeological site high up above the Pha Daeng inflow. The single large daylight chamber has seven teak coffins including some of the biggest yet found in Northern Thailand. Some of the coffins are still on their original supporting beams but most have fallen to the floor and are in a dilapidated condition.

Several coffins have been badly vandalized by graffiti, burning and reuse in temporary shelters. The heads are rather crudely carved and are all of the same style. The back of the cave descends to a lower dark chamber without coffins. The platform outside has a fabulous view and was probably a Stone Age habitation site. Nowadays it is used as a camp site by honey gatherers and hunters.

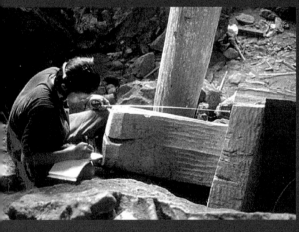

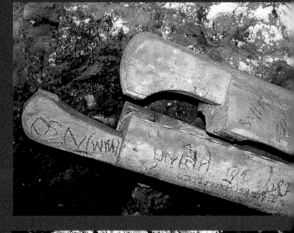

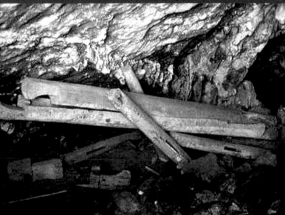

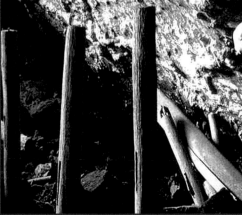

All photos © Pang Ma Pha Caves Project.

Tham Srisopon   ถ้ำศรีโสภณ

A remote cave with a very impressive main cavern south of Tham
Pha Daeng inflow, not far from Tham Long Yak. Two entrances
on opposite sides slope down to a dirt floor and a pile of boulders
where the ceiling is over 50m high. The site has some interesting
examples of rock art, although they are probably not of
prehistoric origin. Some of the drawings may relate to the
Japanese occupation in World War II. A bamboo platform near
the main entrance has a sign in Thai indicating that it was built
during the same period. There are a few log coffin fragments
scattered throughout the cave with 4 styles of head but they are all
in a very poor condition.

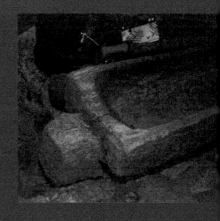

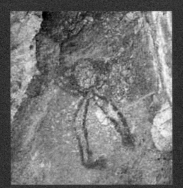
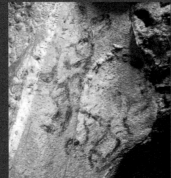
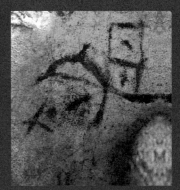

Spirit Cave (Chester Gorman)   ถ้ำผี

A remote dry cave set high in the hillside south of Tham Pha
Daeng with a commanding view over the Nam Khong valley.
This site was made internationally famous by the seminal
excavations of Chester Gorman in 1965-66. His work
confirmed human occupation of the area from about 12,000
to 7,500 years B.P. which at that time was the oldest known
site in Northern Thailand and greatly boosted interest in the
prehistory of the region (see p 90).

Spirit cave is a complex multi-level cave with three major
chambers. The northern chamber is 20 m wide and 3-10 m
high, only accessible by a small hole and completely dark.
The main central chamber has several levels and is daylit. A
half-supported large coffin piece is in the centre of this
chamber and more coffins are higher up on the north and
south sides. The south side has a small room with stratified
floor deposits. This was Gorman's main excavation site.
Further south, a small ledge leads to Gorman's upper cave
with semi-supported coffins. This cavern has three entrances
and two rooms with flat dirt floors.

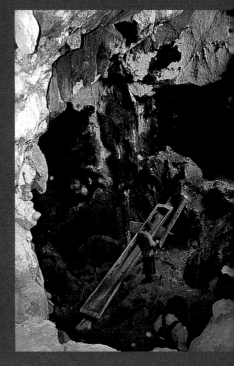

All photos © Pang Ma Pha Caves Project.

Tham Pang Kham ถ้ำปางคาม

An impressive through cave with a dramatic northern entrance which plunges downwards for 40 m to the stream level. The first 200 m of the cave is formed along a vertical fault line and is spectacularly lit by a narrow shaft through which a tributary stream enters. The main stream enters through a large upper level chamber parallel to the shaft and linked to it by a steep scree slope. The cave has many active formations, including impressive orange flowstones. The 550 m long stream passage becomes progressively narrower towards the downstream exit, which is only 4 m wide and high. Below the cave outflow the stream drops down several waterfalls over black andesite rock.

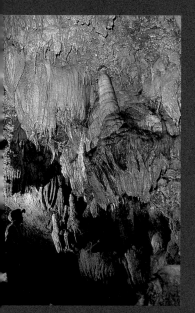
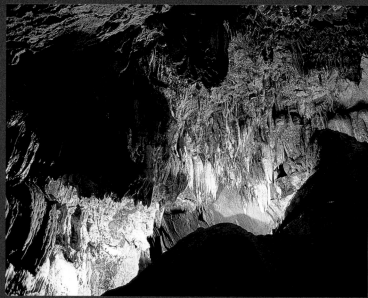
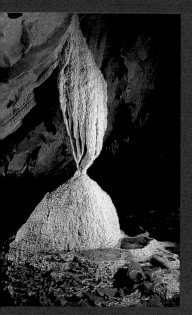
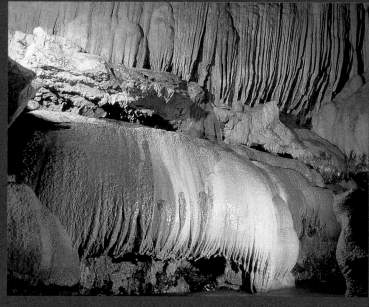

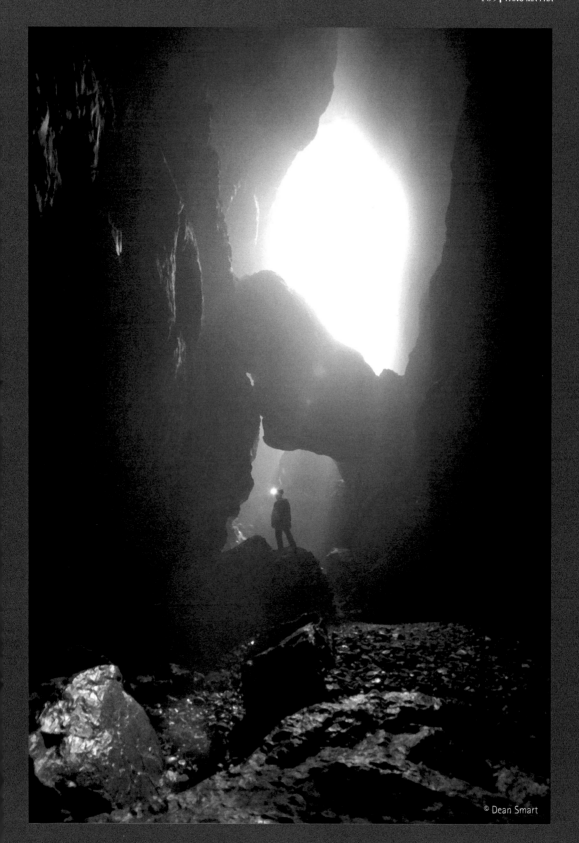

© Dean Smart

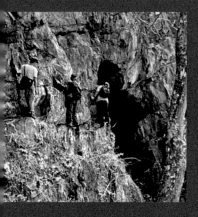

Pang Kham Piman Caves ปางคามผีแมน

At least seven small dry caves in the Pang Kham area are log coffin sites – there are 4 on the cliffs above Pang Kham outflow alone. All of the sites have been vandalized to some degree but still contain a unique collection of prehistoric objects including human bones, pottery, iron tools and bronze artifacts as well as the coffins themselves.

Pang Kham Piman 1 (west) has an 8 m wide entrance leading into a narrow chamber over 100 m long. Only the first section is day lit - most of the coffins are in the dark zone near the back of the cave. The coffin heads are carved in a variety of unique styles, including one shaped like a pig's head which has now been removed to Bangkok for safekeeping. A particularly interesting feature is a 1,350 year old human skull , encrusted with calcite crystals and half-buried in the flowstone.

Pang Kham Piman 2 (east) has a small entrance high up in the cliff face and can only be reached by a nerve-racking narrow ledge. A wide variety of coffin styles and unique artifacts were found in the first survey, including a socketed bronze axe and a marble bangle which has since disappeared.

Pang Kham Piman 3 (north) in the hills north of Pang Kham village is only accessible via a steep narrow path. Two entrances lead into a large partially daylit chamber, 30 m wide and 50 m long . There are many coffin fragments and heads scattered across the sloping breakdown floor. Bat guano and dirt are found all over the cave floor and coffins. The cave is above the natural altitudinal limit of teak trees and is one of the few sites where the coffins are made from another local hardwood rather than from teak.

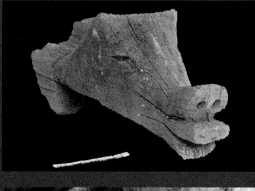

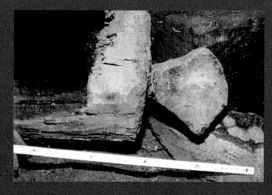

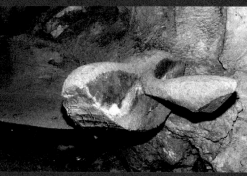

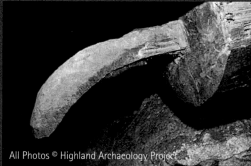

All Photos © Highland Archaeology Project

Some of the log coffins in the Pang Kham area were recently removed from their original sites by the Thai Military and placed in more accessible locations. A couple of coffins were displayed at the inflow entrance to Tham Pang Kham, fully exposed to rain and sunlight.

A small cavern near the outflow entrance has also been used to display four coffins on top of flowstone and rimstone pools where they would never originally have been placed because the pools can fill up with water in the wet season, greatly shortening the life span of the coffins. It is to be hoped that these coffins will soon be returned to their original locations or moved to a museum with proper storage.

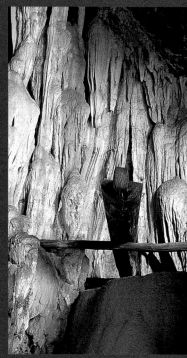

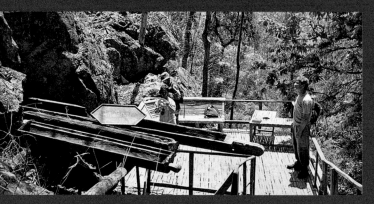

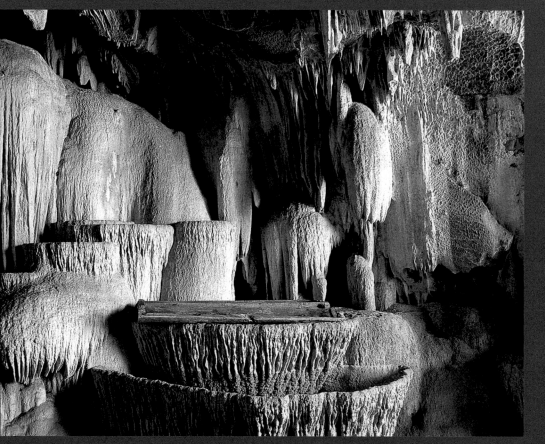

Tham Pha Phuak   ถ้ำผาเผือก

One of the deepest caves in Thailand with a magnificent entrance set at the base of an overhanging white cliff. A small permanent stream runs off the surrounding impermeable rocks and enters a vast entrance chamber over 100 m high. The cave plunges downwards at an angle of 40-50 degrees parallel to the rock bedding planes, gradually becoming steeper and narrower until it becomes impassable at a point about 170 m below the level of the entrance. The stream probably connects with Tham Pung Hung (p 196) and re-emerges at Tham Huai Khun 2 km to the north west.

The main chamber is floored by enormous blocks of fallen limestone and muddy scree. The steeply dipping angle suggests that Pha Pluak is a relatively young cave that has not yet reached equilibrium. Massive stalactites adorn the roof in the lower section and there are a good number of exposed fossils in the rocks. Natural light reaches deep into the cave and supports a wide variety of ferns and other shade tolerant plants.

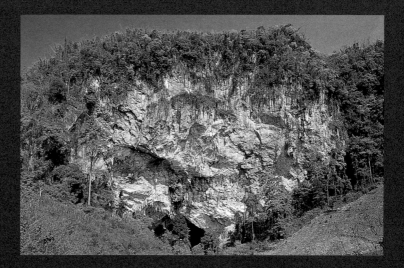

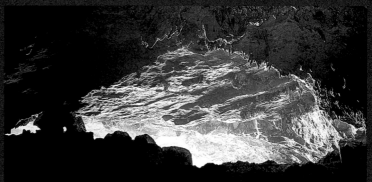

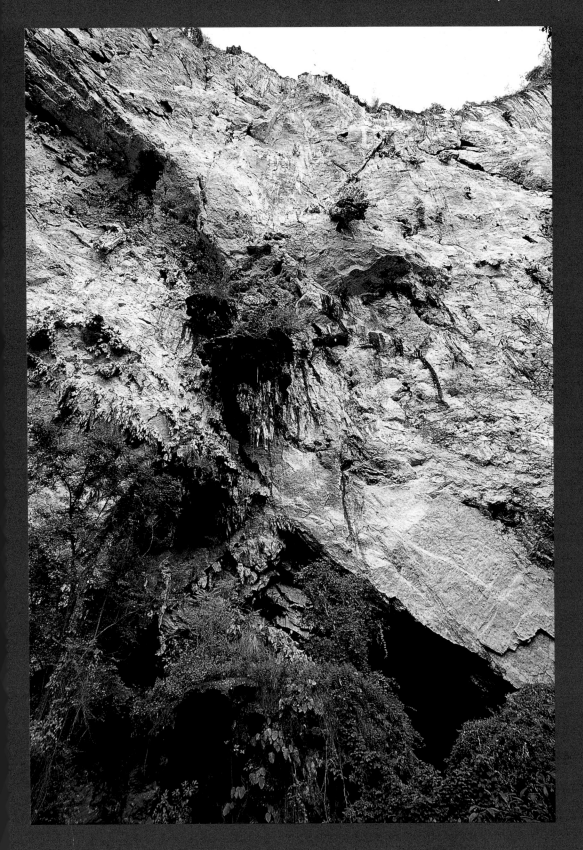

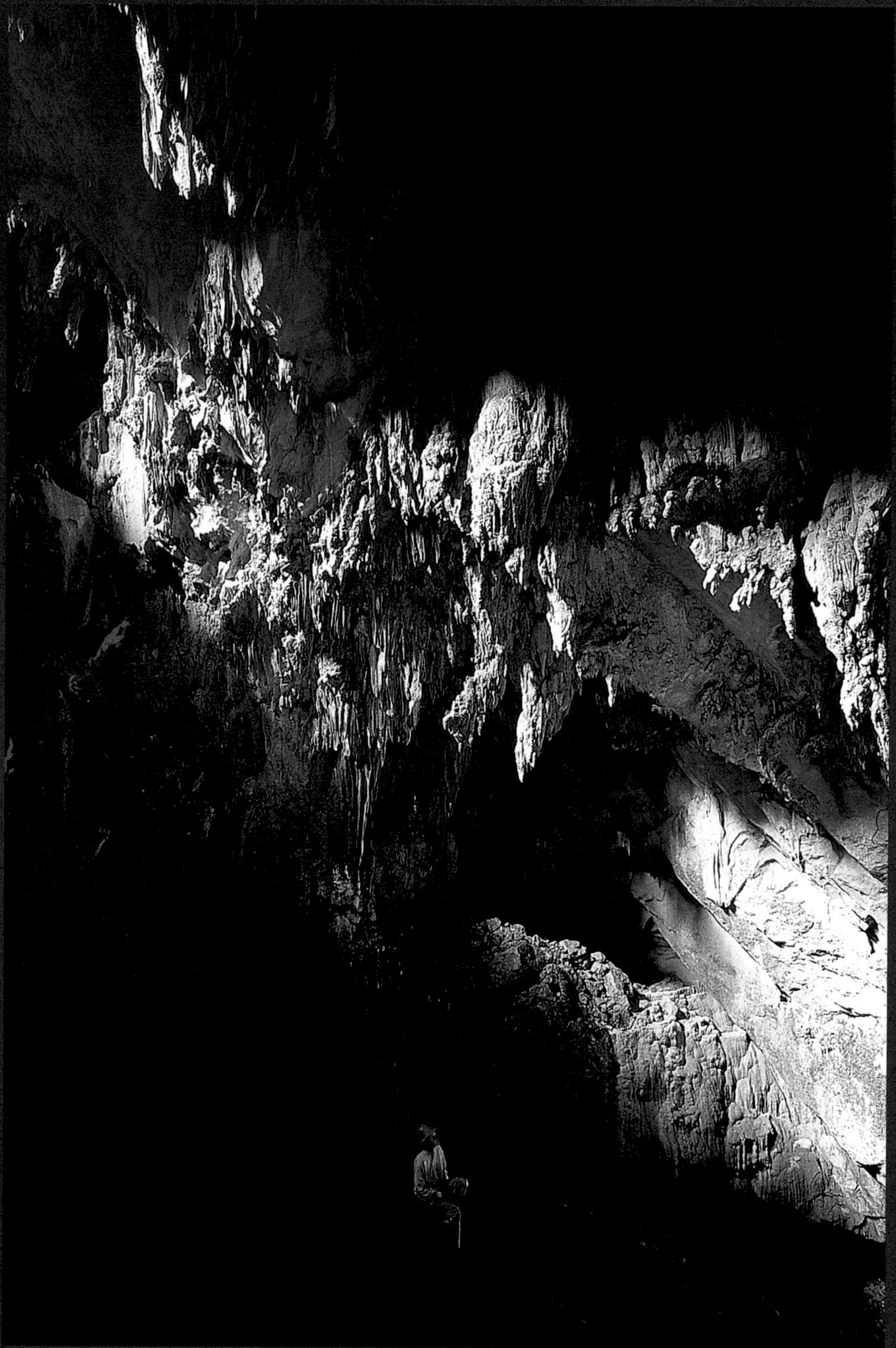

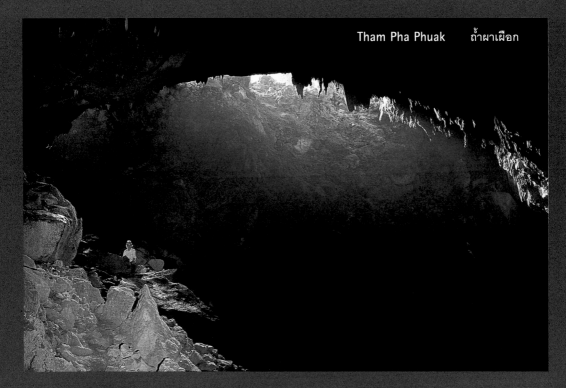

Tham Pha Phuak   ถ้ำผาเผือก

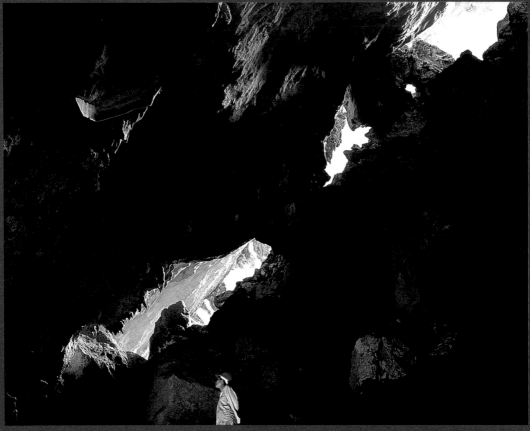

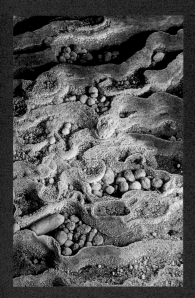

Tham Pung Hung   ถ้ำปุงฮุง

The third longest cave in Pang Ma Pha after Tham Mae Lana and
Tham Nam Lang. Tham Pung Hung is a master cave draining streams
from at least four other caves. A large dry entrance chamber can be
used to bypass the first 100 m of the stream passage. Otherwise
almost the whole cave is an active stream system but in the dry season
the water sinks about 500 m from the entrance. There are no large
chambers or upper levels and most of the stream passage is only 1-4 m
high, which indicates that this is a relatively young cave. The passage
is decorated with many beautiful formations, including the biggest
known cave pearls in the region. After 2 km a 16 m vertical section
has to be negotiated and the cave roof gradually lowers until it is only
10 cm from the floor. The stream probably re-emerges at Tham Huai
Khan 3 km to the north.

The large dry entrance chamber is a prehistoric habitation site
which was made internationally famous by the excavations of Chester
Gorman in 1972 ('Banyan Valley Cave' - see page 88).

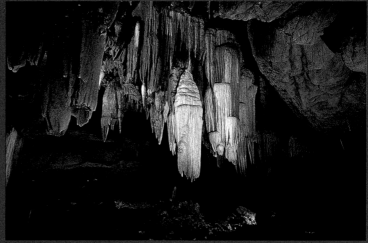

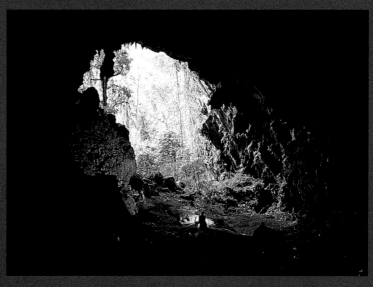

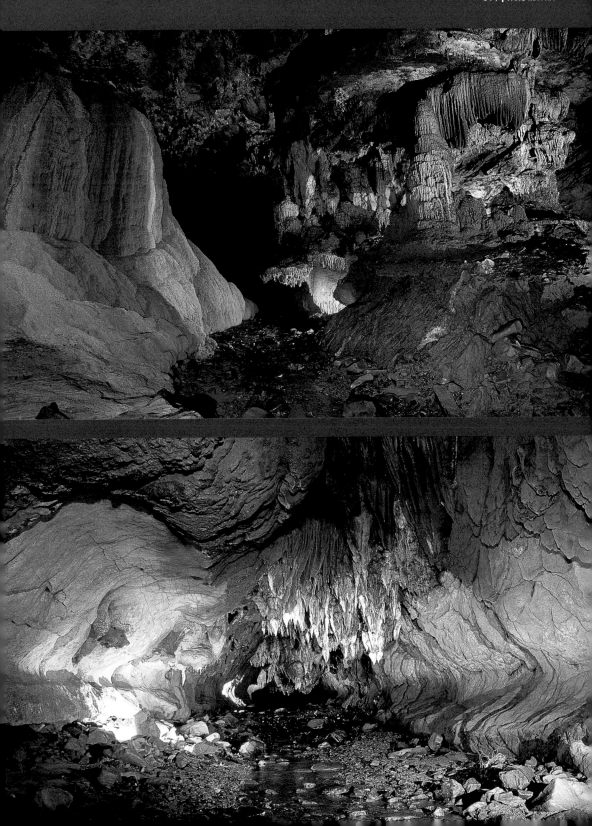

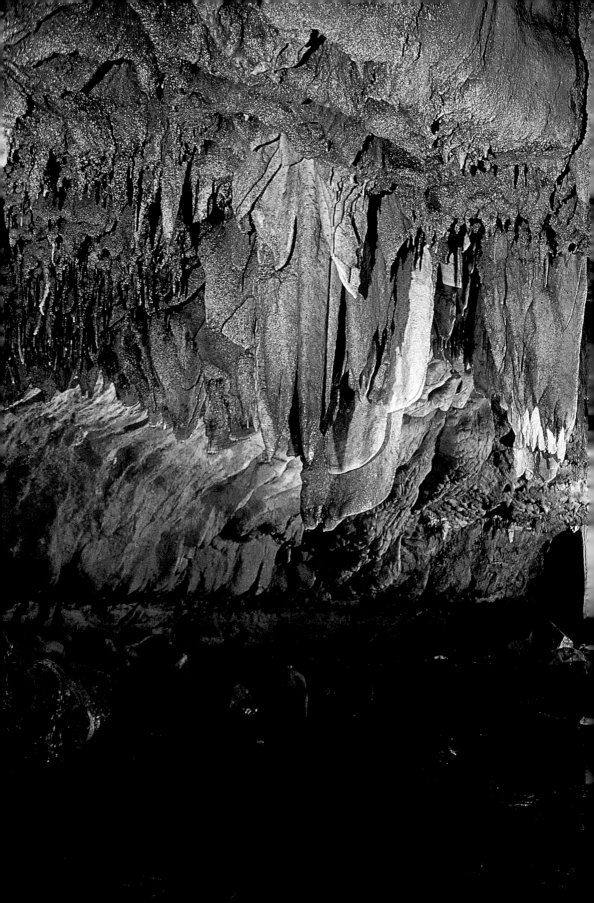

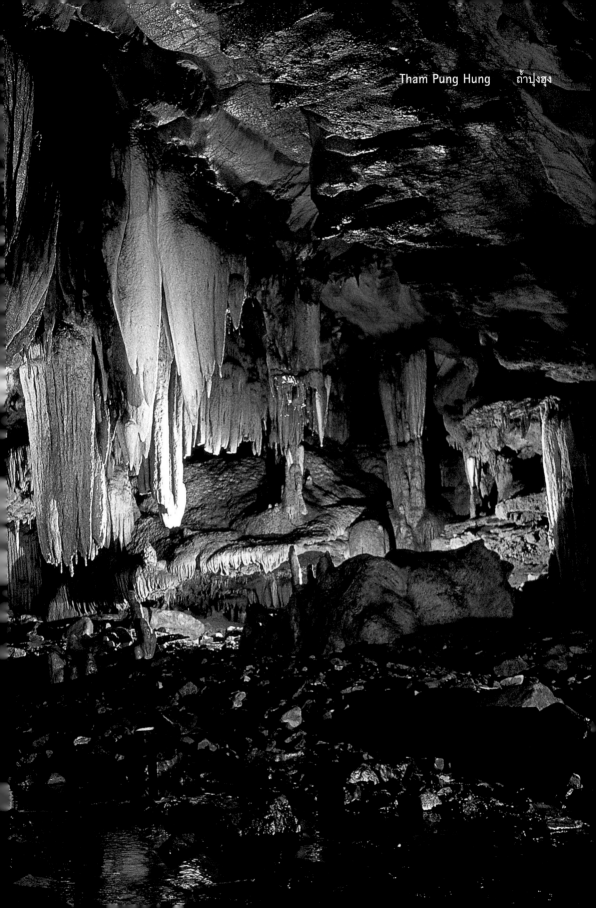

Tham Pung Hung   ถ้ำปุงฮุง

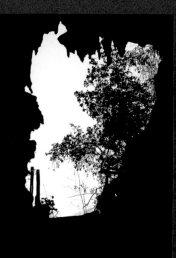

Tham Jabo   ถ้ำจะโบ่

An impressive log coffin site in a large cavern with a commanding view over the upper Mae Lana Valley. The 30 m wide entrance is set 60 m up an imposing cliff face and is only accessible by a steep scramble. The cave consists of a single large chamber which descends steeply towards the rear of the cave but has no apparent passages. One large coffin on supports is in the full daylight on the south side of the main chamber but most of the coffins are tucked away in a series of alcoves on the north side. A number of the coffins are in pristine condition and include some of the best examples in Pang Ma Pha.

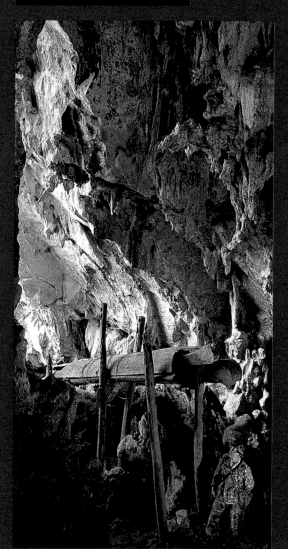

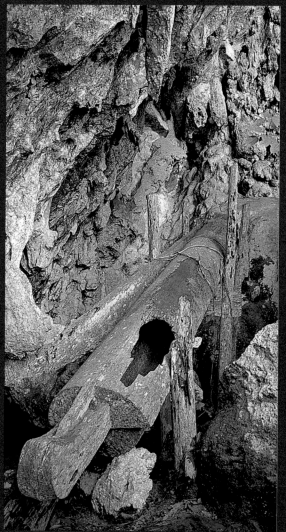

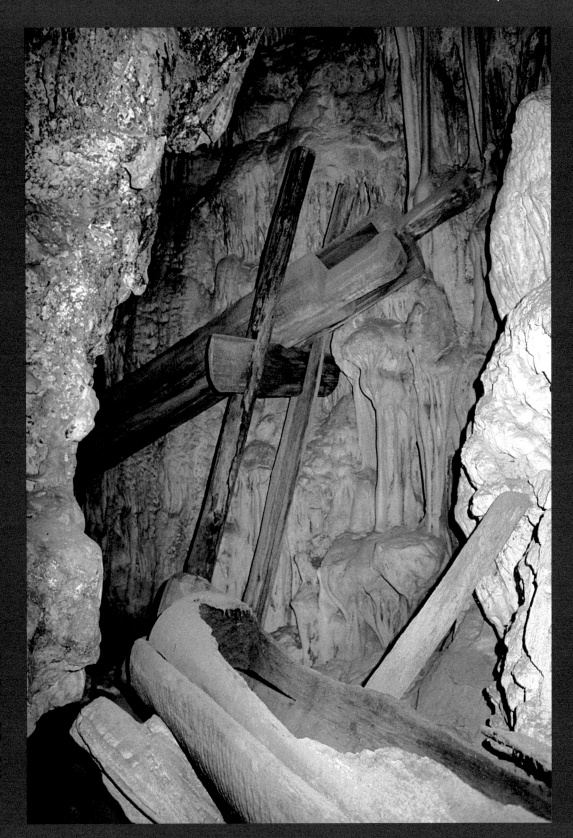

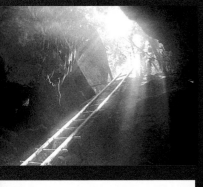

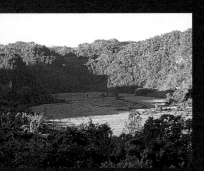

Tham Mae Lana   ถ้ำแม่ละนา

Tham Mae Lana is one of the finest caves in Thailand and of international importance for conservation. When it was discovered by an Australian team in 1986, Tham Mae Lana was the longest recorded cave in mainland S.E.Asia. Since then, several longer caves have been discovered in Laos and Vietnam, as well as Tham Phra Wa Daeng in Phitsanolok which is 13 km long. However, with a total length of 12,600 m Mae Lana is still the longest through cave in Thailand.

The Mae Lana stream sinks in an impenetrable hole 500 m south of Ban Mae Lana, re-emerging in an enclosed depression 2 km to the south which also contains a number of other caves (page 206-209). The stream flows on the surface for 300 m before plunging into the upstream entrance of Tham Mae Lana. A rough track has recently been built from Mae Lana village all the way to the cave entrance but is currently only driveable as far as the edge of the depression, from where it is a half-hour walk to the cave.

The main river passage is 7.2 km long and makes for an exhilarating trip but should only be attempted by experienced cavers as the full through trip takes at least 11 hours and requires considerable physical stamina. Most of the first half of the cave is very spacious with passages up to 40 m high, extensive rock breakdown and many rapids. Two substantial tributary streams join the main river about 2 km and 3 km from the entrance, the first of which is can be followed for over 2 km 3.8 km from the inflow entrance the main river enters a large sump at the 40 m wide Green Lake, re-emerging again after 100 m. This section can be bypassed at a higher level by a passage which branches off on the right 400 m. before the lake. The lower half of the river passage below the sump is mainly narrow canyons with many rapids, finally emerging above the Pong Saen Pik river.

The extent and variety of speleothems in Tham Mae Lana is quite exceptional, including good examples of almost all cave formations known from Northern Thailand. The main river passage has some massive decorations including 20 m high columns but the majority of the unusual speleothems are in the side passages and upper level where air movement is more restricted. The collection of cave pearls (oolites) is particularly outstanding, with several thousand in one chamber alone. Other unusual features include calcite spar crystals (the only known site in Thailand), anthodites, pendulites, mud stalagmites, birdbath conulites and cave shields. Most of these rare formations are extremely fragile and easily damaged even by well-intentioned cavers. Fortunately, the most beautiful sections are several km deep into the cave, which helps to protect them from casual visitors.

The ecology of Tham Mae Lana has not been well studied but is certainly of international importance with large numbers of species found nowhere else in the world. It is the only cave in Northern Thailand with both species of cave fish known of the region. This priceless heritage is in danger of being damaged by inappropriate development both inside and outside the cave.

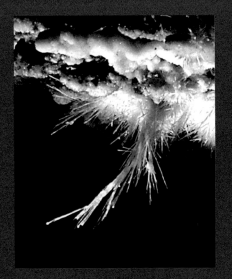

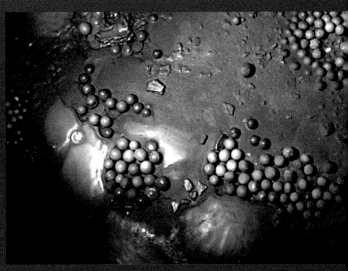

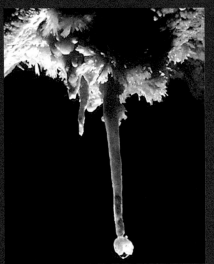

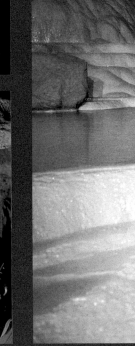

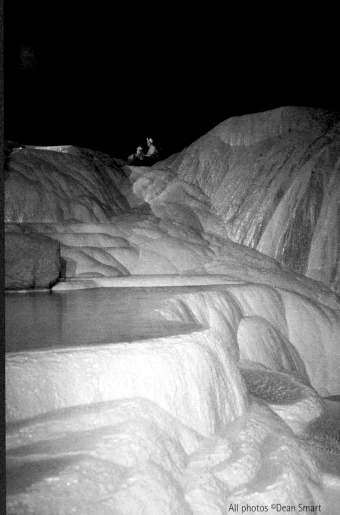

All photos ©Dean Smart

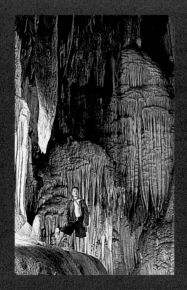

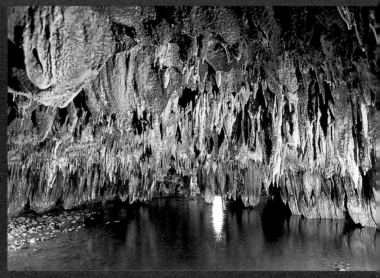

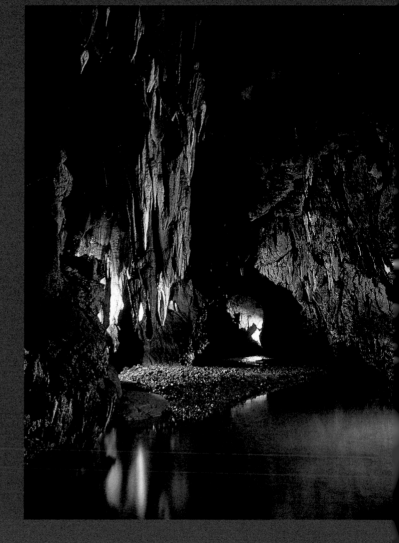

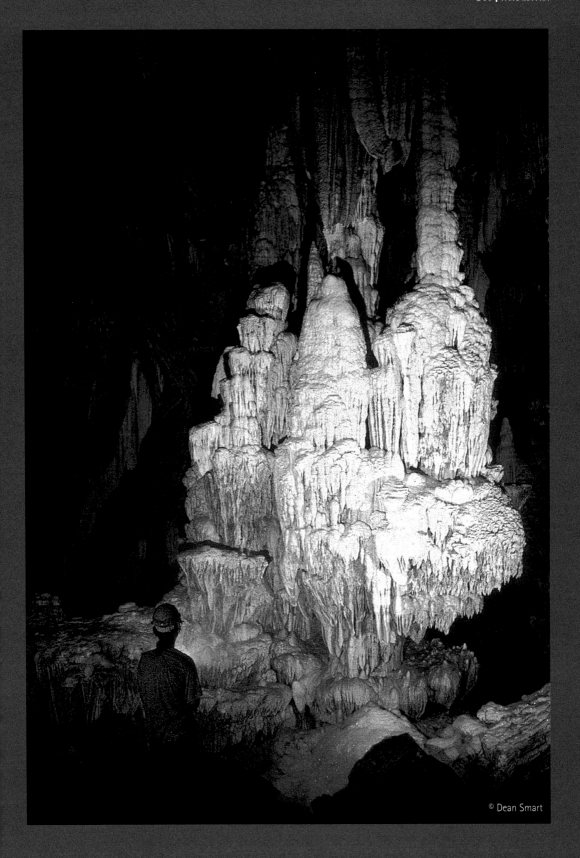

Tham Pet (Diamond Cave)   ถ้ำเพชร

A small but superbly decorated cave in the same doline as Tham Mae
Lana. A narrow entrance right next to the track opens out after 30 m
into a medium-sized chamber with almost every available surface
covered by active speleothems. A path leads straight ahead to another
larger but less decorated chamber behind.

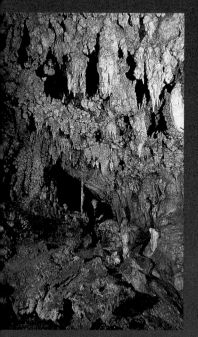

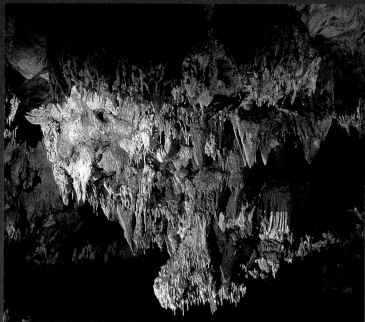

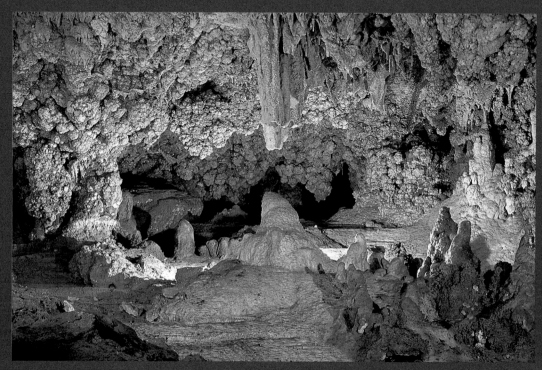

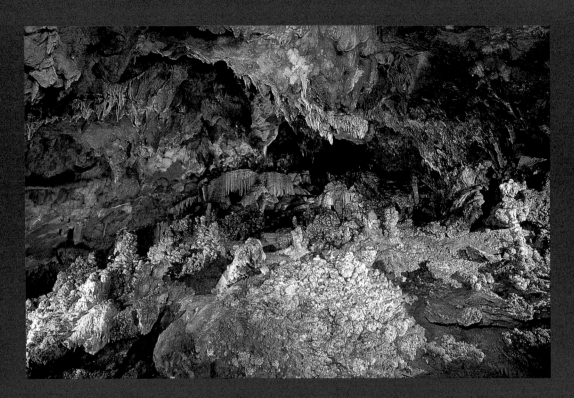

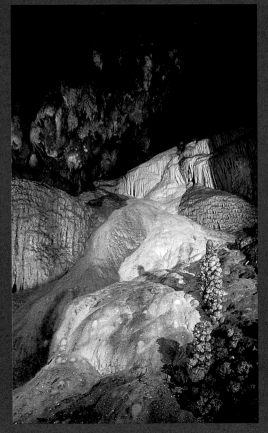

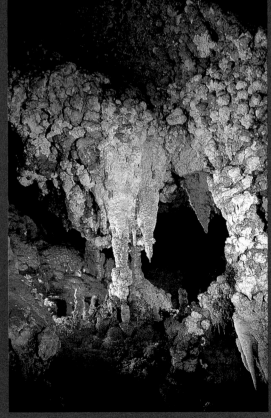

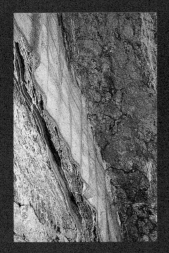

Two other caves in the same doline as Tham Mae Lana are actively promoted for tourism. Local guides are readily available from Ban Mae Lana for a fixed fee.

Tham Kai Mook (Pearl Cave)   ถ้ำไข่มุก
The large entrance chamber slopes steeply downwards over large boulders. The lower sections have some interesting erratic forms tucked away in moist crevices but no obvious cave pearls, despite the cave's name. A narrow passage at the back leads to a fine crystal flow but quickly becomes impassable.

Tham Pakarang (Coral Cave)   ถ้ำปะการัง
A small entrance with a 5 m ladder down to a spacious dry passage which has some active formations and good examples of cave popcorn but no cave coral. After 200 m the main level section of the cave comes to an end at an attractive chamber with no exit. A side passage further back leads to a 60 m vertical shaft which drops down to lower levels and eventually to the Mae Lana stream 100 m below.

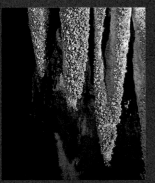

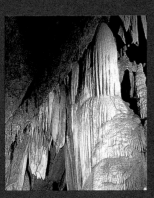

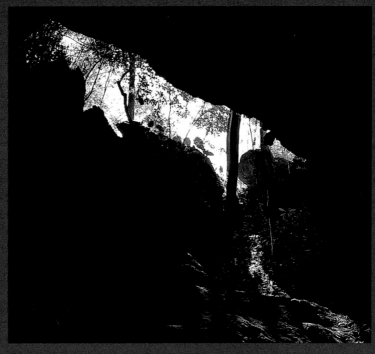

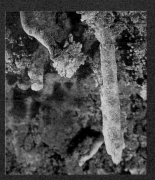

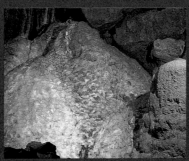

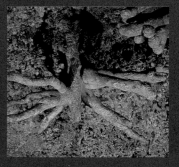

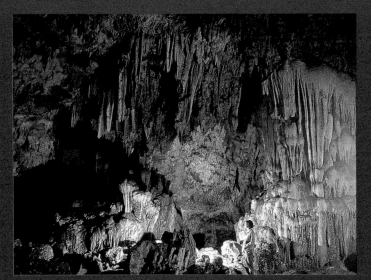
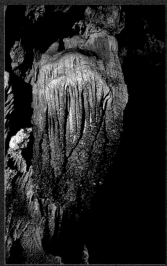
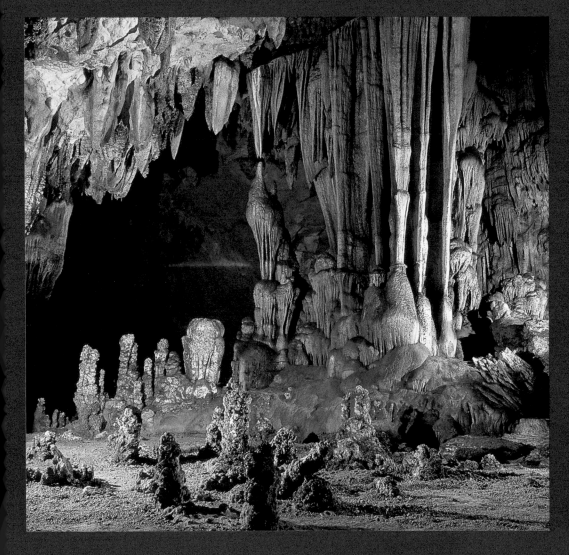

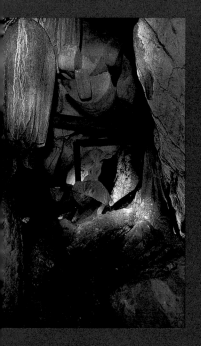

Tham Bor Krai   ถ้ำบ่อไคร้

Tham Bor Krai is one of the finest surviving log coffin sites, with many coffins in excellent condition and a greater range of coffin head styles than any other known site. The site is unusual because the 3 x 3 m entrance is tucked away in the forest rather than high up on a cliff face as is more typical for log coffin sites in Pang Ma Pha. The cave is on two levels separated by a 5 m vertical drop. The upper level consists of a single chamber which is partially day-lit and has only a few coffin fragments in poor condition.

The lower level consists of two chambers which are absolutely dark and tomb-like. The first chamber is dominated by two magnificent coffins complete with lids and still on their supporting columns. The floor is littered with a profusion of broken coffins carved in a wide variety of unique styles, including some with heads carved in the shape of an animal. The inner chamber is long and narrow and packed with coffins on 2 tiers. Some coffins are supported by posts hewn from older coffins. The 2 cm holes near the end of almost all of the coffins are not part of the original design but were made in 2001 by a research team from Mahidol University who were trying to study the prehistorical environment from the tree rings.

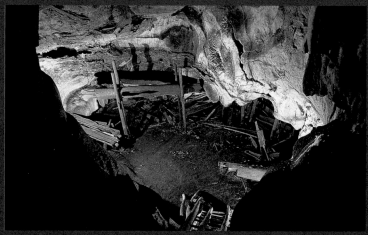

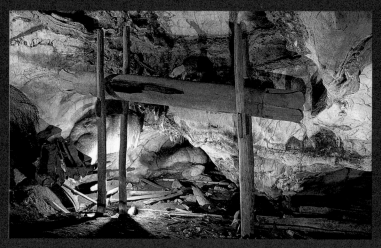

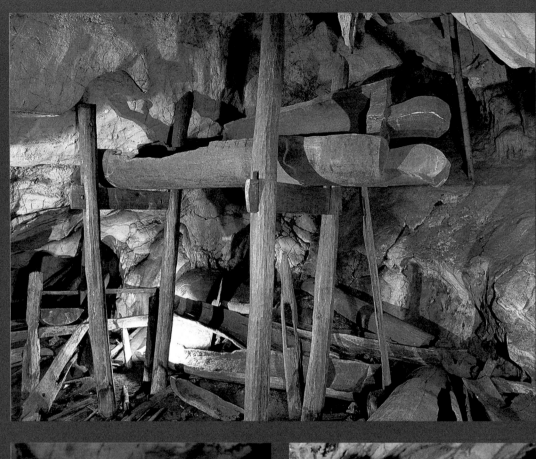

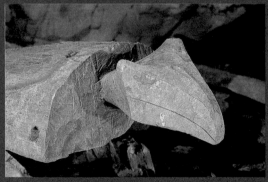

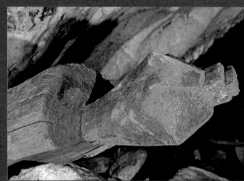

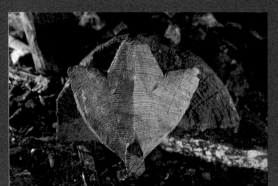

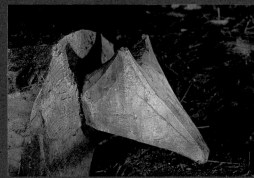

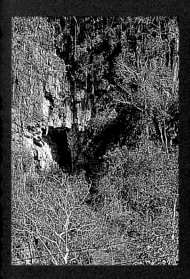

Lang Jan Caves   ถ้ำลางจันทร์

The most well-known and visited coffin caves in Pang Ma Pha, clearly visible in the cliffs close to the main road 2 km west of Soppong. A short steep path leads up from a sign by the road about 150 m to the foot of the cliff. A complex of at least seven small dry caverns containing a large number of coffins, some still on their original supports. Many of the coffins have deteriorated and only fragments remain, but there are still a few in reasonable condition. One cave has a large flat area with a gnarled old tamarind tree and a marvellous view. Stone tools and other signs of prehistoric habitation have been found at this site but have since been removed. Another small interconnected cave to the west has a dark inner chamber with two well-preserved coffins, 7 m long and complete with lids. One of the support posts has been C14 dated as 1,800 years old.

The inner sections of some of the more remote caverns have dangerously high levels of CO_2.

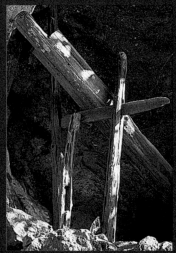

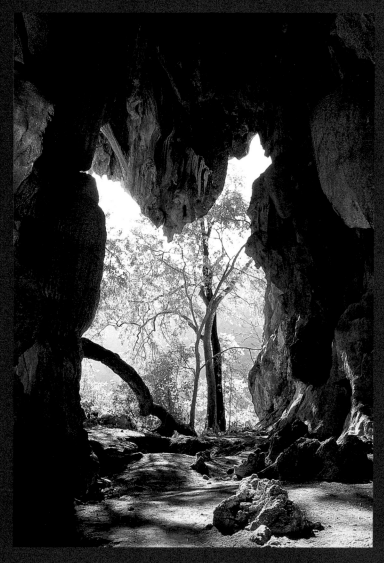

Tham Yappanae 1 ถ้ำยาป่าแหน 1

An inconspicuous entrance between boulders leads downs to two dry moderately-sized chambers. The outer chamber receives a small amount of daylight and has some semi-active formations, including an impressive wall of algae-stained flowstone. The larger inner chamber is completely dark and contains the remains of 8 or 9 log coffins in rather poor condition, including some still on their original support columns. Archaeologists have found a variety of artifacts including iron & bronze tools, glass beads, lacquerware and spindle whorls. Some of the coffins have been burnt by vandals.

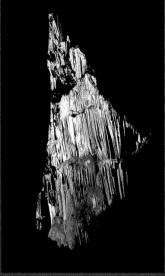

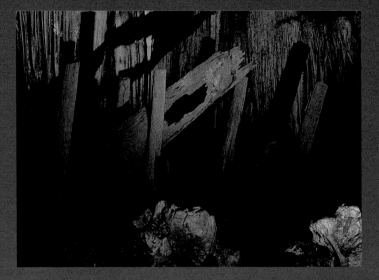

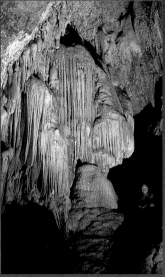

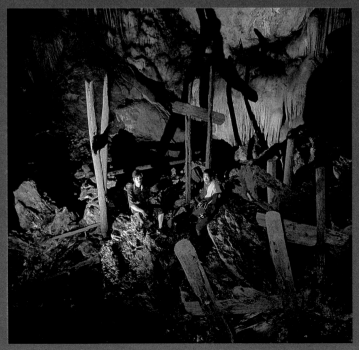

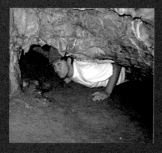

Fossil Cave   ถ้ำฟอสซิล

A triangular-shaped entrance leads to a narrow squeeze opening into a
beautifully decorated chamber with very clear rock bedding planes. At
the further end of this chamber the cave narrows to a seasonal stream
passage with large angular boulders. After 100 m the cave reaches a
T-junction. The right branch contains high CO_2 levels, but the left
branch can be followed along an easy gravel bed for another 100 m
to some excellent examples of shell fossils (brachiopods).

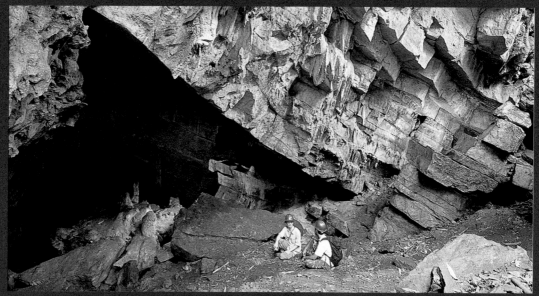

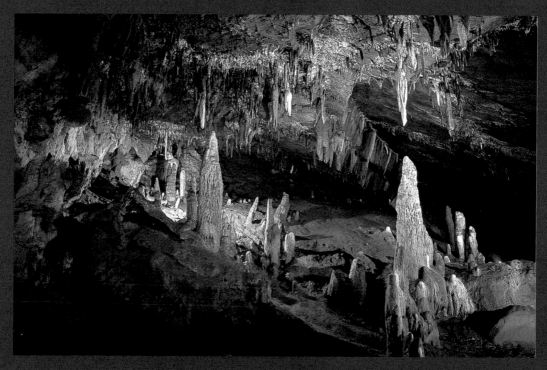

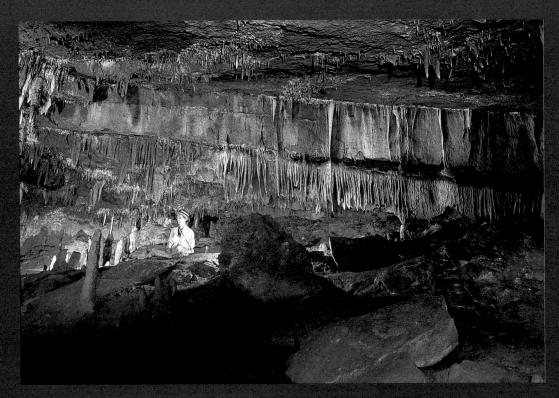

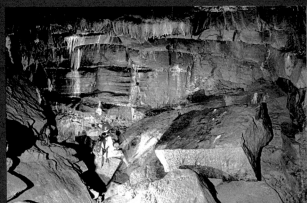

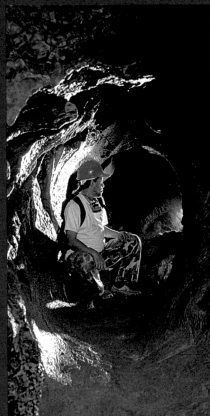

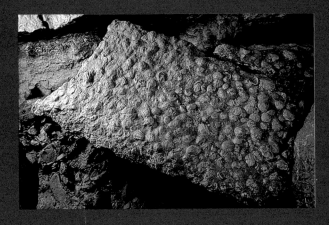

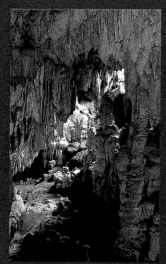

Christmas Cave ถ้ำคริสต์มาส

A superbly decorated main chamber with 3 entrance windows, beautifully lit by the afternoon sun. The main chamber is V-shaped, involving a steep scramble down over boulders to reach the inner passage. A 5 m vertical drop leads down to the stream, which is usually dry except during the rainy season. The stream passage is only passable for a short distance , but the stream probably connects with Tham Hued and eventually re-emerges at Tham Nam Hu (page 220).

Christmas Cave is one of the easiest places to see active cave formations in good condition, including good examples of drapery, rimstone dams, crystal-lined pools and cave pearls. In the upper daylight sections many of the formations and cave walls are tinged green with algae, but in the darker areas the algae cannot survive and the surface of the formations is coated with glittering white calcite crystals.

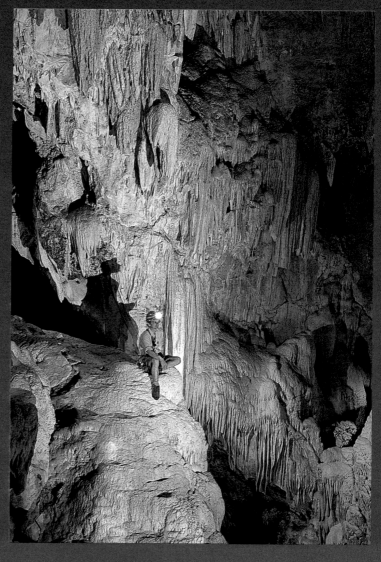

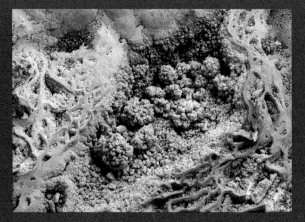

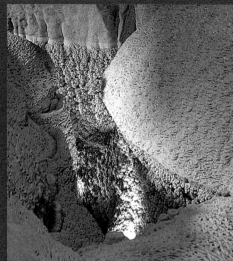

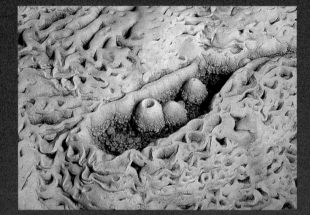

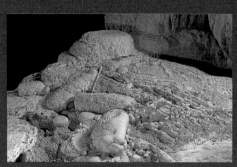

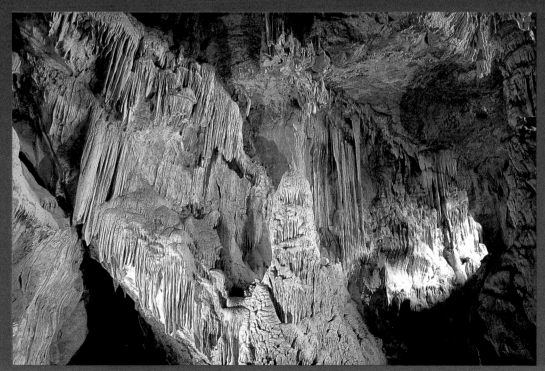

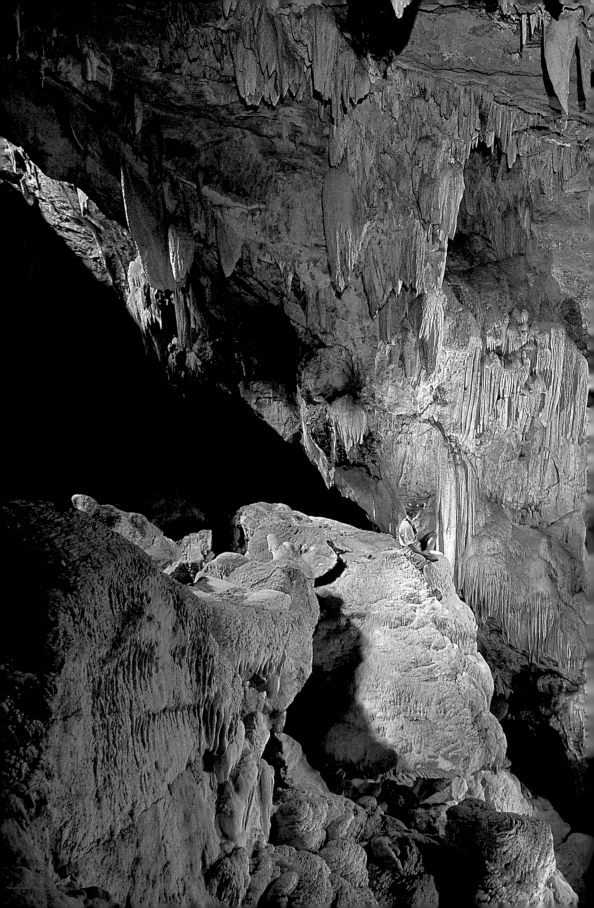

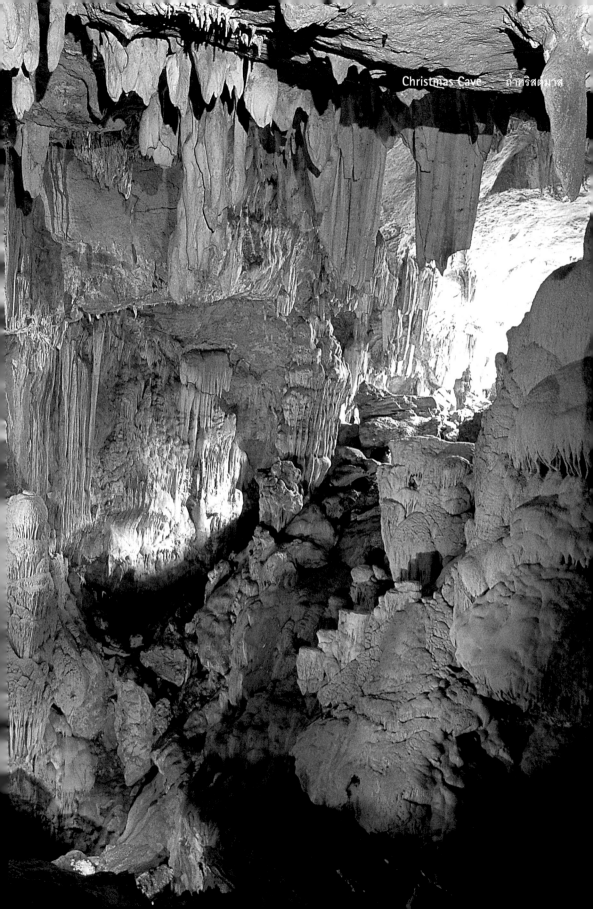

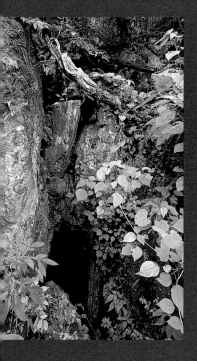

Tham Nam Hu   ถ้ำน้ำฮู

A resurgence with a small permanent stream next to the road 3 km west of Soppong market. The stream is an important local water source. The entrance passage is usually full of water 10 m inside the cave and cannot be visited . In exceptionally dry periods the passage beyond can be followed for at least 1,200 m but normally contains lethal levels of CO_2 and should on no account be visited except by experienced speleologists. The further reaches of the cave have not been explored but it is likely that the stream drains a large area of karst as far north as Ya Pa Nae village over 10 km away. Tham Nam Hu may eventually prove to be one of the longest caves in Thailand.

Tham Hued   ถ้ำฮืด

Tham Hued is probably one of the longest caves in Pang Ma Pha butis very difficult to visit due to the extremely high levels of CO_2. The CO_2 levels vary at different times of the year but can be as high as 5% – ie.100 times normal atmospheric conditions! Tham Hued is easily accessible from Ban Tham and the high CO_2 is usually obvious just at the back of the entrance chamber, so this is a good place to experience this remarkable phenomenon – take a lighter and on no account be tempted to go beyond the entrance chamber. The entrance leads down to a stream which is an important habitat for a species of troglobitic fish endemic to Pang Ma Pha. The stream probably connects with Christmas Cave and re-emerges at Tham Nam Hu.

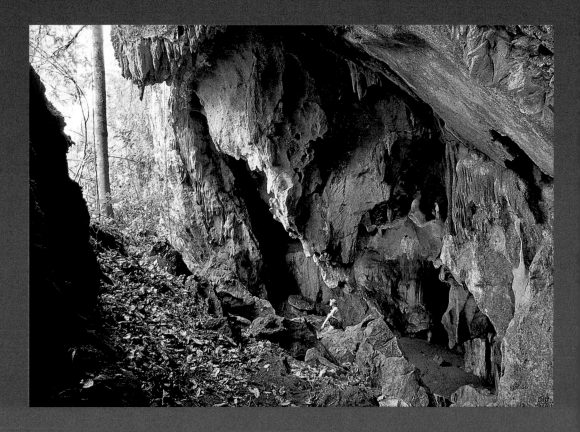

Nong Pha Cham หนองผาแจ่ม

A large descending entrance at the foot of a collapsed doline overgrown with luxuriant vegetation. The main dry chamber is 100 m long and 60 m deep with few active formations, descending steeply over large boulders. The lower sections can contain dangerously high CO_2 levels. In the lowest part of the cave, just beyond the limit of daylight, a short section of stream crosses the floor of the cave, impassable in both directions. The stream supports troglobitic fish and probably connects to an extensive subterranean network.

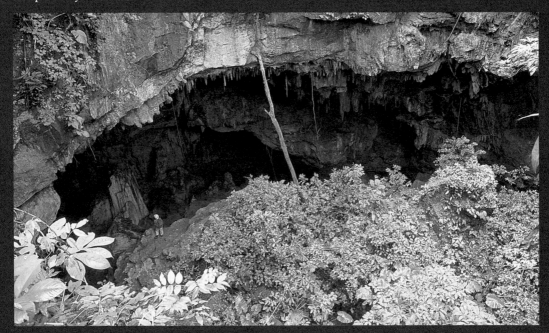

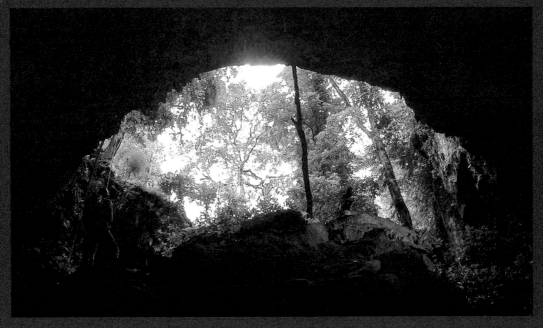

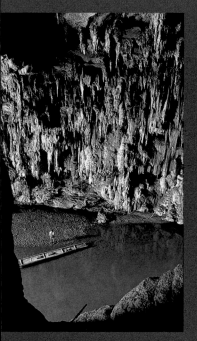

Tham Lod   ถ้ำลอด

The most popular cave in Northern Thailand after Tham Chiang Dao, developed as a tourist cave by the RFD and local villagers over the past 20 years. There is no lighting system but guides with lanterns are readily available.

An active through cave with a permanent large stream and two well-decorated upper levels. The main stream passage is 600 m long and can be negotiated without difficulty for the first 100 m The remainder of the main stream passage is most easily visited by raft which can be hired from within the cave. A concrete path continues on the south side to a large dry chamber with an impressive 22 m column and other mostly inactive decoration. A side route can be followed to a window with a superb view of the central river passage and a chance to see the swifts close. On the opposite side of the river a steep staircase leads up 30 m to a series of dry chambers with excellent decoration including rimstone pools, flowstone and a group of 'doll-like' stalagmites. The roped-off sections lead to constricted passages with dangerously high CO_2 levels.

The 40 m high outflow chamber is the most impressive part of the whole cave and has figured in many popular travel articles. A large colony of Himalayan Swiflets roost in this chamber, producing copious quantities of guano which is a vital food source for the cave ecosystem. A series of small dry chambers perched high up above the river contain about ten log coffins in an advanced state of decay.

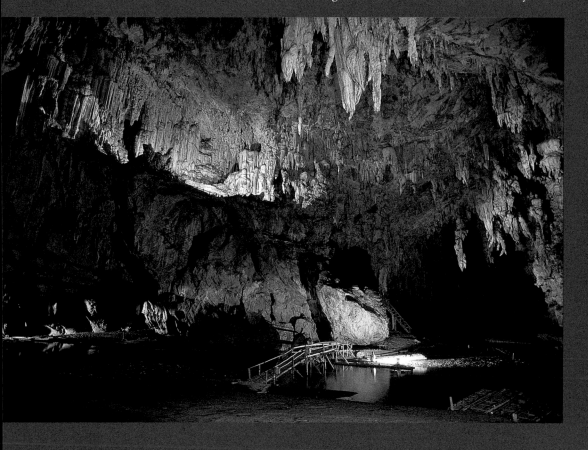

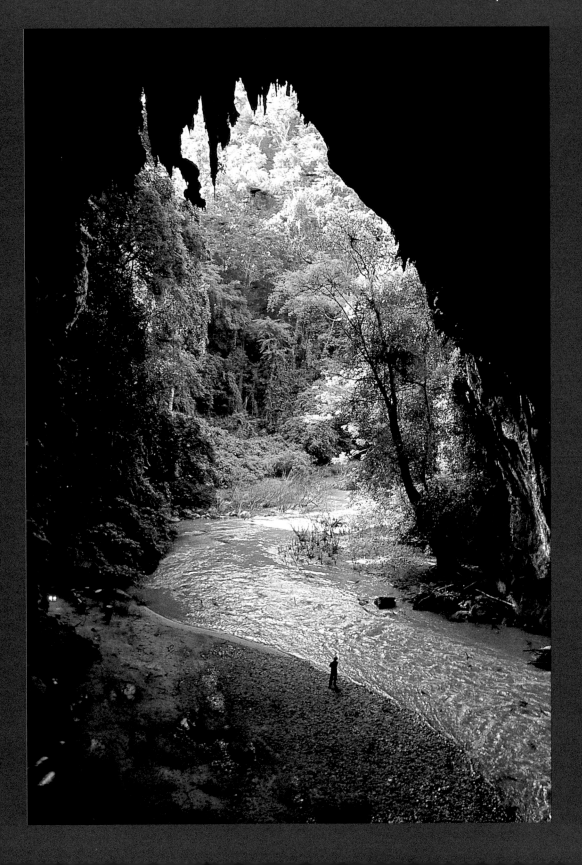

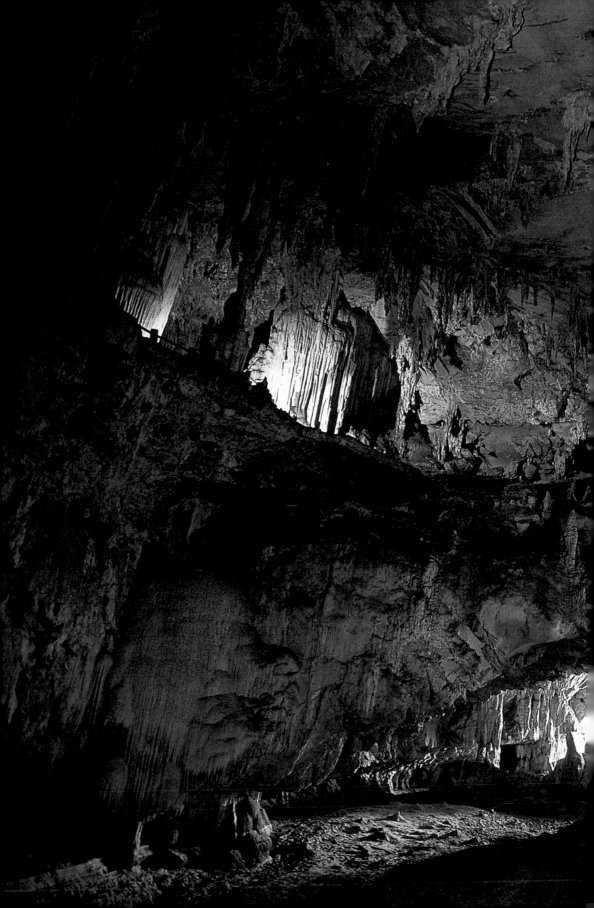

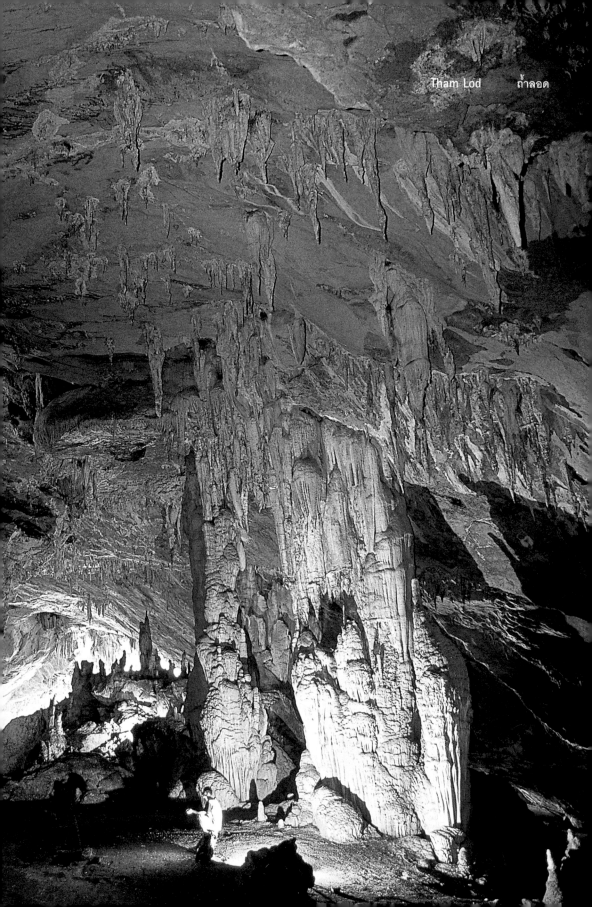

Tham Lod ถ้ำลอด

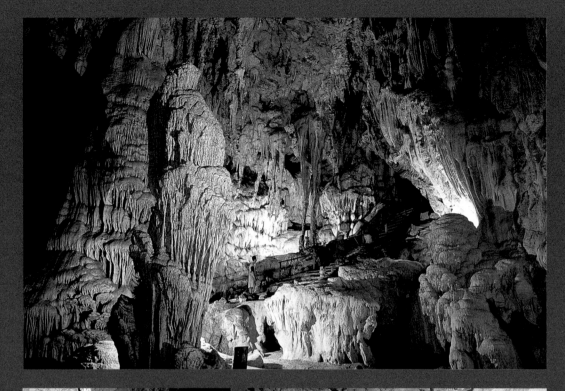

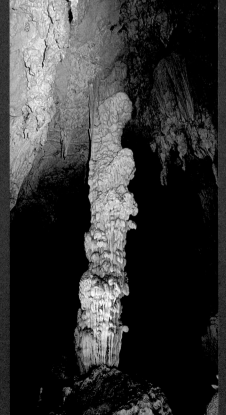

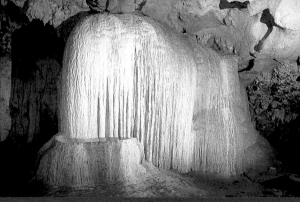

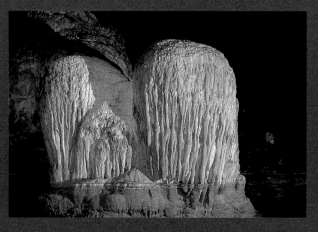

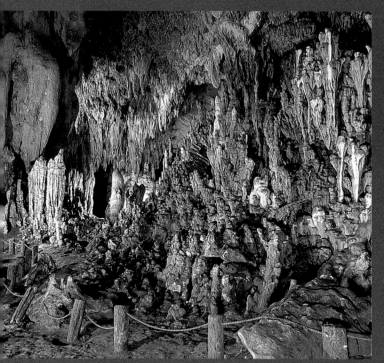

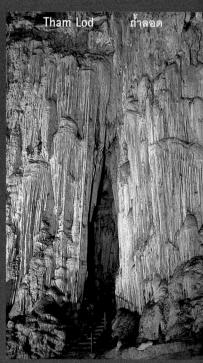

Tham Lod ถ้ำลอด

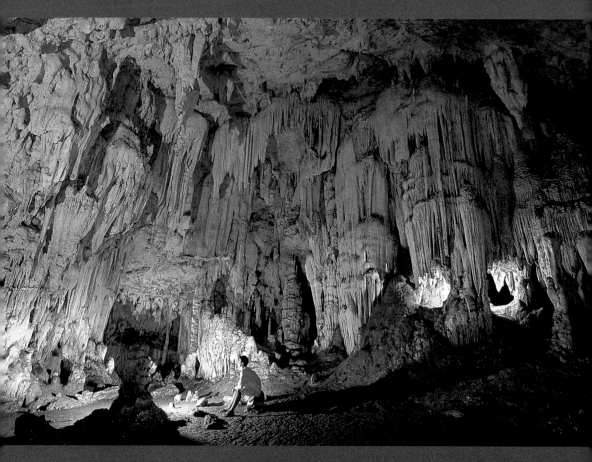

Tham Lod   ถ้ำลอด

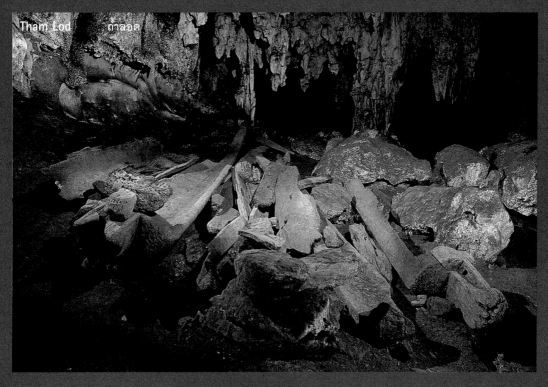

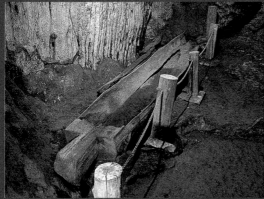

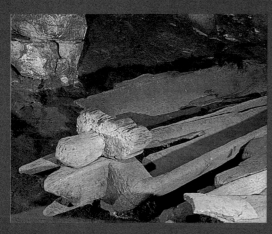

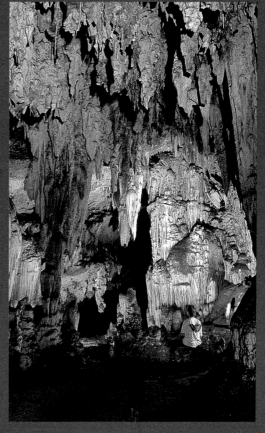

Tham Lod Overhang   เพิงผาถ้ำลอด

A rock overhang 300 m north of Tham Lod inflow, the site of a
major excavation by a team of Thai archeologists from Silpakorn
University in 2002. Their research has uncovered evidence that the
site was used by prehistoric people over a very long time period
stretching back more than 13,000 years BP. The most exciting
discovery was two human skeletons which appear to have been
buried and are 12,000 and 13,640 years old . The first skeleton was
in a lying position 50 cm below the surface, whilst the second older
one was found 30 cm lower in a crouched position, like the skeleton
found in Ban Rai (see page 106).

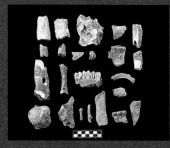

Bone fragments.

An incredible quantity of river pebbles was unearthed in a layer
over 2 m thick. Over 100,000 flakes and stone tools of the
Hoabinhian type have been identified, including sumatraliths, short
axes and choppers. Similar tools from Chester Gorman's excavation
at Spirit cave were dated 7,500-12,000 BP . The dedicated
archeologists are still processing the material and it will take many
years before any definitive conclusions can be made. A large number
of stone flakes were found along with the tools, suggesting that this
was a prehistoric "factory" where the tools were actually
manufactured.

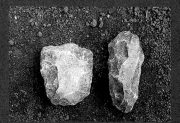

Hoabinhian Stone Tools.

The abundance of shells and animal bones, such as deer and wild
cattle, indicate that the site was also used for habitation at least in the
earlier periods. Pottery, blue glass beads and even a Victorian coin
were found in the more recent upper layers but have not been dated.

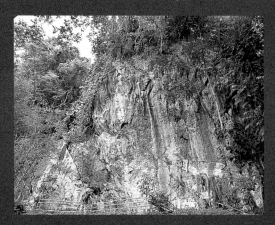

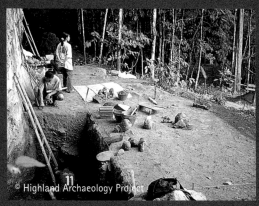

© Highland Archaeology Project

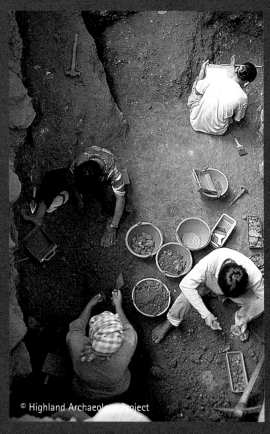

© Highland Archaeology Project

Tham Long Yaow   ถ้ำล่องยาว

An easily accessible cave with attractive formations, one of the safest caves to visit in Pang Ma Pha. The entrance was originally only 60 cm high, but has been widened and gated by local villagers. After a few metres crawl near the entrance, the cave opens out into a level dry passage 4-12 m wide and up to 10 m high. This main passage meanders along an old stream way for 500 m, eventually meeting a small permanent stream after 500 m The stream can be followed further for a short distance, before it disappears beneath a large rockpile.

Long Yaow has good examples of a range of cave formations, including some fine groups of active stalactites and stalagmites in upper side chambers visible from the main passage. The wind currents in the final chamber have caused the stalactites to grow in many different directions. Helictites and other erratic forms can be found in moist crevices with little air movemen – many of them are less than 1 cm long but wonderfully intricate.

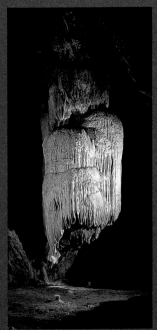

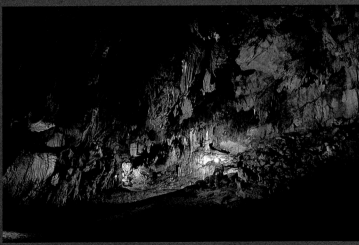

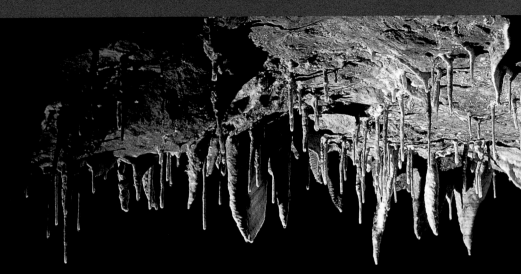

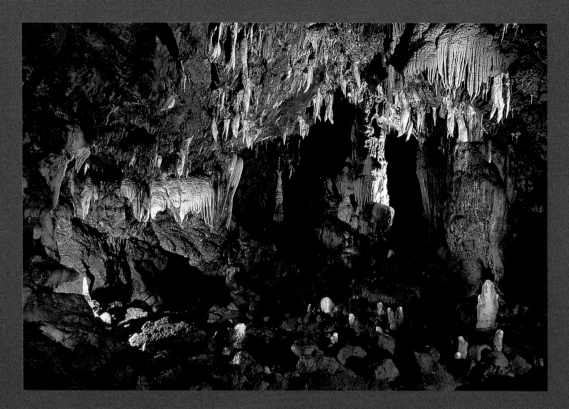

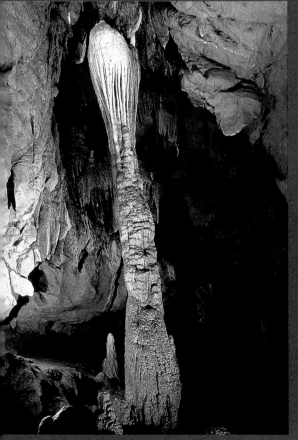

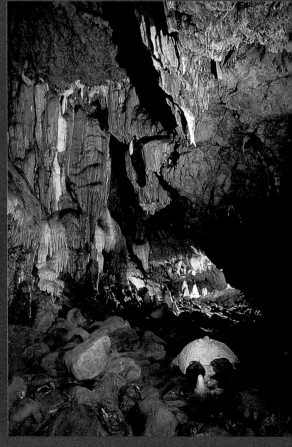

Tham Phra ถ้ำพระ

A rarely visited through cave close to Tham Long Yaow. The upstream entrance follows along a seasonal stream bed day lit by two natural windows. After 50 m the passage becomes very muddy with deep pools of water even in the dry season. The cave is best visited by raft, which can sometimes be hired from Muang Phaem Village. There is reputed to be an image of the Buddha half-buried in the mud somewhere in the middle of the cave, but whether this is natural or man-made we were unable to find out.

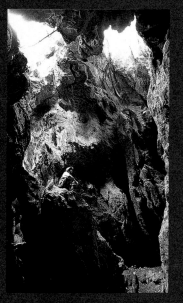
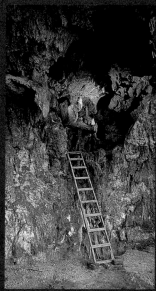
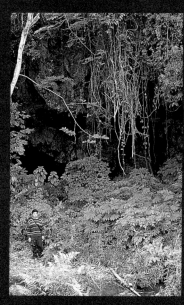

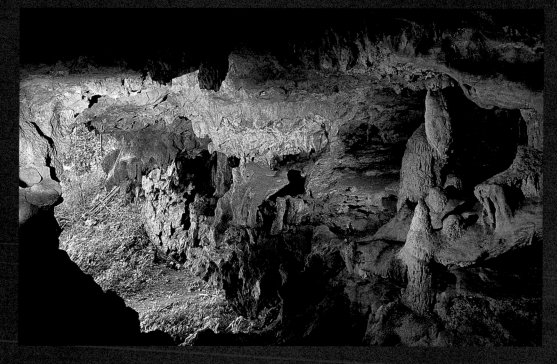

Tham Chedi ถ้ำเจดีย์

A small dry cave high up in a cliff reached by a steep wooden ladder on top of a much older brick base. The cave is used as a retreat by monks and is the only cave with a chedi in Pang Ma Pha. The chedi is probably at least 150 years old but has recently been renovated.

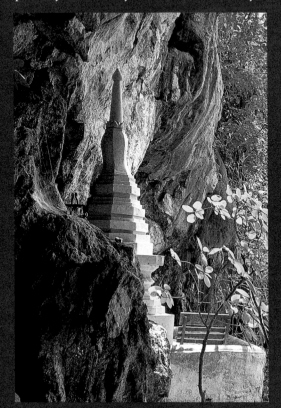
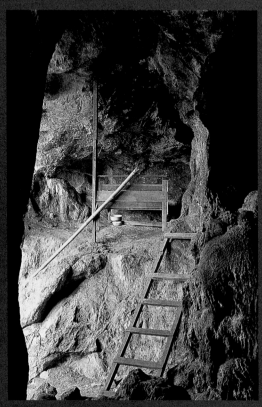
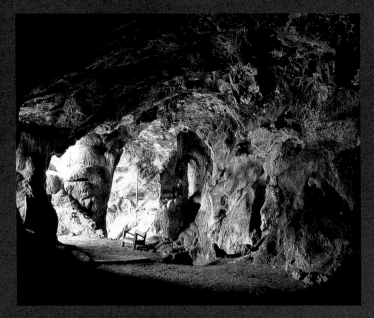
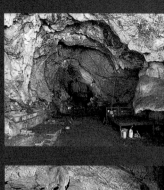
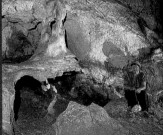

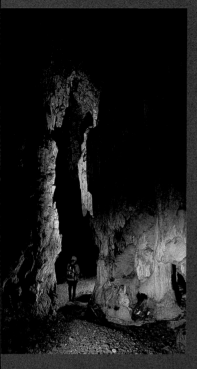

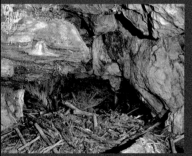

Tham Talu (Huai Rai 1)   ถ้ำทะลุ

A through cave with a seasonally active stream 200 m north of Tham Long Yaow. The 5 x 3 m inflow entrance is almost completely blocked by an impressive log pile caused by a severe flood. The 285 m inner passage is fairly uniform, 3-5 m wide and up to 20 m high, trending northwards after a dogleg bend 60 m from the entrance. The gravel floor is littered with many large logs which have broken some of the speleothems , exposing a cross section of their growth rings. 50 m from the outflow end the stream bed drops 4 m into a series of pools before tumbling down a large pile of boulders at the foot of a dramatic vertical slit in the cliff face. In 1985 there was a large bat colony but only a few individuals remain today.

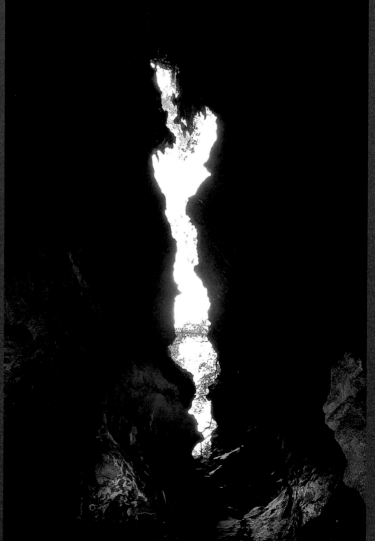

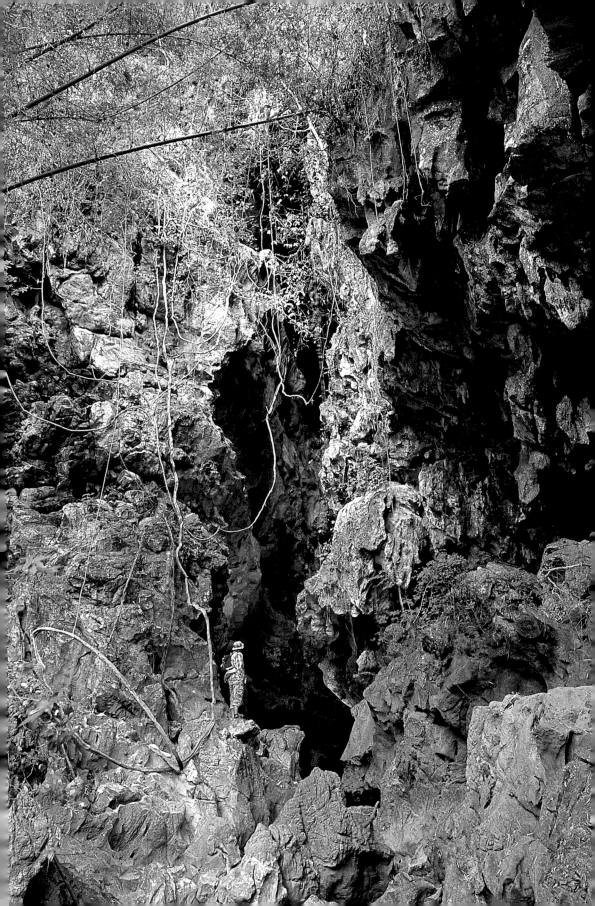

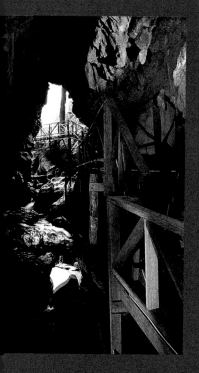

Tham Pha Mon ถ้ำผามอน

One of the most interesting caves in Thailand with an unrivalled collection of outstanding speleothems. The cave was first explored by a French team in 1985 when it was in pristine condition. The upstream section is currently being extensively developed by the military in preparation for a royal visit and is closed to the general public in the meanwhile.

The main through passage is 3,000 m long with a permanent stream and can only be visited during the dry season. The first 1 km from the upstream entrance now has electric lighting and a wooden or bamboo walkway all the way to the well-decorated upper chamber. The remainder of the main passage is still undeveloped and makes a exciting sporting trip of about 3-4 hours, enlivened by some low sections and a narrow canyon where the only way across is to swim. Elsewhere, it is mostly easy walking along a gravel floor in passages up to 30 m high with some fine formations and a couple of cave windows. A pleasant breeze blows throughout the main passage, becoming very strong in places and there are no problems with CO_2. The outflow is close to the main road 2.5 km east of Soppong.

The variety and beauty of speleothems in Pha Mon is unparalleled in northern Thailand, including good examples of most formations known from the region. The upper chamber is especially well-decorated with extensive areas of multi-tiered rimstone pools, pendulites, mud stalagmites, cave pearls, helictites, anthodites and a famous blue stalactite which is the only known example in S.E.Asia. The blue colour is due to a copper compound called azurite, $Cu_2 (CO_3)\text{-}(OH_2)$, a recrystalized form of the more common green malachite.

Tham Pa Mon supports a particularly rich fauna notable for the absence of guano-feeding species. A number of species are true troglobites and found nowhere else in the world, including several species of isopod, Collembola and a new species of beetle, *Ozaenaphaenops deharvengi,* named after the leader of the team who first explored the cave. *Macrobrachium sirindhorn,* a blue shrimp named after the princess was first discovered inside Pha Mon but has since been found in the surrounding karst. The cave also supports a population of *Schistura oedipus,* a troglobitic fish which is endemic to Pang Ma Pha. The development of Pha Mon as a tourist attraction could cause some of these unique species to become extinct.

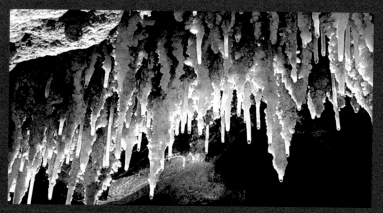

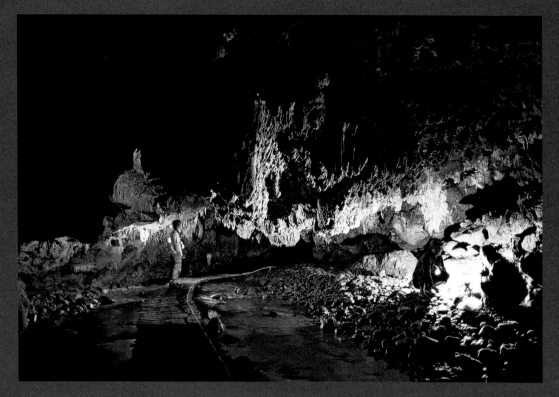

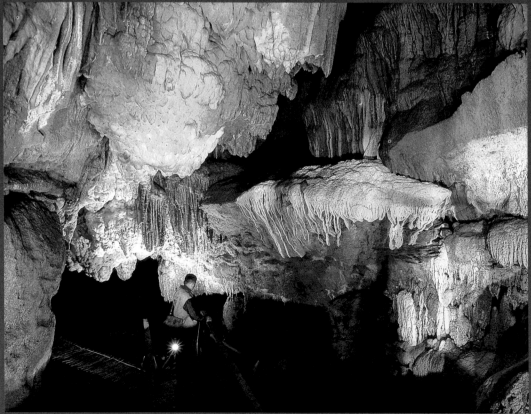

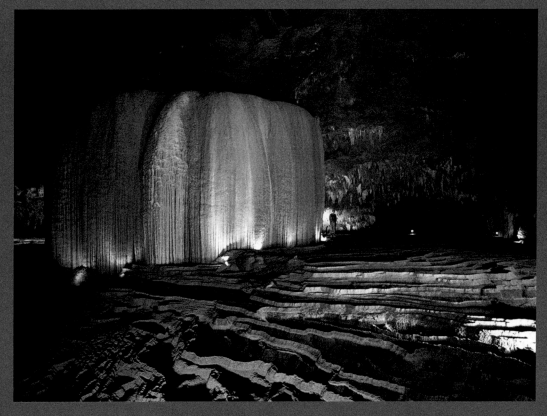

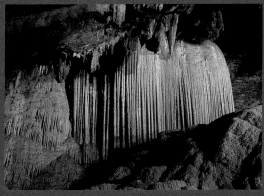

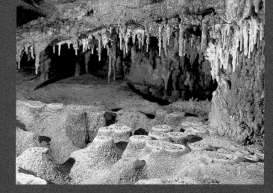

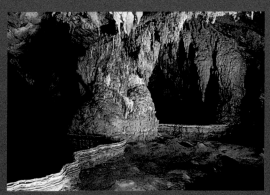

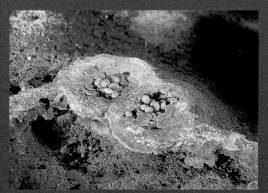

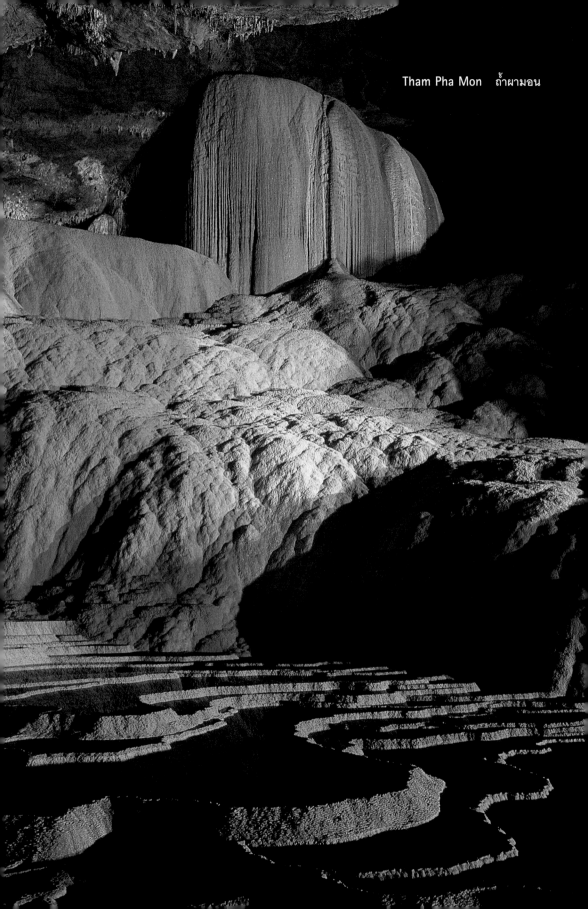

Tham Pha Mon   ถ้ำผามอน

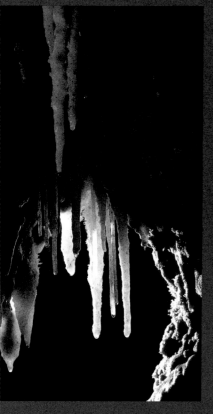

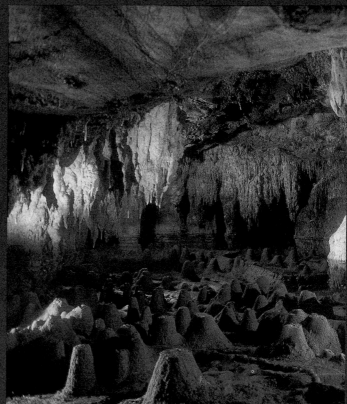

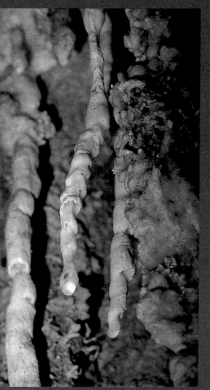

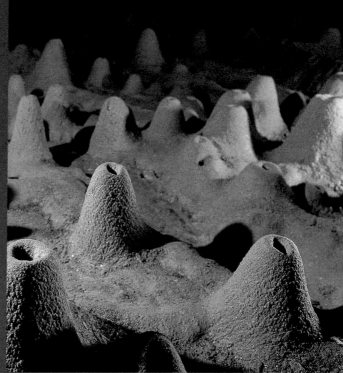

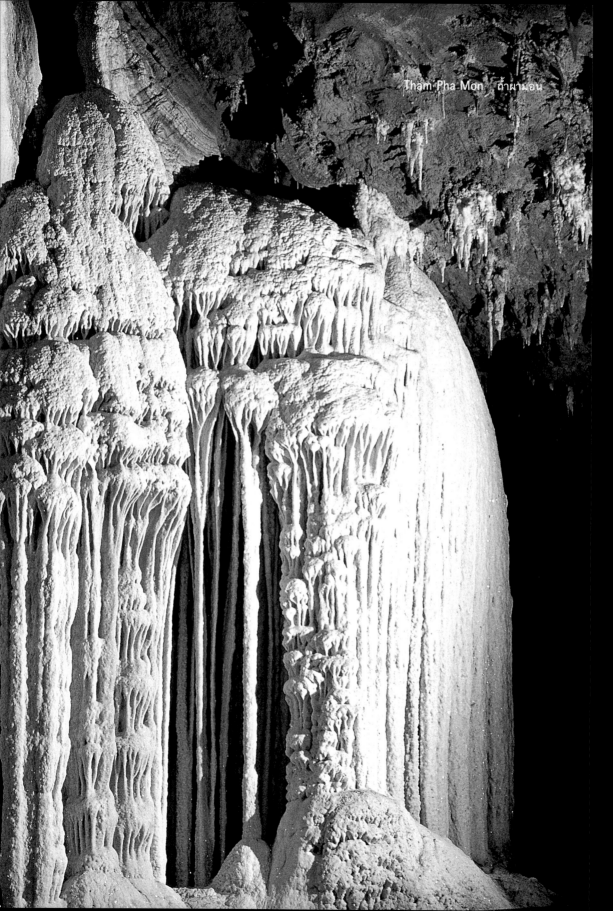

Tham Pha Mon   ถ้ำผามอน

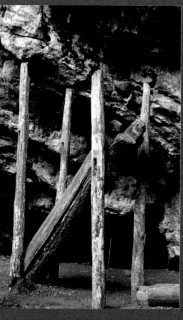

Ban Rai Overhang   เพิงผาบ้านไร่

A magnificent natural amphitheatre high above the Nam Lang River with an unrivalled suite of teak coffins. After a steep 200 m climb, a slight dip leads down into the site itself, which is hemmed in by towering white cliffs and eerily silent.

The log coffins are ranged all around the back of the amphitheatre behind the drip line; some are still on their original supports but there are many more scattered all over the floor. There are at least 15 coffins and 95 supporting columns, one of which is carved at the top. The coffin heads are mostly rather crudely carved compared with sites such as Bor Krai. The biggest coffin is 9.3 m long, which is the largest known example in Pang Ma Pha.

A unique collection of prehistoric rock paintings depicting humans, animals and abstract designs is hidden in sheltered crevices. Ban Rai is the only site where stone tools, iron artifacts, pottery, rock paintings and log coffins have been found in the same location, suggesting human use over a very long time span stretching back more than 10,000 years BP.

Ban Rai was excavated by a team of Thai archaeologists from Silpakorn University in 1999-2002 (see pages 104-107).

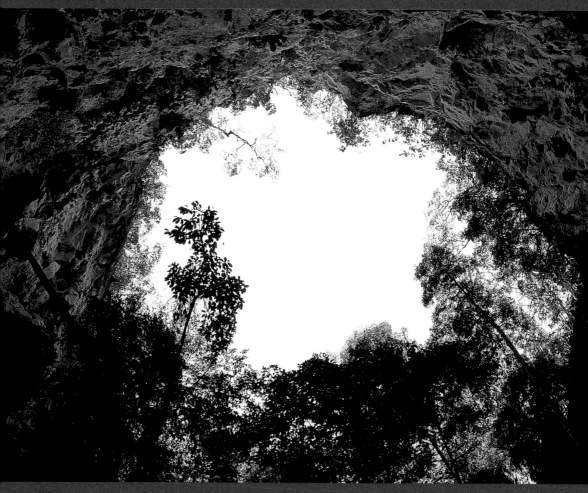

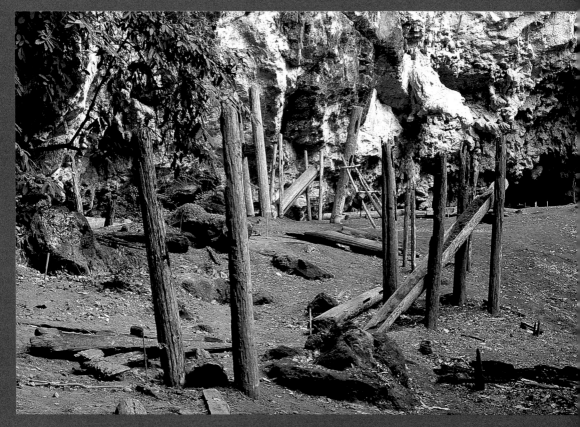

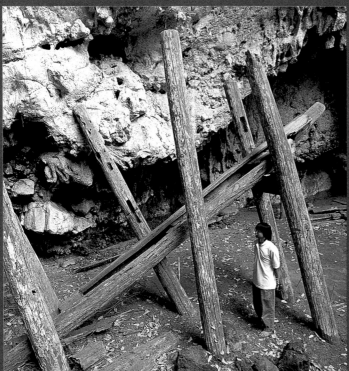

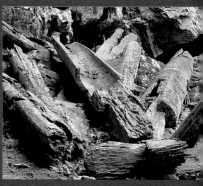

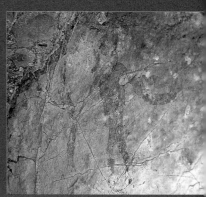

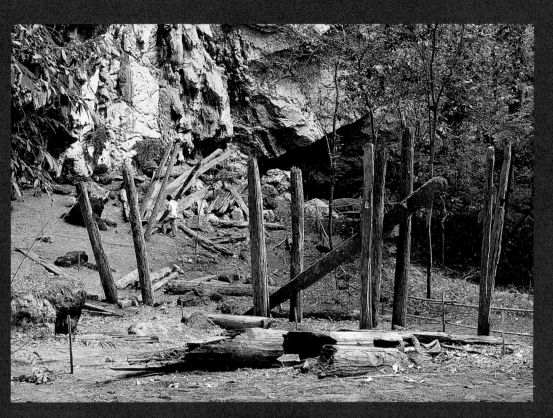

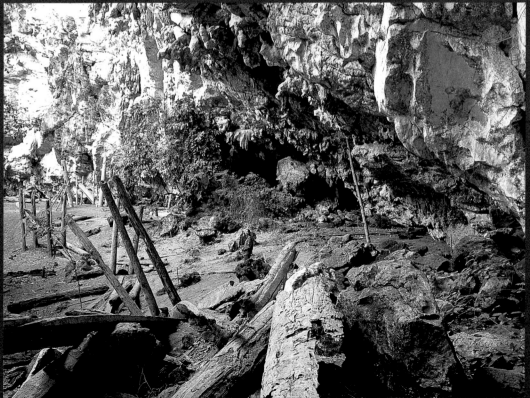

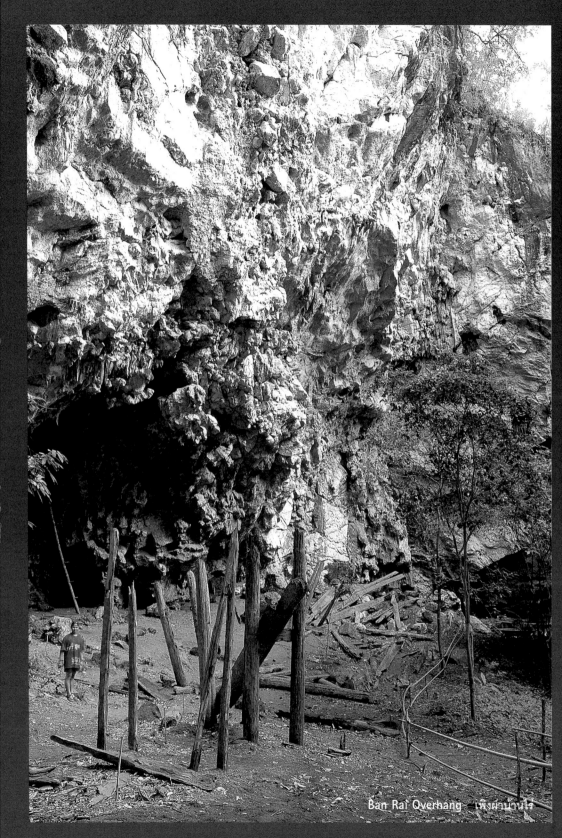

Ban Rai Overhang เพิงผาบ้านไร่

Piman Face Cave   ถ้ำผีแมนหน้าคน

A small dry cave perched 6 m up a vertical cliff face with a stunning view out over the Nam Khong Valley. The outer daylit chamber contains 3 or 4 log coffins of a style common throughout Pang Ma Pha, including two in good condition. A tight squeeze leads to the completely dark tomb-like inner chamber. The coffins in this chamber are only 2 m long, much smaller than average for Pang Ma Pha. Two of the coffins have a unique style of carving on their lids, resembling a human face.

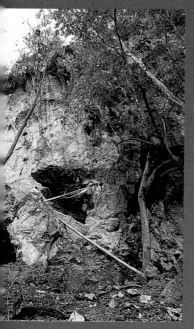

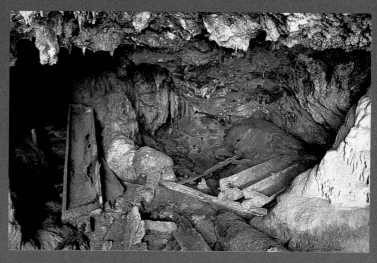

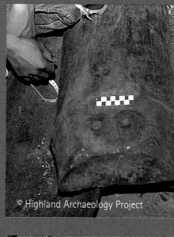

© Highland Archaeology Project

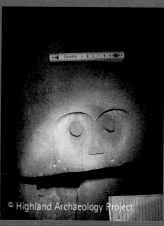

© Highland Archaeology Project

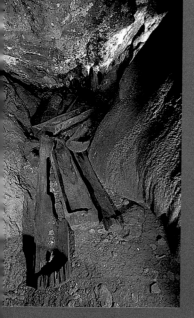

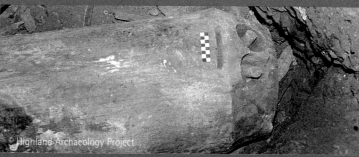

© Highland Archaeology Project

Pha Nam Lang ถ้ำผีแมนน้ำของ

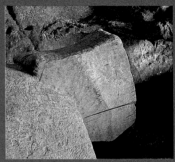

Small dry cave in towering cliffs north of Tham Nam Lang. The main entrance is perched high up the cliff and affords a magnificent view over the Nam Khong Valley. The cave contains a 4 m log coffin in good condition, complete with it's lid and support columns. An archeological survey in 1999 found glass beads and 1724 fragments of pottery associated with this one coffin.

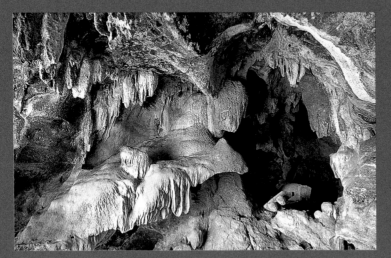

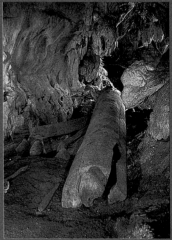

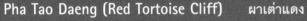

Pha Tao Daeng (Red Tortoise Cliff) ผาเต่าแดง

A remote prehistoric rock art site in a spectacular setting at the base of towering red cliffs north of Tham Nam Lang. The paintings are about 2 m above the ground sheltered by a huge boulder. There are 3 red hand prints and a curious abstract swirling design, all quite faint. The style is similar to the rock art in Pra Thu Pha, Lampang Province (page 98-101) which has been dated at 3,000 years old. Stone tools and pottery were also found underneath the paintings. The site is currently used by intrepid local hunters who build extremely precarious-looking bamboo structures to collect honey from bee hives high up the cliff face.

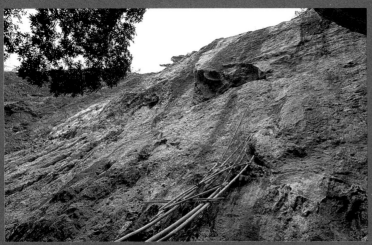

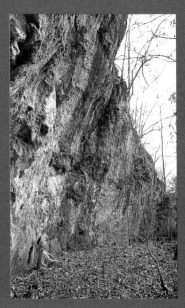

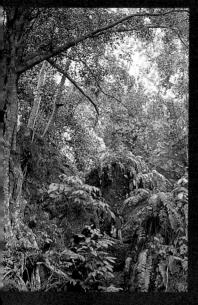

Tham Nam Lang ถ้ำน้ำลาง

The third longest cave in Thailand (8,550 m) and certainly one of the most spectacular. The vast entrance chamber leads to an awesome river passage up to 100 m high which is one of the biggest in the world. The river can be followed continuously for more than 6 km, through wonderful underground vistas in a passage seldom less than 10-20 m high and wide. For the first 4 km the water usually flows gently over gravel beds but can be chest-deep even in the dry season and a couple of 30 m wide lakes have to be negotiated. Massive rockfalls disappearing up into the blackness attest to ongoing cave breakdown. Formations are scarce in the main river passage except for two huge flowstone deposits, the second of which is over 100 m long. 3 km into the cave a side passage leads to an upper chamber with some of the most stunning cave formations in Thailand. Only a few cavers have ever made it further than 4 km since this requires an overnight camp in the cave and the water gets even deeper. The cave ends at an impenetrable sump only 500 m in a direct line from the equally impenetrable inflow.

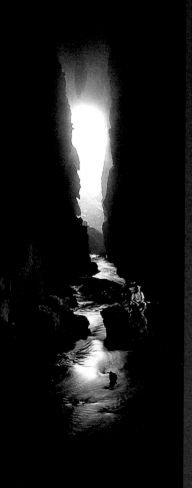

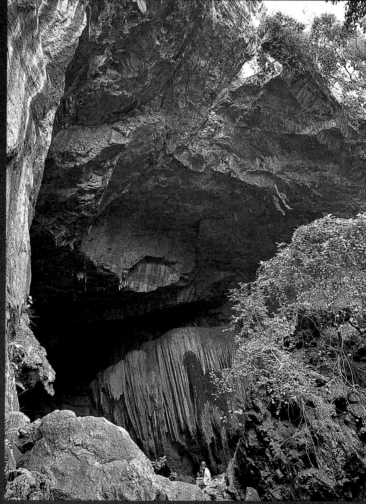

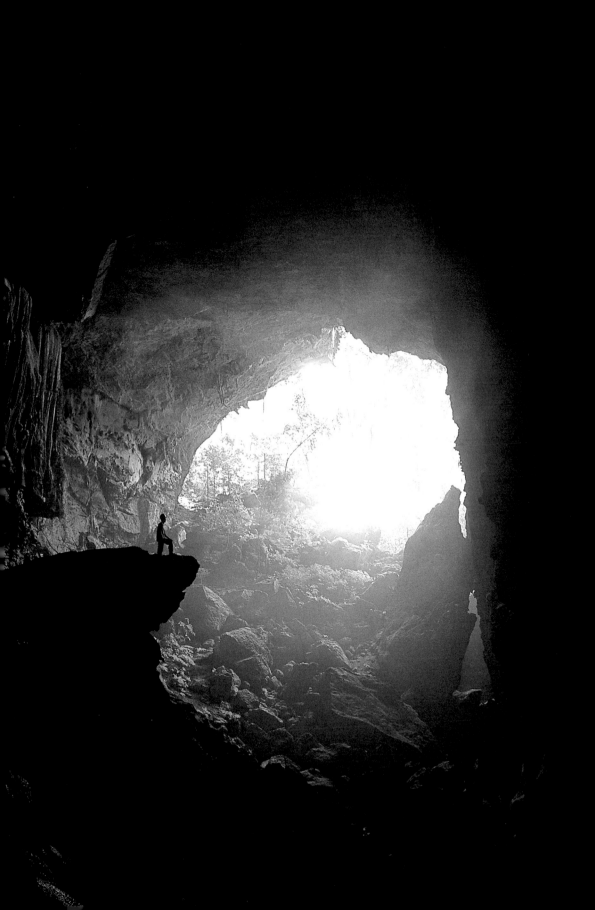

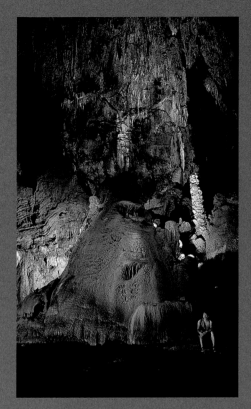

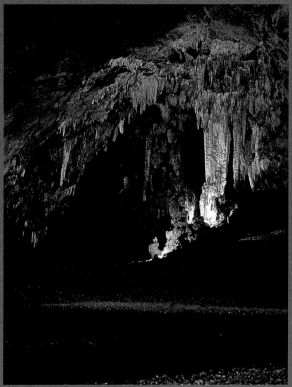

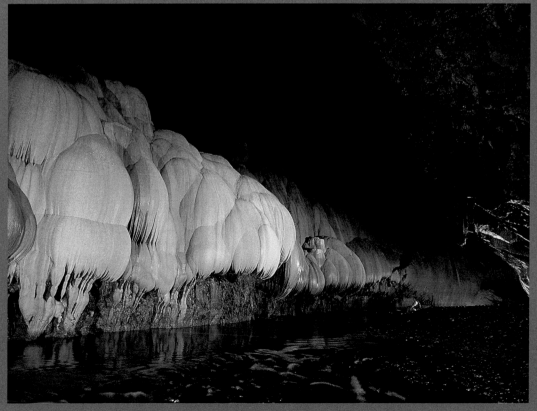

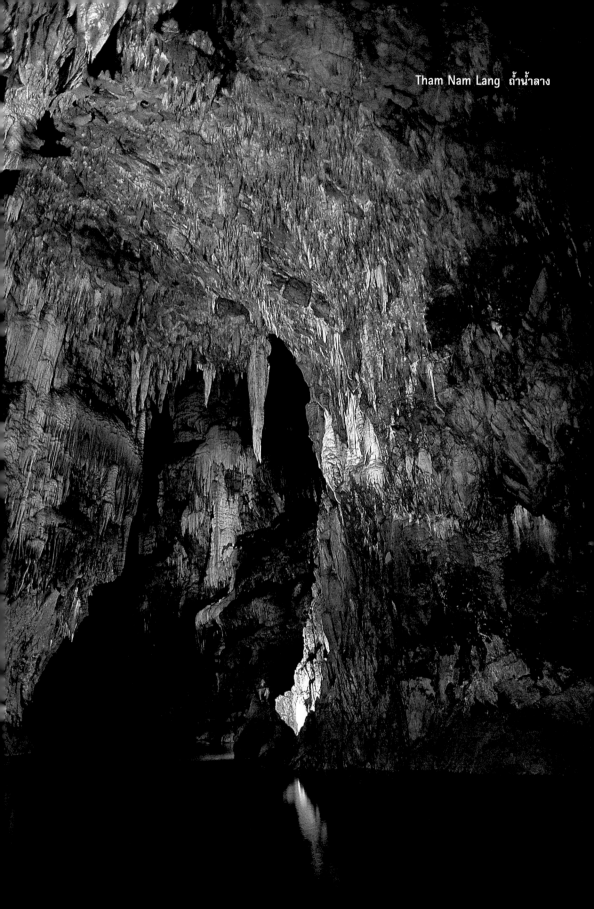

Tham Nam Lang ถ้ำน้ำลาง

Tham Susa   ถ้ำซู่ซ่า

A remote cave at the base of an impressive cliff, the only known stream outflow for the extensive karst plateau between the Nam Lang and the Mae Nam Pai Rivers. Downstream of the spring the stream has built an spectacular travertine delta 400 m wide, falling into the Nam Khong River over attractive waterfalls 3 m high. The delta supports one of the few examples of lowland evergreen forest in the region.

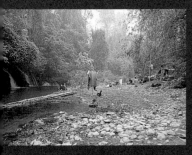

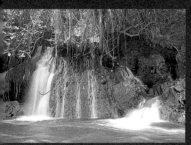

There are two entrances to the cave, both situated above the spring. The higher of the two entrance chambers contains archeological evidence of a prehistoric occupation site. Beyond the entrance chamber the cave descends down a steep and very slippery muddy slope to the stream. The main stream passage is 5-7 m wide with slow-moving waist-deep water in the dry season. Annual flooding of the stream passage has created the massive muddy banks which are a prominent feature of this part of the cave. CO_2 levels are dangerously high throughout the first 500 m. After 500 m a deafening roar signals the proximity of an impressive 5 m waterfall formed by a band of very hard basalt rock. A second waterfall lies just around the corner. A further 1,500 m has been surveyed above the waterfalls, but there is no known inflow entrance.

Tham Susa is the type location for *Cryptotora thamicola*, the "waterfall-climbing" fish.

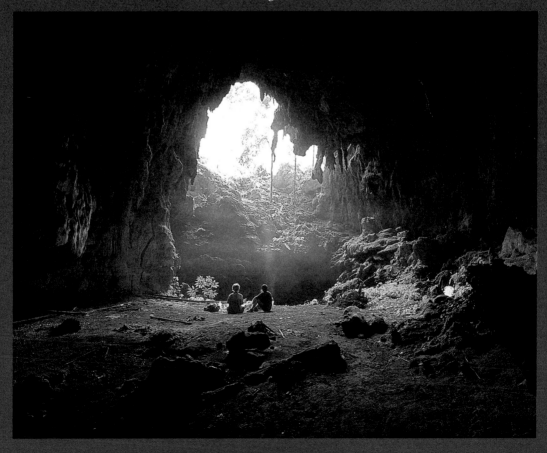

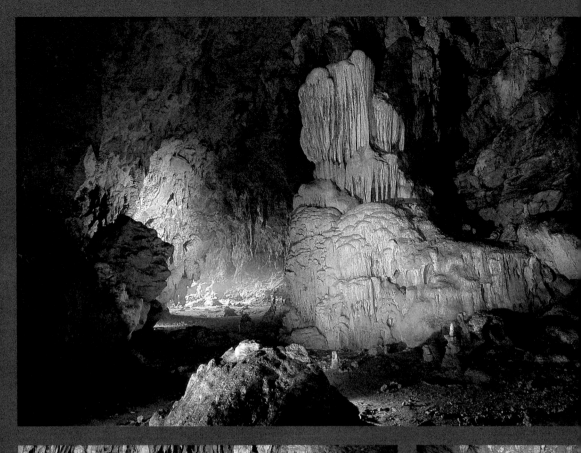

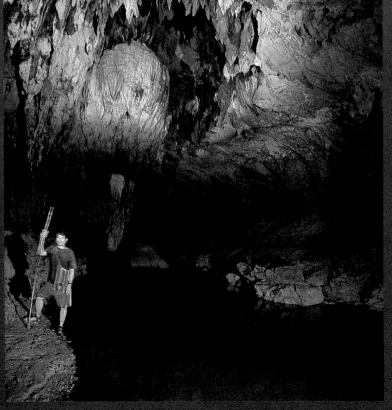

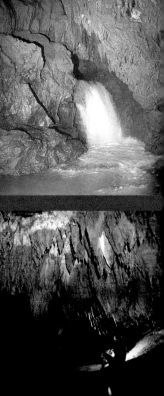

Seua and Lom Caves   ถ้ำเสือ ถ้ำลม

A network of interconnecting caves on many levels with at least five known entrances, one of the longest surveyed cave systems in Thailand (3,100 m). The system is characterized by seasonally flooded stream passages which become very constricted in several places. There are very few formations but the elegant smooth-walled passages with distinct scallop marks have a beauty of their own and are an excellent illustration of cave formation processes. Some sections of the cave cut through marblized limestone, a much harder rock formed under high pressure.

A distinct breeze can be felt throughout the cave but is particularly noticeable at the Wind Cave outflow (see page 53). Although the system has been extensively surveyed by both British and Australian teams, many passages are still unexplored.

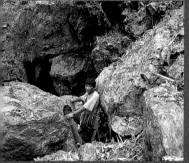

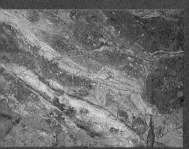

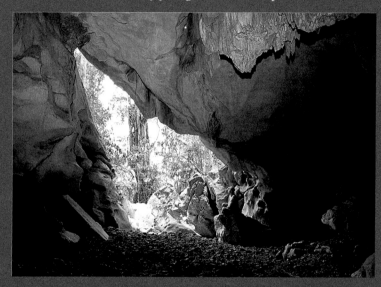

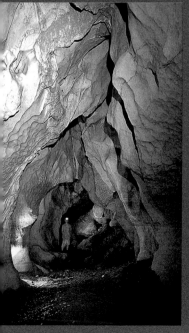

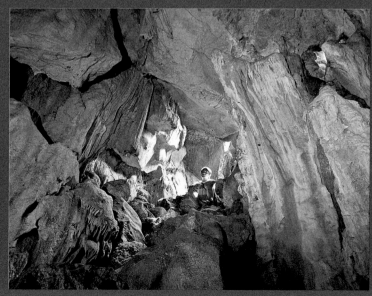

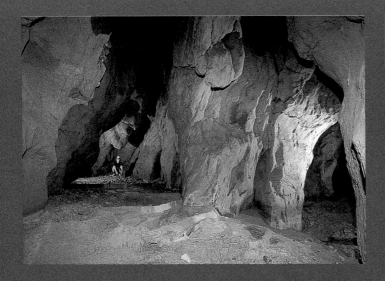

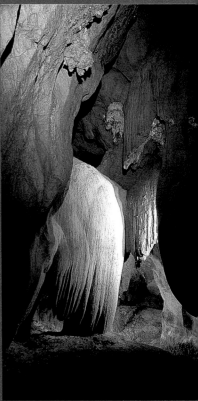

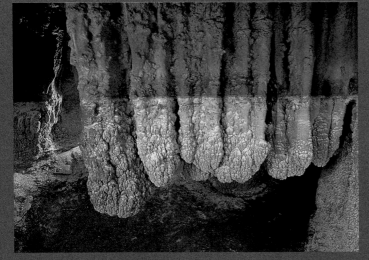

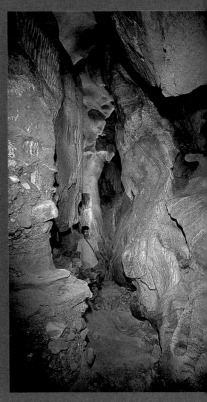

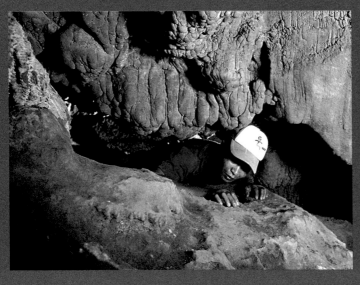

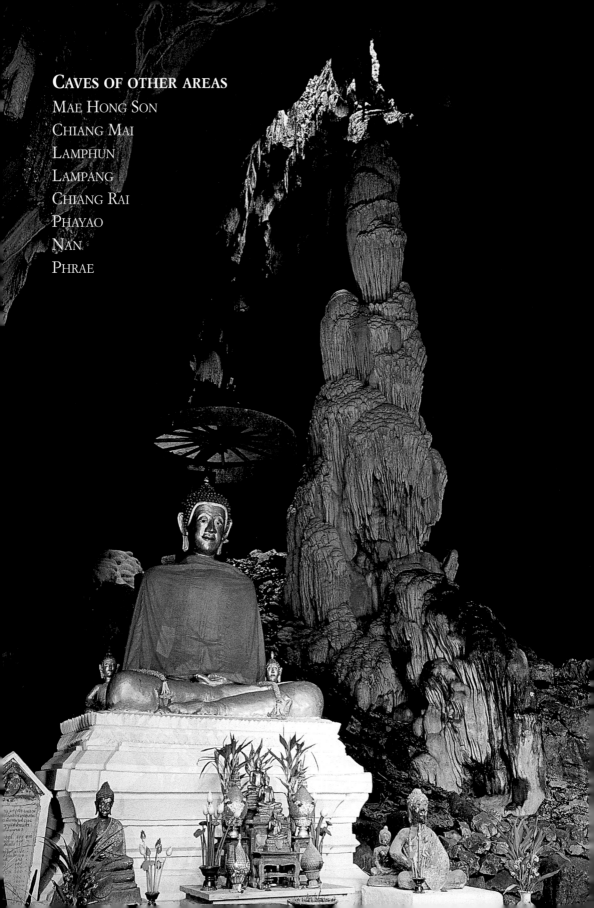

CAVES OF OTHER AREAS
MAE HONG SON
CHIANG MAI
LAMPHUN
LAMPANG
CHIANG RAI
PHAYAO
NAN
PHRAE

Woa Overhang   ถ้ำวัว

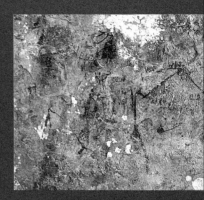

A shallow overhang hidden in the forest behind the tranquil temple
of Wat Tham Wua. Several platforms have been built along a 100 m
stretch of cliff face, housing several Buddhas and other religious
imagery.

   6 to 8 m above the largest platform is a remarkable series of rock
paintings that are probably of prehistoric origin, although there is
some doubt as to their authenticity. Fragments of bone can be seen
embedded in sediments cemented to the bedrock up to 4 m from the
ground. The concrete floor is built on top of thick layers of
prehistoric deposits, now well-protected from the ravages of time.

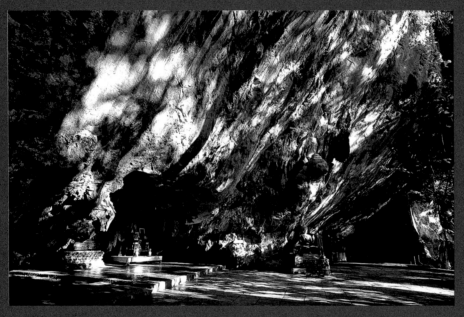

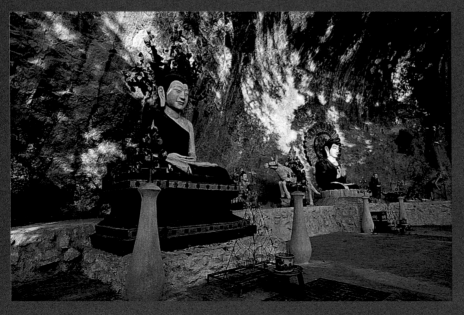

Tham Pla   ถ้ำปลา

A well known tourist attraction 18 km north of Mae Hong Son.
A pleasant 200 m trail along a crystal clear stream leads to the
resurgence, which harbours a seething mass of enormous well-fed
Brook Carp. The cave itself cannot be entered here, although it may
be possible to reach the inner passage via a small entrance further up
the hill. A circular nature trail starting near the bridge takes in Tham
Russi, a small dry cavern formerly used by monks.

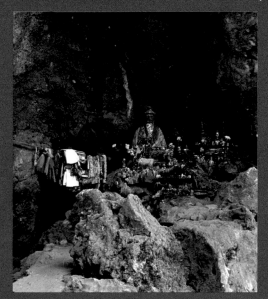
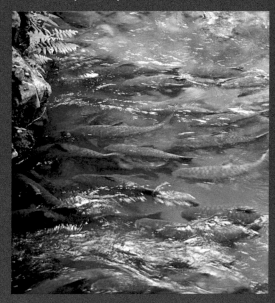

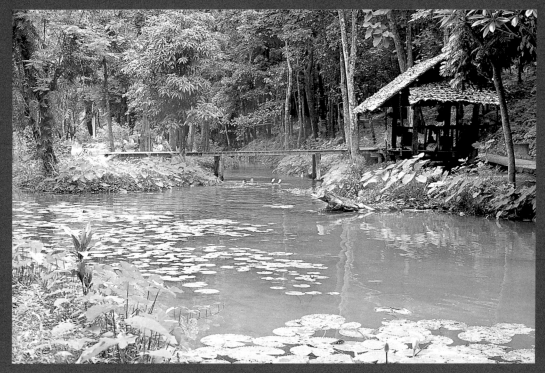

Tham Khang Khao (Ru Hoa Koa)   ถ้ำค้างคาว

An easily visited natural cave close to the main road 500 m west of
the Tham Pla turn-off. The entrance is hidden at the head of a dry
valley behind a group of Buddha statues under black canopies.
The main passage trends south along a seasonally flooded stream
bed and is straight-forward except for a few low points and
scrambles around flowstone blockages. After 200 m, a moderately-
sized chamber is home to a colony of bats which have left
distinctive dark markings all over the ceiling. The passage is well-
decorated in places with massive flowstone and attractive rimstone
pools. The cave is passable for at least 1,400 m but CO_2 levels
become noticeably higher further in and may force a retreat long
before this point. Tham Khang Khao supports a rich cave life
including two recently discovered species of Isopod (see page 65).
A small cave 50 m to the west is worth visiting for its wonderfully
tranquil atmosphere.

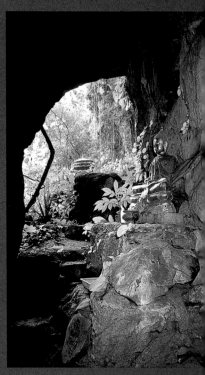

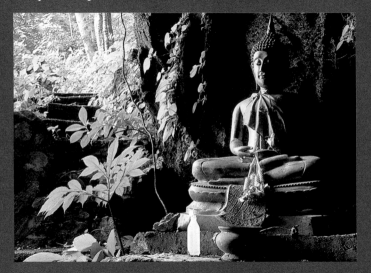

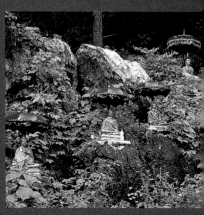

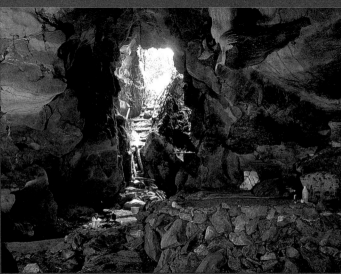

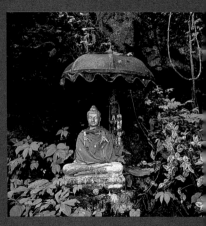

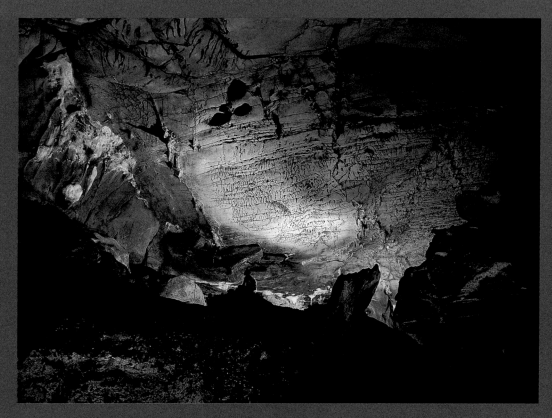

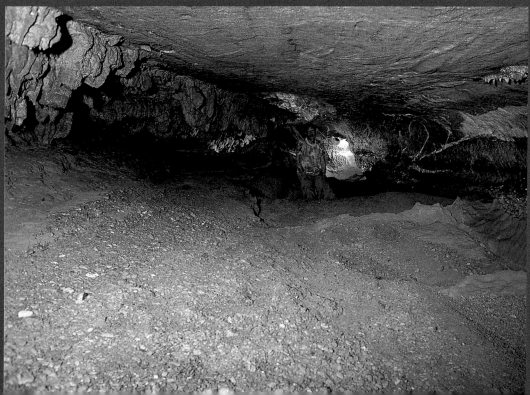

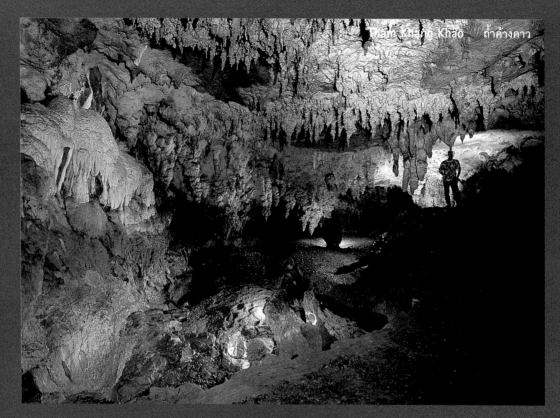

Tham Khang Khao   ถ้ำค้างคาว

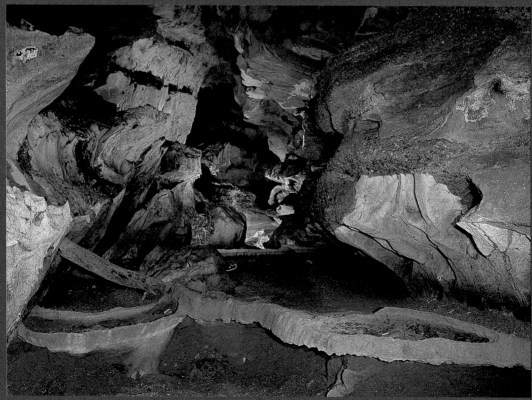

Pha Phung   ผาผึ้ง

An impressive overhang close to the road 2 km south of Tham Pla.
The site is several hundred metres long and riddled with a maze of
precarious bamboo scaffolding used by local villagers to collect honey
from the numerous bee hives high up on the cliff face. One of the
more sheltered sections harbours a remarkable collection of prehistoric rock art, including one drawing which seems to be a
representation of a bee, suggesting that the site has been used for
the same purpose since time immemorial (page 96-97).

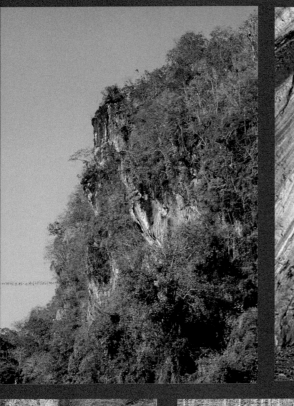
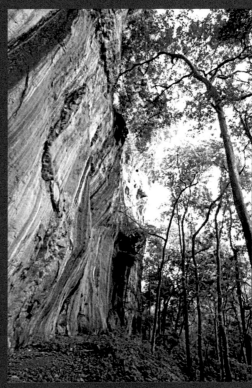

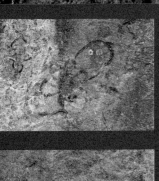
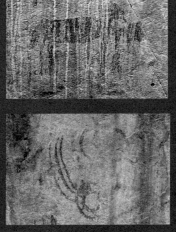
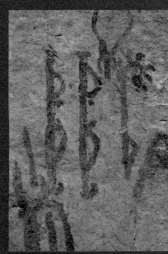

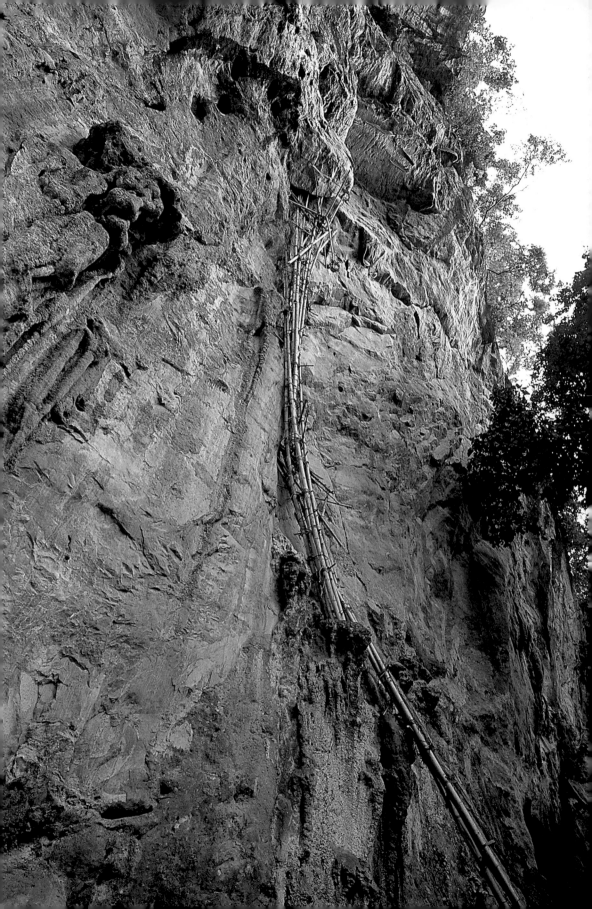

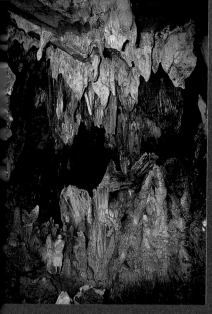

Tham Jak Dtor (Phra Pbat Ku)   ถ้ำจ้ากต่อ (ถ้ำพระบาทคู่)

A locally famous cave 1 km northwest of Tham Plaa and probably one of its major water sources. The cave serves as a forest temple in the traditional fashion, with at least one monk in permanent residence. It is considered sacred on account of a hand-like impression on a rock and a rimstone pool said to represent both left and right footprints of the Buddha. The 20 m wide entrance has a large platform with a series of shrines and an ornate canopy over the 'handprint'. The rock-strewn main passage descends steeply behind the shrines with only occasional signs of running water. After 150 m a slight climb up to a narrow side passage leads to the upper end of a vast hall which plunges downwards into the murky darkness. A delicate descent over a morass of unstable boulders coated with guano leads to a curious collection of large bones encrusted with calcite, said to have belonged to a hermit. The lower section of the cave is unexplored and could connect with Tham Pla over 200 m below, which would make it the deepest cave in Thailand.

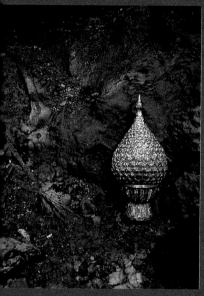

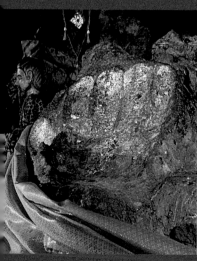

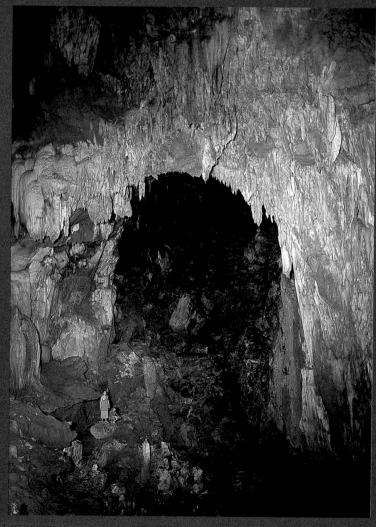

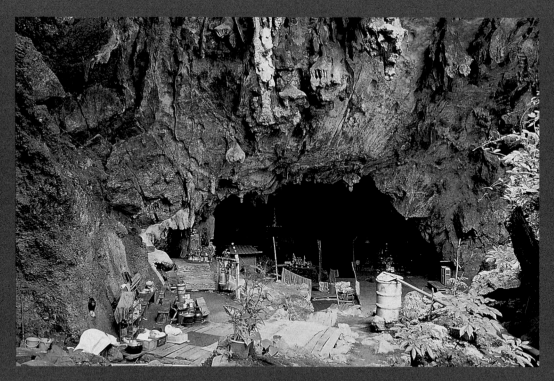

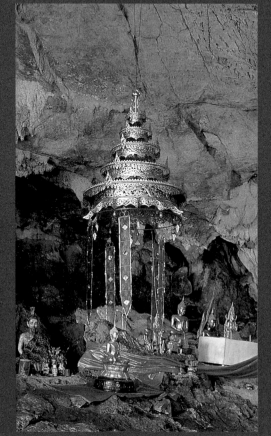

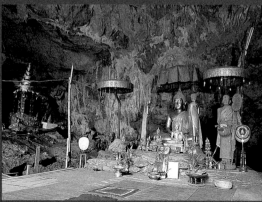

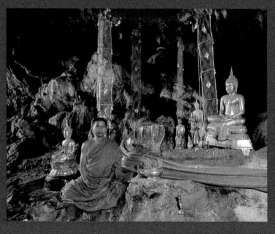

Pha Phung ผาผึ้ง

A shallow overhang at the base of an imposing 100 m cliff next to a stream. The site was occupied by a group of monks until fairly recently but has now been abandoned, although a couple of the shrines still show signs of care. An interesting feature is a wall embedded with (amulets) and pictures of Thai kings. The cliff is split by a high vertical rift which provides some shelter but is only 15 m deep with no apparent extensions. At the western end a tiny chamber has been converted into monks' sleeping quarters.

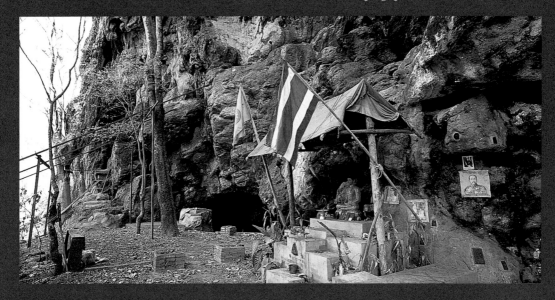

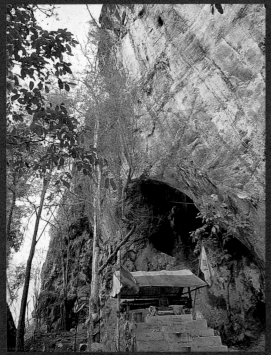

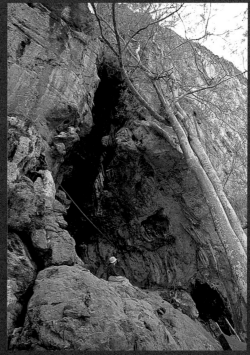

Tham Mae Ga Noi   ถ้ำแม่กาน้อย

50 km south of Khun Yuam, planned to be a major tourist
attraction with a special 3 km trail built right up to the cave
entrance. The trail is now in a dilapidated condition and the cave
itself seems to have been abandoned to its fate. An inconspicuous
hole in the hillside leads down into a surprisingly spacious
chamber 50 m across with a large gilt Buddha which is
nevertheless dwarfed by the grandeur of the natural setting.
The far end of the chamber rises up over large boulders to a short
passage leading into a completely dark inner chamber 10 m across
with some active speleothems but no obvious exit.

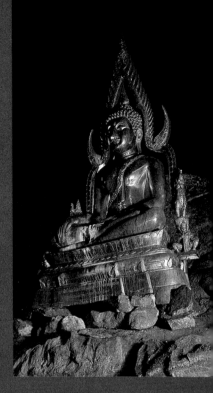

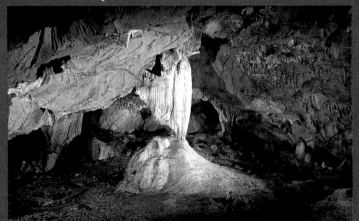

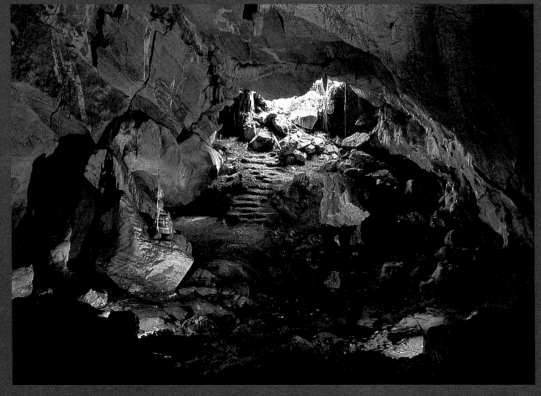

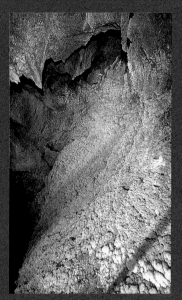

Unnamed Cave   ถ้ำกลางเหมือง

A newly discovered cave in the middle of an active mine, uncovered by accident during blasting. The small entrance descends steeply for 15 m before dividing into two small chambers lined with crystals similar to those in Tham Kaew Komol. The left branch continues downwards to unexplored depths. The cave is not safe to visit as sections of the roof weakened by blasting are in danger of imminent collapse.

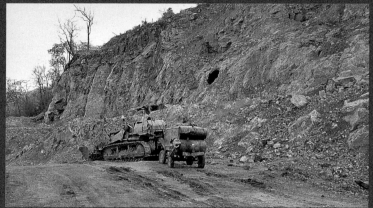

Tham Mae Hu   ถ้ำแม่ฮุ

A small temple cave maintained by a solitary dedicated monk, 2 km SE of Tham Kaew Komol. The 15 m wide entrance chamber has been partially enclosed and a level concrete floor constructed. There is a small library and a shrine with statues of the Buddha and a hermit.

   A 20 m passage behind leads past a spartan bedroom cell to a larger inner sanctuary with a tiled floor and a collection of gilt statues on a raised platform. Another less-developed cave in the same temple is 100 m down the hill and can only be reached by a sturdy ladder built by the enterprising monk. A medium-sized chamber with active speleothems and no obvious exit.

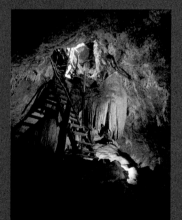

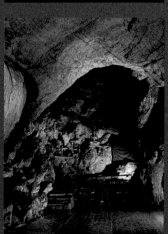

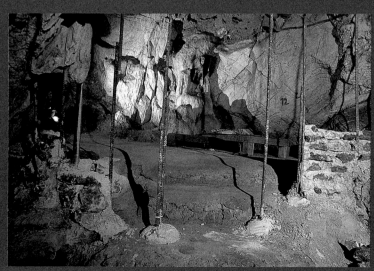

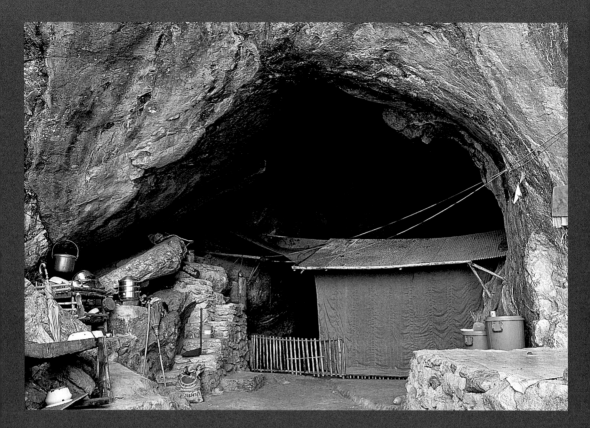

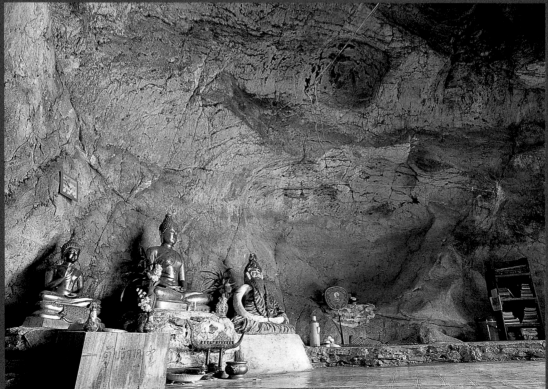

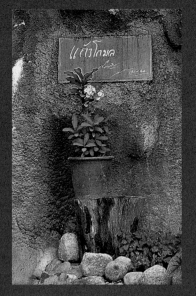

Tham Kaew Komol (Calcite Cave) ถ้ำแก้วโกมล

This small cave has the best examples of calcite crystals in Northern Thailand and is the only known example of a hydrothermal cave in the region. The site was discovered over 10 years ago by mining operations but was not opened to the general public because of its fragile condition. A major effort has recently been made to develop this unique cave as an important tourist attraction, culminating in the Queens's visit in the year 2000. A specially constructed tarmac road leads right to the cave entrance, where there is a car park and a visitor information centre. The cave itself has been equipped with a permanent walkway, halogen lights and an air circulation system, which is only used for VIPs due to concerns of the adverse effect on the cave environment. All visitors must be accompanied by an RFD ranger.

Tham Kaew Komol was formed by superheated water from deep underground forcing its way upwards along the line of weakness at the limestone/shale boundary and dissolving the limestone in the process. The crystals developed slowly under conditions of high temperatures and very low oxygen levels, probably when the cave was completely filled with water. These special conditions have created a fabulous growth of crystals quite unlike those seen in normal caves. Since the cave was opened up to the outside air these hydrothermal crystals have stopped growing and are now being smothered by stalactites and flowstone.

The cave consists of a series of five small chambers on a steep gradient, the final chamber being more than 50 m below the surface. Crystals originally covered every available surface and even now are very impressive despite vandalism in the early days after it was discovered. Please be very careful not to touch anything as these natural wonders are an irreplaceable part of our heritage.

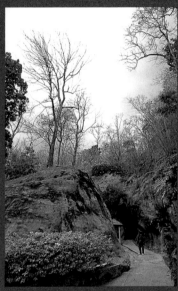

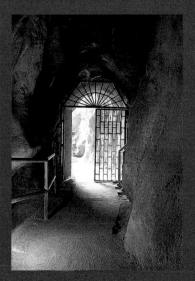

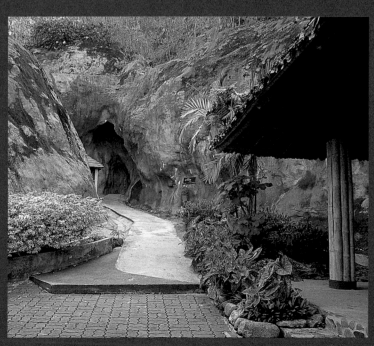

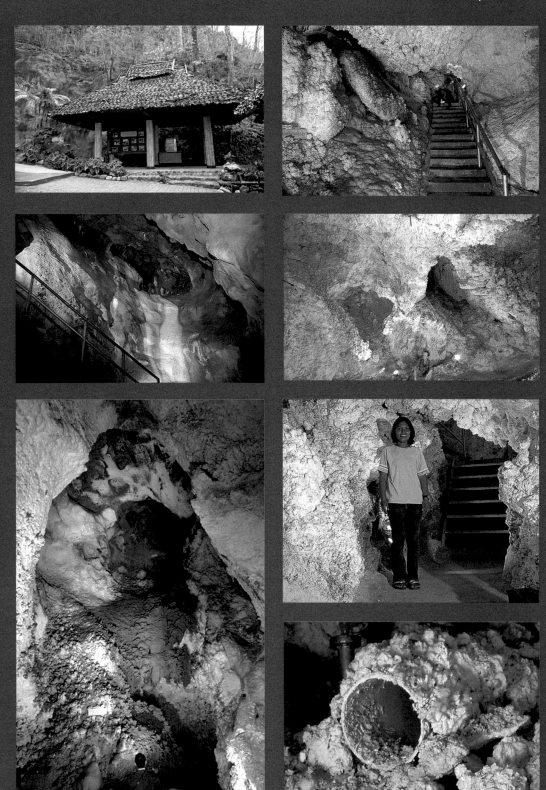

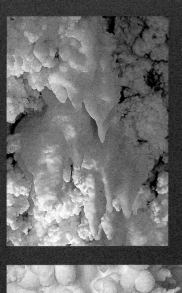

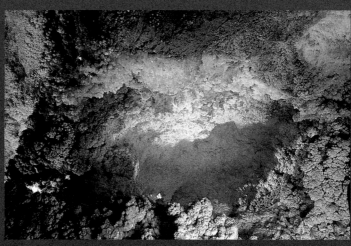

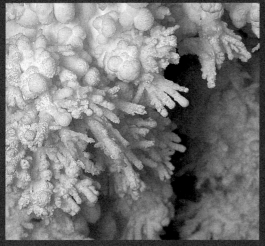

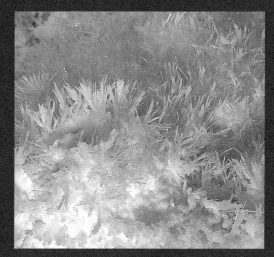

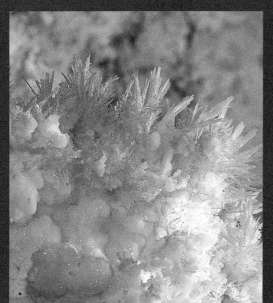

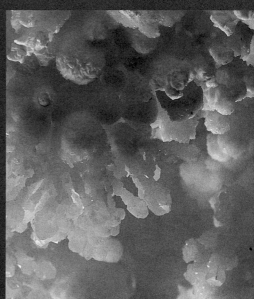

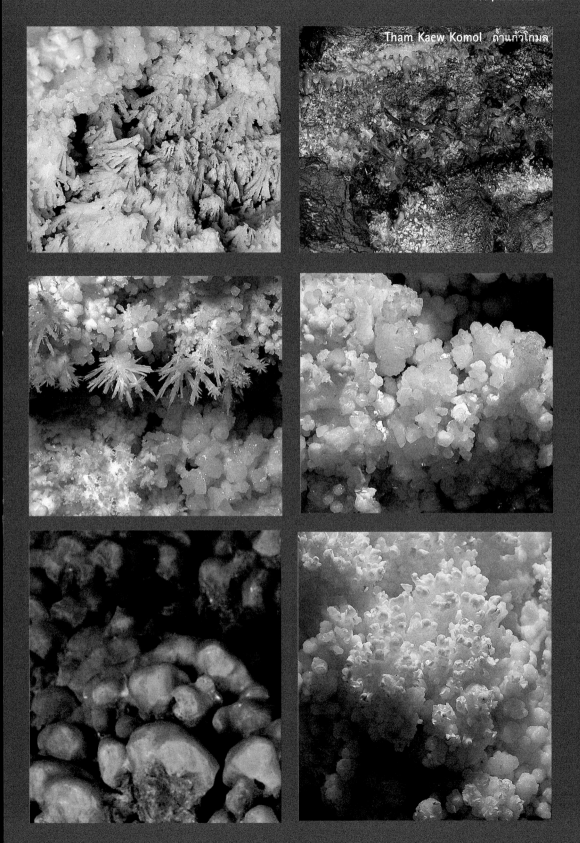

Tham Kaew Komol ถ้ำแก้วโกมล

Tham Chiang Dao   ถ้ำหลวงเชียงดาว

The most famous and revered cave in Northern Thailand located at the base of Doi Chiang Dao, Thailand's third highest mountain at 2,220 m. According to local legend Tham Chiang Dao is several hundred km long and stretches all the way to Burma, but scientific surveys have shown the total length to be 5,190 m, which still places Tham Chiang Dao as the (seventh) longest cave in Thailand. Despite the great length the difference in elevation between the highest and lowest points is only 75 m.

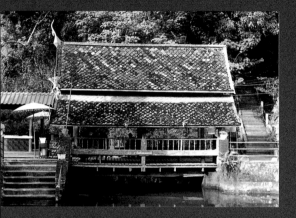

HISTORY AND CULTURE The cave has probably been known for several thousand years as there is archeological evidence that Chiang Dao Town less than 5 km away has been important settlement since prehistoric times. There is an extensive folklore associated with the cave and a well recorded written history stretching back over 300 years (page 120).

VISITING THE CAVE The stream resurgence cannot be entered but a series of mostly dry passages is reached through a man-made entrance at the top of a covered stairway next to a crystal-clear fish pond. The most important religious site is a raised platform with a large collection of Buddha images under a natural cave window known as "Plong Sawan" (chimney to heaven) which is situated about 20 m from the entrance. Until the 1880s this window was the only entrance to the cave, presumably equipped with a precarious series of ladders to negotiate the 20 m vertical drop. The steps up to the platform are guarded by a singha, kinnari and a naga. 10 m further on, a 345 year old bronze bell carried by two soldiers is the oldest surviving artifact in the whole cave.

At this point the passage opens out into a medium sized chamber with several more shrines. From here, there are four choices of route on two clearly defined passage levels, three of which require a guide with lantern which can be hired from this point. Tham Chiang Dao is a complex cave with a bewildering network of interconnected passages, many of which are dead ends – do not consider exploring unmarked sections on your own as you are highly likely to get lost.

1. **Tham Phra Non** (Sleeping Buddha Cave) is the main tourist cave, with a path and electric lighting throughout and the only one that can be visited without a guide. The dry fossil passage is sparsely decorated and is more of cultural than geological interest. Several tight side tunnels on the right lead to four other small exits in the cliff face to the north. A natural rock concretion half way has been sculpted to represent the demon Phanthurat who was turned to stone by a hermit whose elephant he had just eaten. After 350 m the main attraction is revealed – a statue of a reclining Buddha built in 1913 at the top of a flight of steps guarded by singhas, one of which is feeding a human baby. The statue is called Phra Phutthachao Kassapa and unusual because he is depicted lying on his back rather than on his side as is customary.

2. **Tham Nam** (Water cave) is the continuation of Tham Phra Non beyond the electrically lit section. A level 5 x 5 m passage leads to a series of permanent water pools up to 40 m across. Further on, the passage becomes progressively more constricted with numerous side tunnels and eventually ends at an impassable sump 900 m from the entrance.

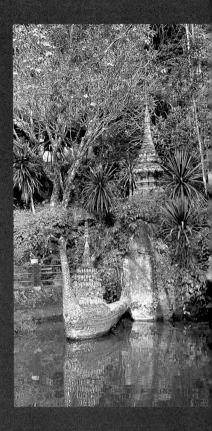

3. **Tham Maa** (Horse cave) is a popular side diversion and has more natural features than Tham Phra Non. Take the left-hand passage at the guide-hire point, leading to a dry upper level with a series of medium-sized chambers decorated with rock formations many of which have picturesque local names. The large scallop marks on the walls indicate that this passage was formed underwater (page 26). After 200 m a T-junction is reached – take the left branch along a spacious level passage for a further 150 m ending at a spectacular flowstone next to a vertical shaft which plunges down into the darkness. Retrace your steps to the T-junction and follow a small passage on the right which leads back to the lower level 50 m from the electrically lit area.

4. **Tham Kaew** (Crystal Cave) is less well-visited than Tham Maa but supports a good variety of cave life including a colony of bats. The left branch 80 m beyond the guide-hire point leads down slightly to a series of spacious level passages, ending after 500 m at a siphon below a steep rock-filled chute.

ECOLOGY Chiang Dao is by far the best studied cave in Northern Thailand, notably by a series of French expeditions 1985-88. Several new species have been discovered which are known nowhere else in the world including a crab *Potamon doichiangdao*, a blind millipede *Dyomerothrix gremialis*, a blind spider *Spermophora* sp. and 8 species of springtails (Collembola). Over 70% of all aquatic fauna sampled were microcrustaceans whilst the terrestrial fauna was dominated by Diptera (fly) larvae. There are great differences in population densities in different sections of the cave, possibly due to visitor impacts.

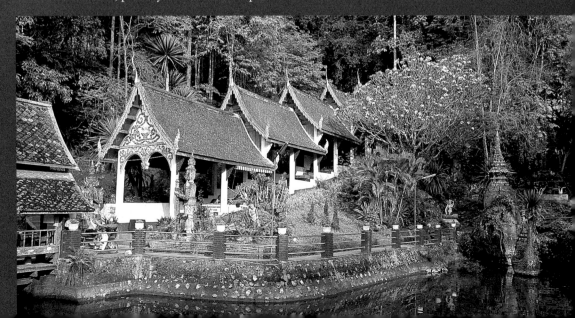

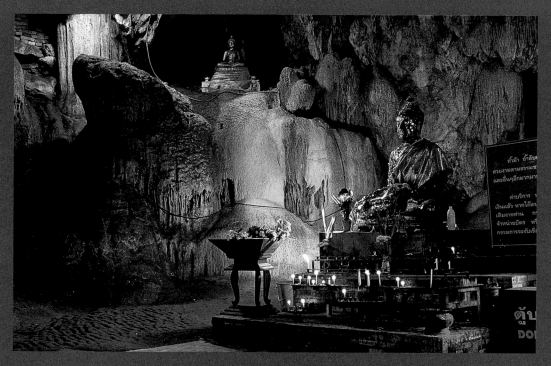

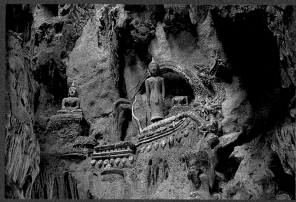

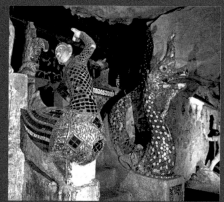

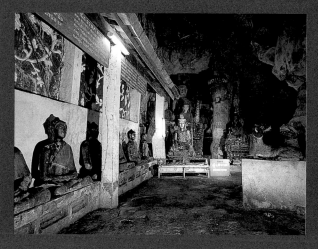

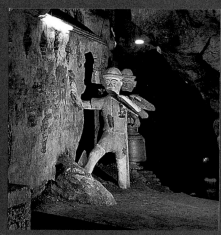

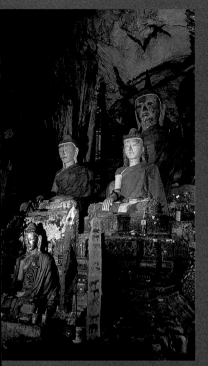

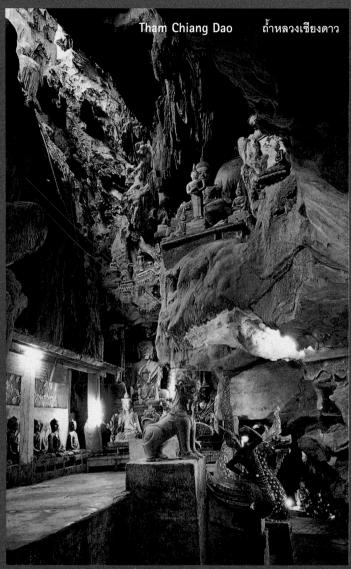

Tham Chiang Dao ถ้ำหลวงเชียงดาว

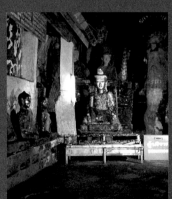

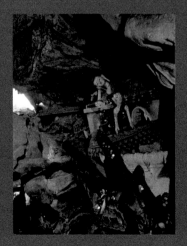

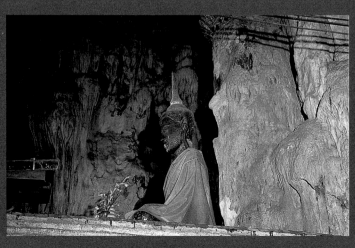

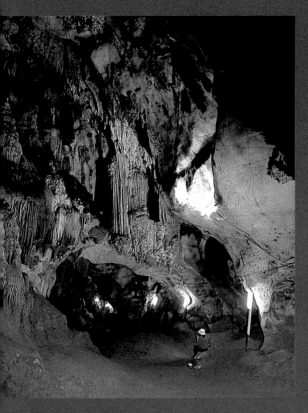

Tham Phra Non (Chiang Dao Cave)

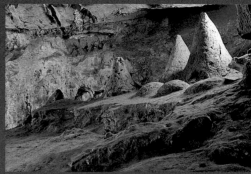

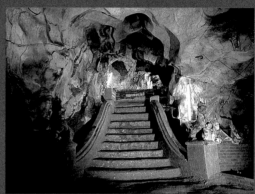

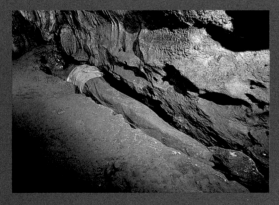

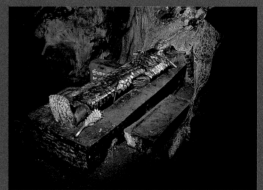

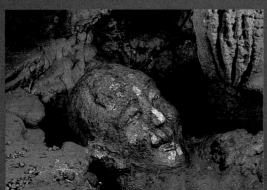

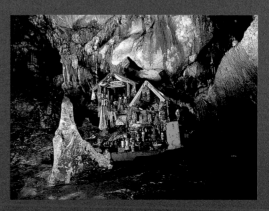

Tham Maa (Chiang Dao Cave)

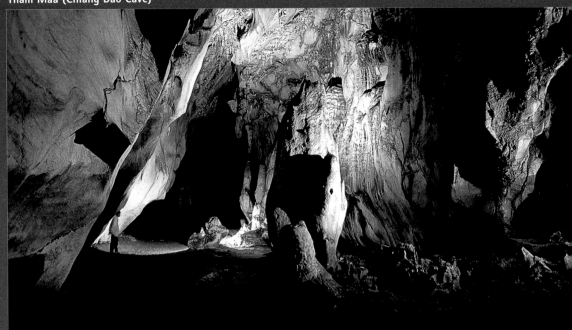

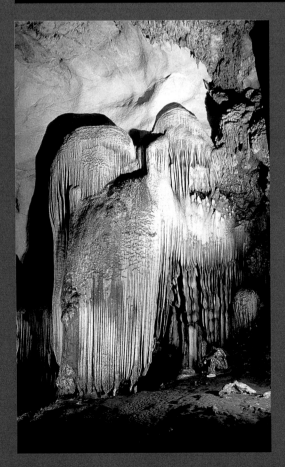

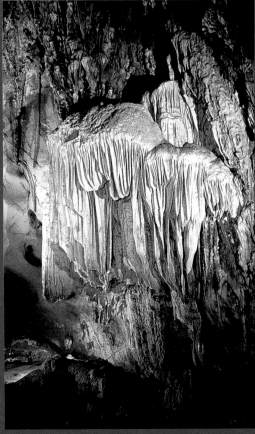

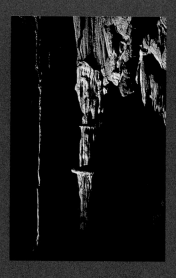

Tham Ki Nok ถ้ำขี้นก

A steep slippery path leads up from the back of Tham Chiang Dao to Tham Ki Nok ("bat guano" cave), 200 m above the main Chiang Dao Cave. A narrow bamboo ramp skirts along the edge of towering cliffs to reach the small entrance flanked by good examples of eucladioliths (stalactites that curve outwards towards the light – see page 37).

The cave itself ascends upwards for 55 m through a series of four chambers becoming increasingly grander in scale until the last one which is over 60 m long and divided by a gigantic stone island encrusted with flowstone. A side tunnel on the left leads to two vertical shafts 30 m deep choked with rockfall. The second chamber is naturally lit by a large cave window and houses a sacred stalagmite wrapped in an orange cloth ('Ji Won').

The entire cave is coated with copious quantities of bat guano up to 1.5 m thick which is actively mined by local people and supports a diverse range of species not found in Tham Chiang Dao.

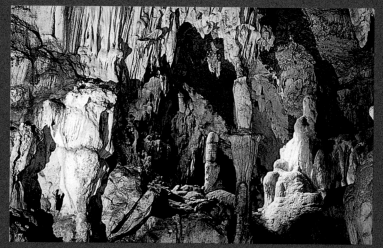

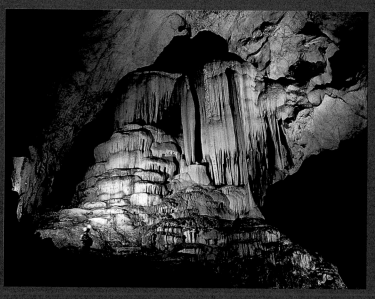

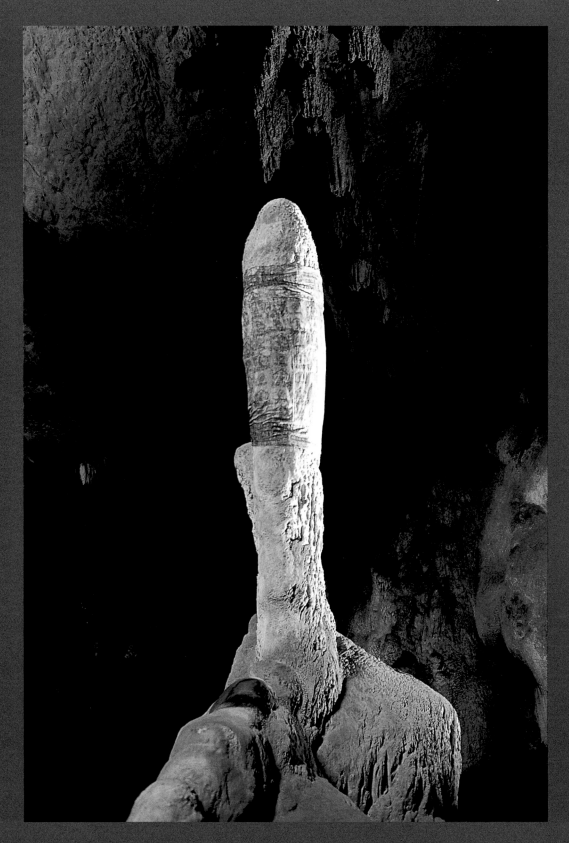

Tham Pbak Sung   ถ้ำปากสูง

An exceptionally well-decorated chamber in a temple complex 1 km
northeast of Tham Chiang Dao. Originally established as a meditation
cave by a yogi over 30 years ago but only recently refurbished to its
current level of comfort. A magnificent teak floor completely disguises
the chasm underneth and gives the whole cave a feeling of domesticity.
Electric lighting, a library and several shrines including one to the
Chinese goddess Guan Yin complete the atmosphere. An elaborate
system of corrugated iron sheeting has been erected to capture water
dripping from stalactites which is believed to have miraculous
properties.

Pbak Sung is said to connect with Tham Pbak Piang, another
small meditation cave on the other side of the same rock outcrop with
an active shrine.

Another cave in the same temple complex, "White Buddha Cave"
seems to have been abandoned and no longer has a white Buddha.
A 20 x 15 m entrance chamber leads to a maze of narrow passages on
two levels which all seem to be dead ends.

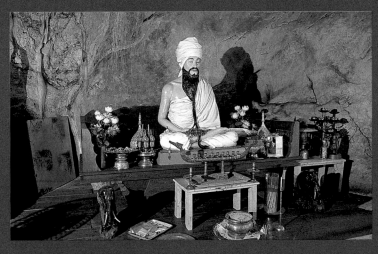

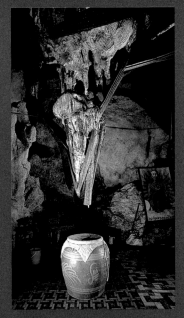

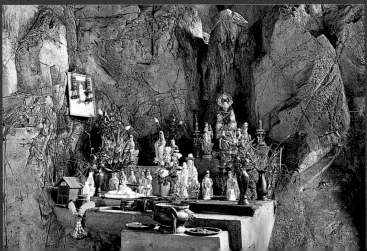

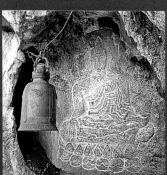

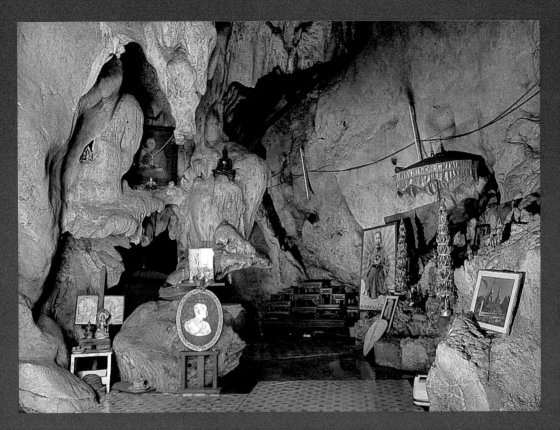

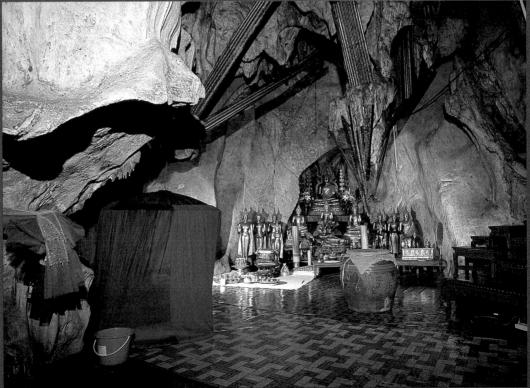

Tham Ki Mi   ถ้ำขี้หมี

A fine through cave, suitable for parties with limited experience
and already included on the itineraries of several tour companies.
Situated on the western edge of Chiang Dao Wildlife Sanctuary, a
pleasant 2 km walk through fields and secondary forest leads to the
stream outflow in a strip of lush evergreen forest. The outflow itself is
impassable but an entrance is provided by a dry fossil passage 50 m
up the hill at the end of a fabulous natural canyon festooned with
creepers and fig tree roots. A precarious 6 m ladder must be
descended to reach a high narrow passage with a seasonal stream
which can be followed up stream for 100 m to a large chamber full of
gigantic boulders. A narrow gap at the upper end leads back to the
main passage which remains mostly level and spacious right up to the
inflow entrance several hundred metres further on. There are several
sections with active speleothems and some areas where the formations
have been deeply re-eroded, indicating that the stream level has
fluctuated considerably over time. A good track around the side of
the hill takes you back to the starting point.

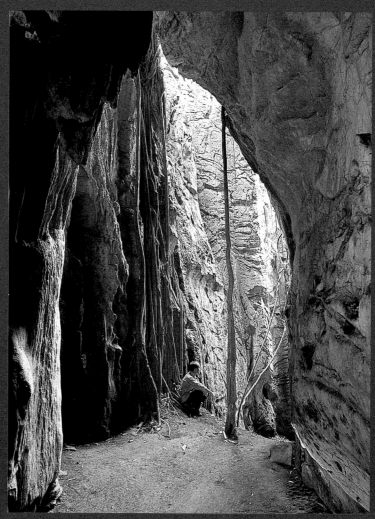

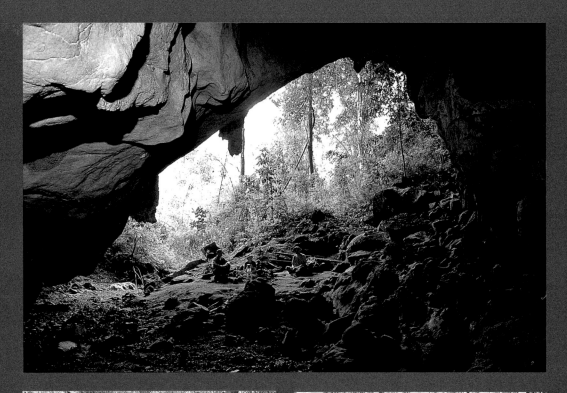

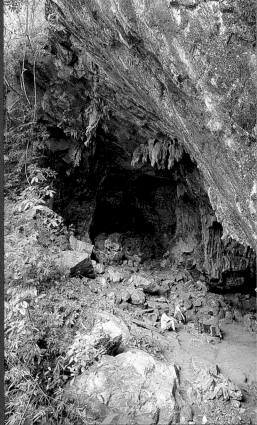

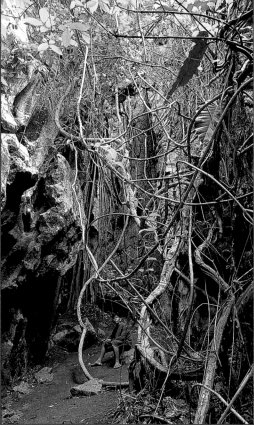

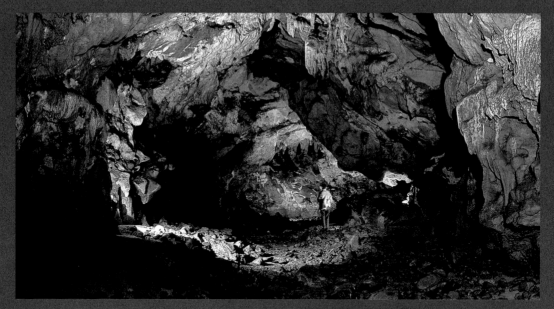

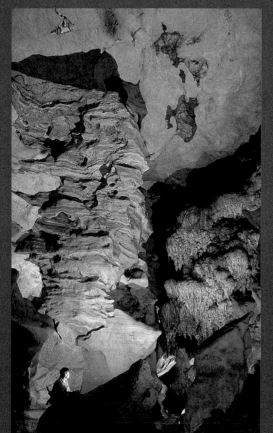

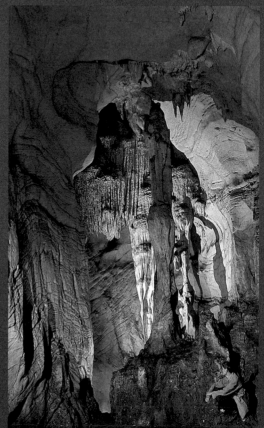

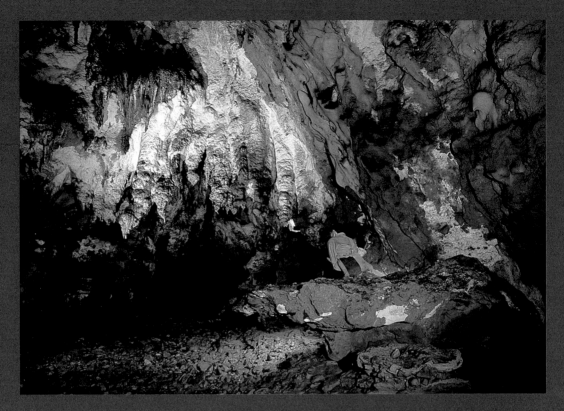

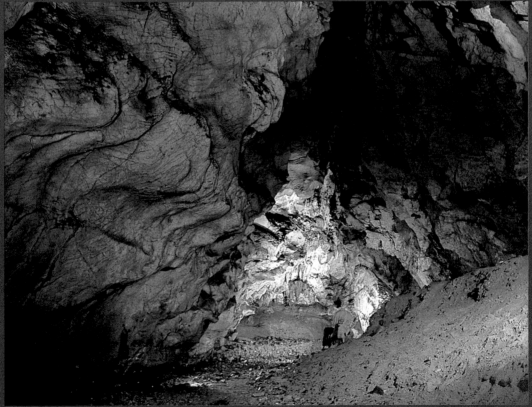

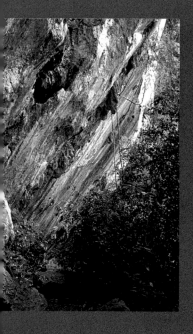

Tham Klaeb Yai   ถ้ำแกลบใหญ่

A seasonally active stream cave skirting the edge of Doi Phu Daeng in Chiang Dao National Park. The inflow has not been located but the main passage can be reached by a 15 m scramble down through a gap between boulders at the base of an overhanging cliff. An excellent example of a paragenetic cave (formed from the bottom upwards – see page 28) with the passage becoming noticeably wider towards the ceiling. The upstream section is a real treat for geologists with gracefully meandering stream passages and beautifully scalloped walls. A good variety of cave life is supported by the guano from a colony of bats. The main passage is mostly spacious and fairly easy going with a level floor but there are some narrow sections (0.5-1 x 4-8 m) as well as low crawls (2-4 x 0.3-0.6 m). Further downstream there are several fossil galleries visible above the active stream level. The outflow is at the head of a dry stream bed next to the road 800 m south of the upper entrance but may no longer be passable.

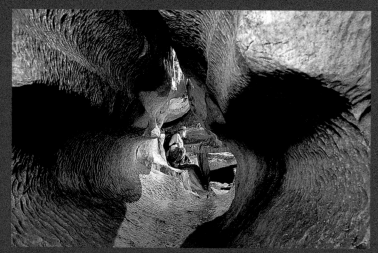

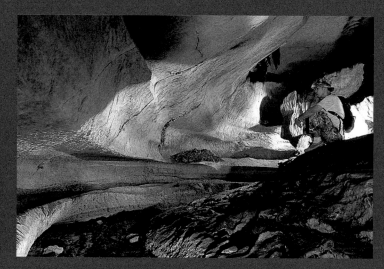

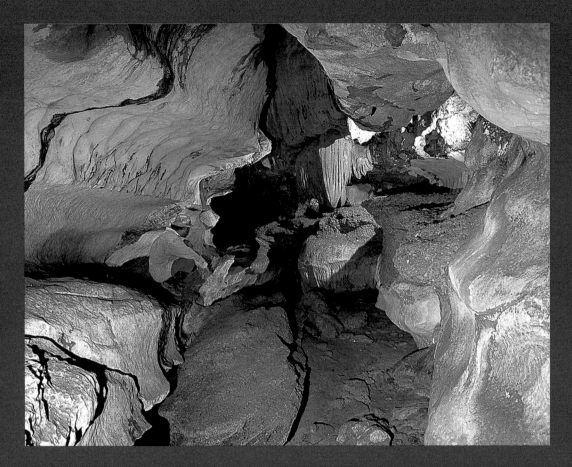

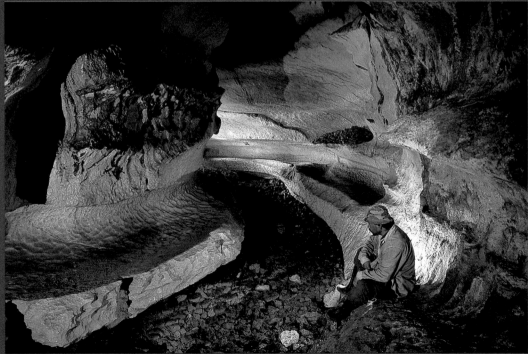

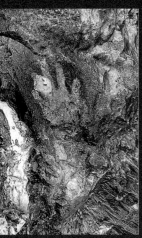

Tham Long ถ้ำโลง

One of the best examples of a 'log coffin' cave outside of Pang Ma Pha (see p 108). A newly constructed nature trail leads precariously up the steep mountainside to an imposing entrance set high in the cliff face with a marvellous view over the valley. A ladder made partly of coffin fragments leads up to the cave itself. The spacious entrance chamber 30 m wide contains a group of at least a dozen teak coffins said to have been placed in a large central pile by Forestry officials. Wood fragments in further recesses suggest that the coffins would originally have been positioned throughout the cave. Most of the coffin heads are T-shaped and many are still in excellent condition.

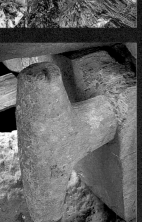

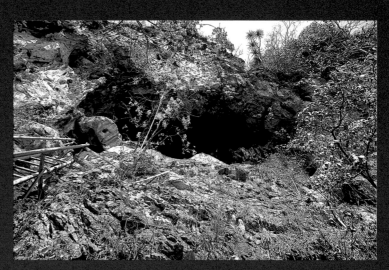

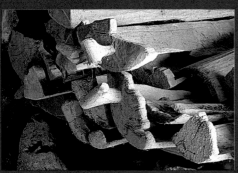

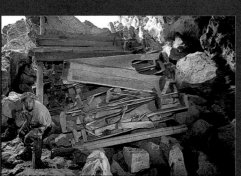

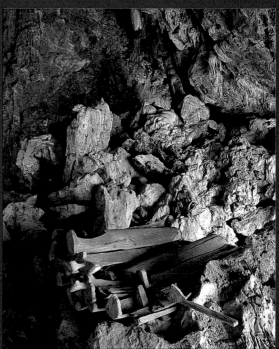

Tham Klaeb   ถ้ำแกลบ

Situated in an isolated limestone outcrop next to the main road. A narrow entrance with a shrine leads obliquely downwards for 10 m to a single chamber 15 m across and 10 m high. Several tight dead-end tunnels lead off in several different directions, including one that can be followed for over 60 m.

   The central chamber was used until quite recently to store rice husks ('Klaeb'), the remains of which can still be seen on the floor. The very dark walls make an ideal canvas for some contemporary yellow hand prints curiously reminiscent of prehistoric paintings in other sites.

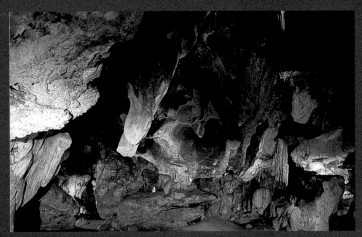

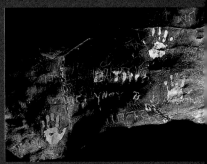

Tham Ngung Chang   ถ้ำวงช้าง

Rock overhang in a small temple with a remarkable natural arch like an elephant's trunk ('Nguang Chang"). A narrow electrically lit passage behind can be followed for 30 m The overhang used to be completely enclosed by a temple building until a few years ago when the new abbot had a dream that the giant of the cave would continue to claim more lives until the structure was removed. The entrance was the site of an excavation by the Thai-British Archeological Expedition 1965-66, which found two buried skeletons, burial urns and metal implements.

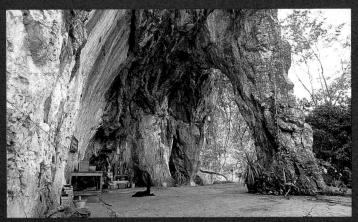

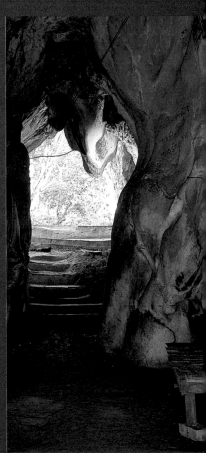

Klang Muang   ถ้ำกลางเมือง

Doi Klang Muang is a large limestone outcrop close to Muang Na with many caves, none of which have individual names. **Cave 1** is visible from a considerable distance 30 m up a huge cliff face, reached by a series of nerve-racking bamboo ladders. The small entrance opens out into a level dry passage with very few active speleothems. The floors and ceilings have collapsed in several sections, indicating that the cave is in an advanced stage of development. Local villagers have installed a series of bamboo ladders up to the higher levels, probably for collecting bats or guano. Our team only managed to explore the first 200 m but the passage is said to continue for another 800 m without major obstacles.

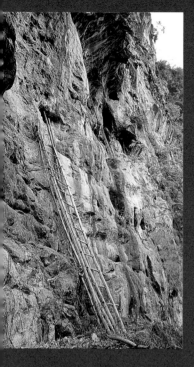

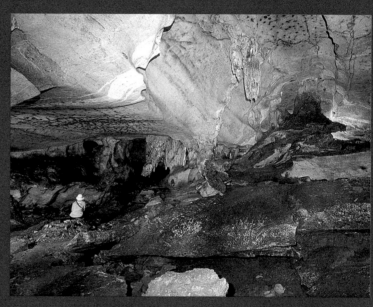

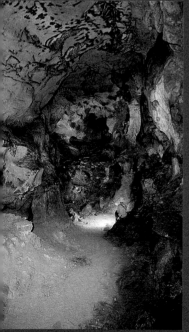

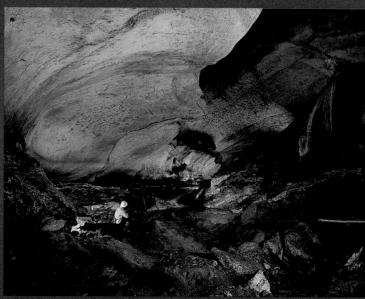

Cave 2 is at the base of the cliff below cave 1. Two small entrances lead down to a complex chamber on several levels with several small side passages none of which extend for more than 20 m Fragments of teak coffins in an advanced state of decay are scattered throughout the cave, including some carved coffin heads similar to those found in Pang Ma Pha but smaller.

A curious feature are the Y-shaped carved teak timbers which used to be standing upright with a horizontal piece between them, as can be seen from the negative imprint of algae growth on the cave wall behind them.

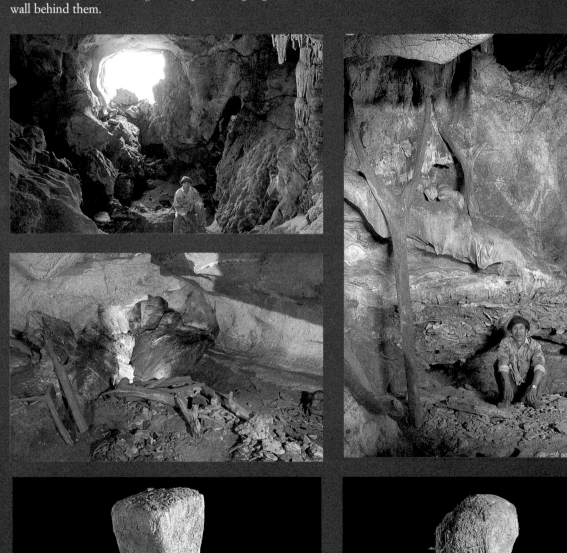

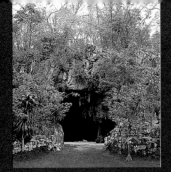

Tham Naresuan (Jhang–Pha Hok) ถ้ำนเรศวร

A large well-decorated cave in an isolated forest temple less than 2 km from the Burmese border, said to have been the last place King Naresuan stayed on Siamese soil in 1605 AD before meeting his death at Muang Hang, 30 km to the north. The vast entrance hall would be certainly be large enough to hold a sizeable army and now provides a dramatic setting for a Buddhist shrine. A short climb up over a rockfall leads to a completely dark inner chamber with very fine cave formations, including some magnificent active flowstone.

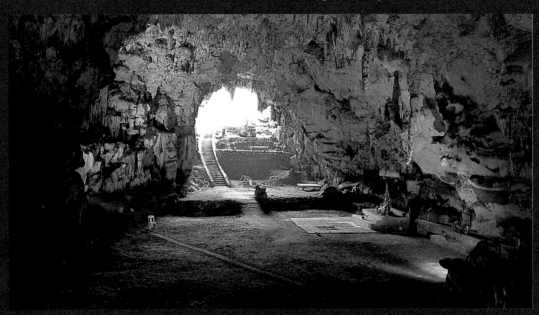

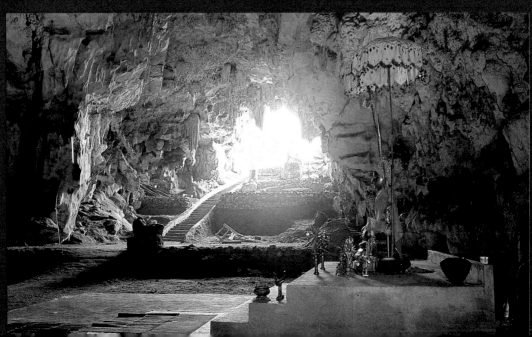

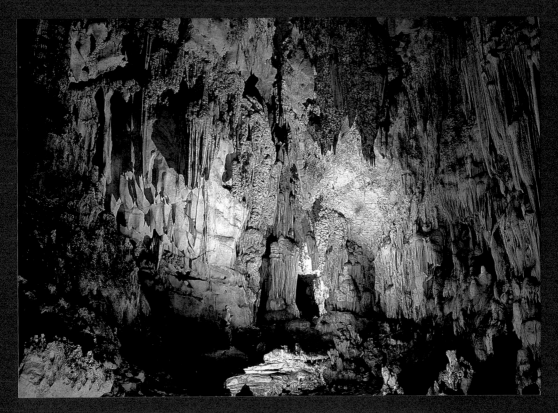

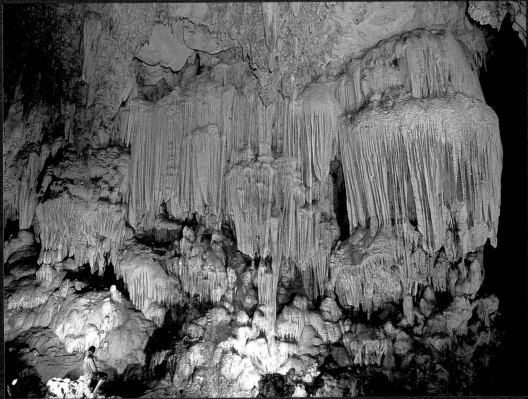

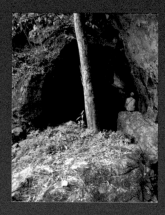

Tham Luek   ถ้ำลึก

The largest of a series of at least four caves in a collapsed doline less than 1 km from the Burmese border. A narrow entrance hidden behind boulders descends steeply to the main level 15 m below. The main passage is level and spacious, following a seasonal stream for at least 400 m. The sides of the stream bed are lined with calcite crystals caused by the evaporation of stagnant pools of water, which has also created pendulites and other interesting cave formations. There are a series of large chambers and extensive unstable rock falls on higher slopes, possibly leading to upper levels. The cave trends northwest directly towards the border and almost certainly crosses into Burma although no exit on the Burmese side has been reported.

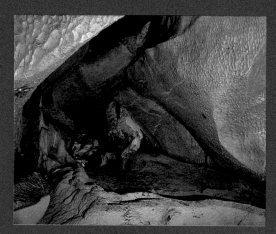

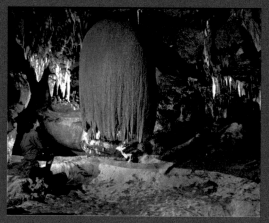

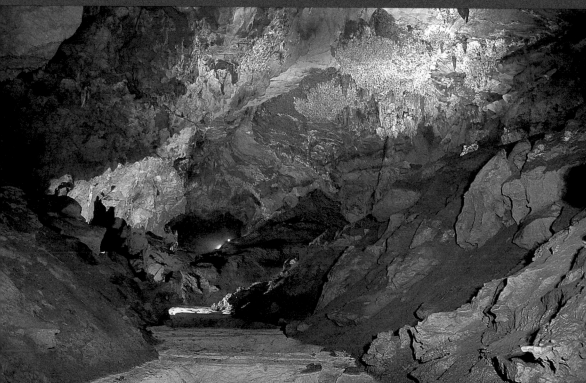

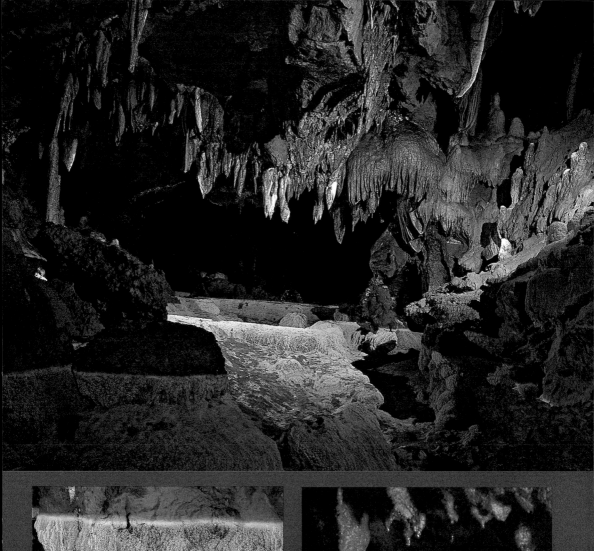
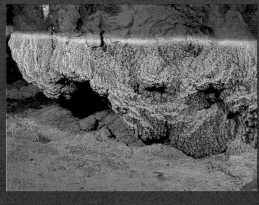
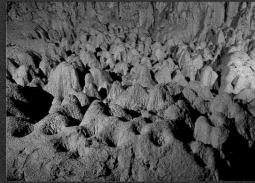
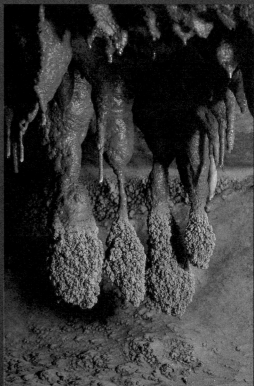

Tham Nam, Tham Mai   ถ้ำน้ำ, ถ้ำใหม่

Two other unexplored caves in the same doline as Tham Luek. **Tham Nam** can be reached via a spectacular cave window down which the enterprising monks have constructed a hair-raising ladder. The cave seems to be the eastwards extension of Tham Luek and may connect with it through a 100 m long rock-filled passage skirting the edge of the doline. **Tham Mai** is 30 m further up the hill with an impressive descending entrance chamber.

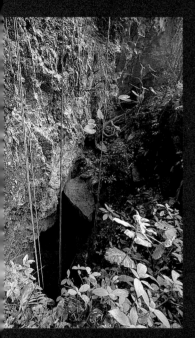
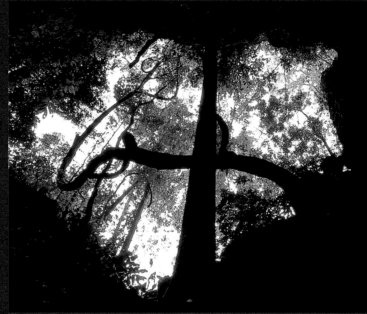
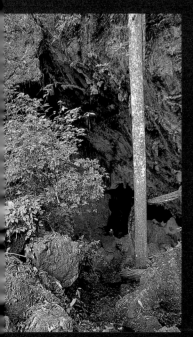
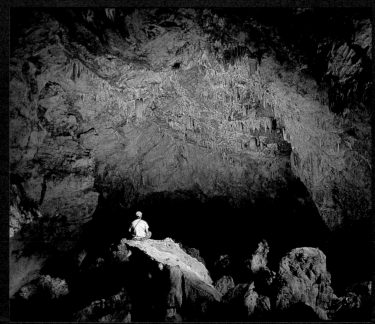

Tham Wat Huai Bon   ถ้ำวัดห้วยบอน

A 40 x 20 m rock shelter in Huai Bon Temple on the edge of Fang National Park, halfway up a 60 m cliff and reached by a very steep staircase. The shelter is divided into two sections by a natural rock pillar with a gilded Buddha in an alcove behind. The right section contains sleeping platforms made of wood from coffins, a potent reminder of the immanence of death. The message is further reinforced by macabre wall paintings of a skeleton and of the Hand of Awareness. The association with death probably comes from the deep layer of animal bones embedded in the floor sediments. The shelter appears to be a prehistoric site as pottery and other artifacts can be seen along with the bones.

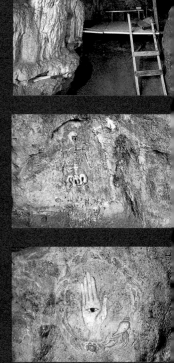

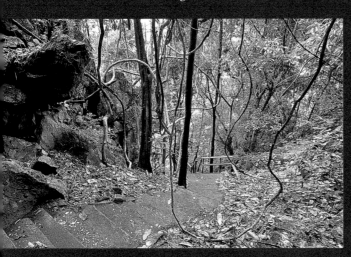

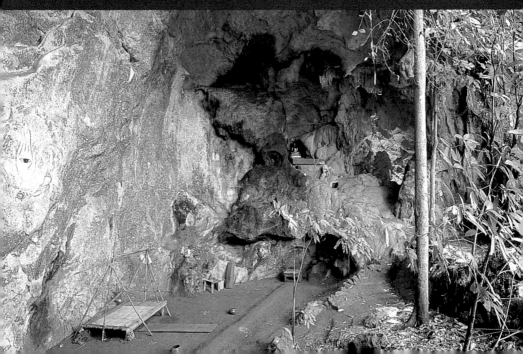

Other caves near Huai Bon

There are at least nine other caves near Wat Huai Bon.

Tham Haeng is a dry through cave close to the temple. The small entrance des-cends steeply into a medium-sized chamber with two Buddhist shrines and more macabre paintings. At the far end of the cave a narrow passage opens into a well-decorated inner chamber, behind which is another larger but sparsely decorated chamber with a shrine. 100 m further on the passage splits into two. The left branch leads to a high vertical rift and an exit window. The right branch is said to continue steeply upwards to another exit.

Tham Nam is close to the left exit of Tham Haeng. An inconspicuous entrance leads steeply downwards to a large complex chamber on many interconnected levels. The far end of the chamber contains many active speleothems and there is a permanent stream in the lowest level. The stream passage is said to continue on for a considerable distance.

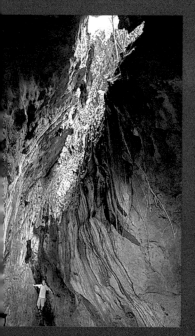

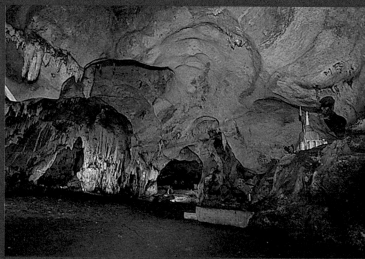

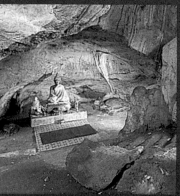

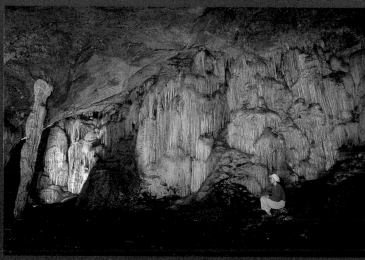

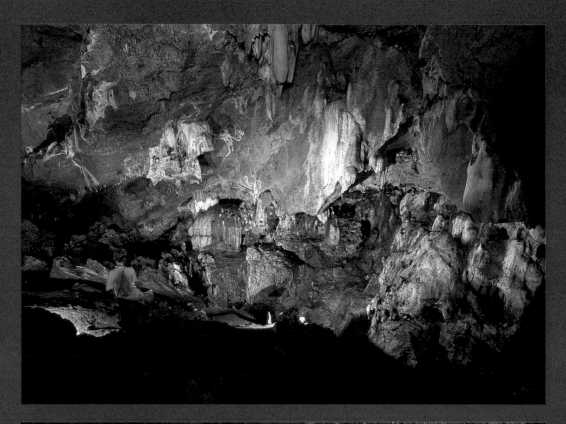

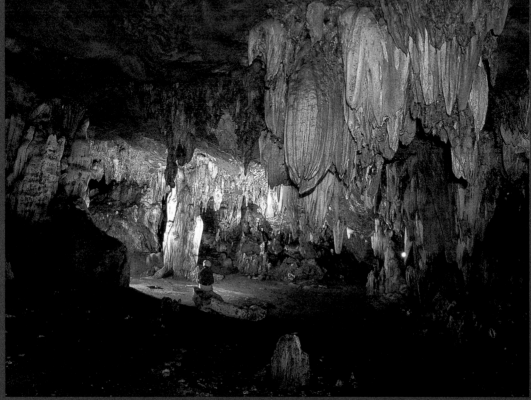

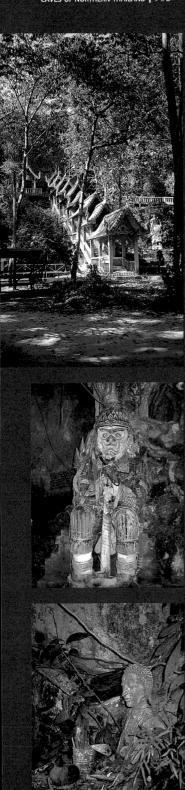

Tham Tab Tao   ถ้ำตับเต่า, ทับเถ้า

A famous temple cave reputed to be over 1,000 years old, set in a tranquil monastery with large shady trees and a moat. Covered steps lead up to two separate entrances connected by a walkway. The left-hand entrance, known as Tham Pha Jik (Light Cave), consists of a single chamber with two very large Buddha figures, one in the reclining position surrounded by his disciples, the other in the seated position framed by an alcove. Shafts of light from the collapsed vault give the whole chamber an air of mystery and sanctity.

The right-hand entrance descends down a naga stairway to a dry fossil passage with a number of small shrines but no active speleothems. After 100 m the passage opens out into a medium sized chamber with a small gap in the far right hand corner leading to another smaller chamber full of bats. Another tight section leads to a larger chamber with an enormous rockfall, then up to a very large chamber over 100 m long known as Tham Pha Khaw (Dark Cave). At the end of this huge gloomy chamber there is an extraordinary shrine with a chedi and a multitude of mud statues in an advanced stage of decay, an eerie reminder of the impermanence of our lives. A concrete path and lighting system up to his shrine have been installed since 1982. The vertical rift beyond the shrine has not been surveyed. Below the steps leading up to the Dark Cave a ladder can be descended to a mostly spacious passage which can be followed north and west for over 500 m until high CO_2 levels are reached.

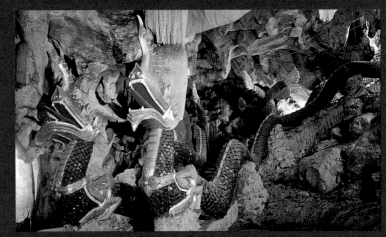

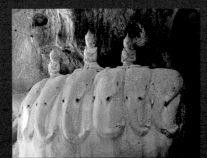

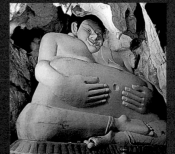

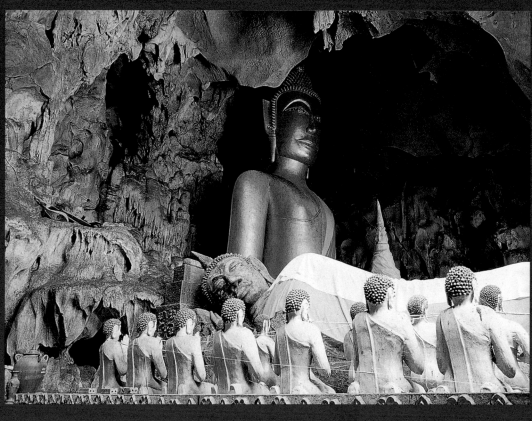

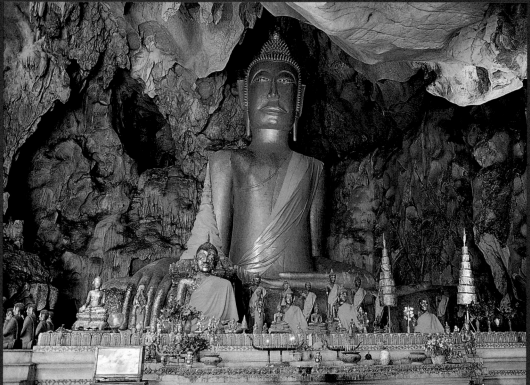

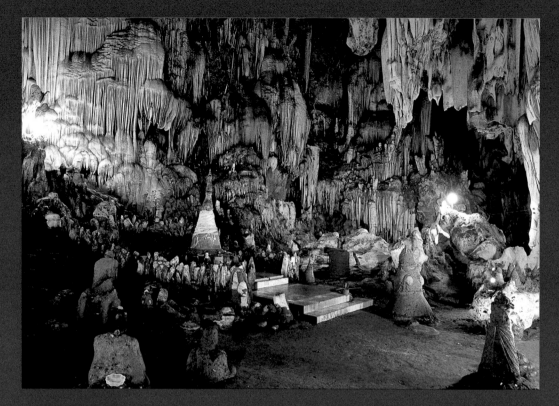

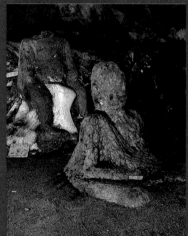

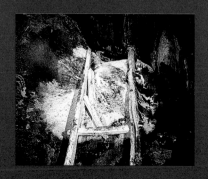

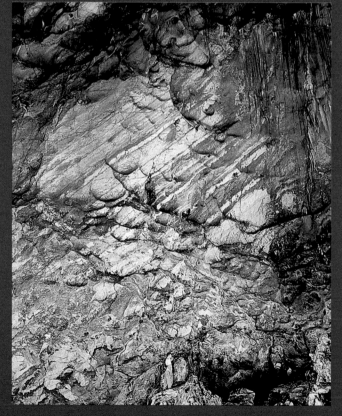

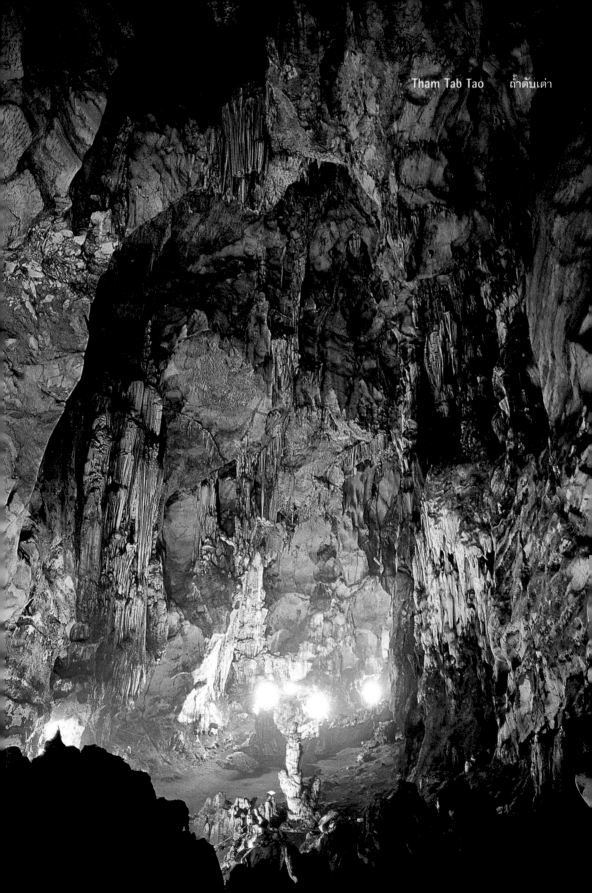

Tham Ngam ถ้ำงาม

2 km south-west of Tham Tab Tao, a 5 minute walk from the end
of the road up a flight of steps to a monastery which is actively
undertaking reconstruction of the natural surroundings. The cave
itself is approached through a landscaped garden of exotic plants.
The single large chamber is dominated by a vast red tiled platform
recently constructed directly on top of the stalagmites. A shrine
houses a group of gilded statues including a translucent green
Buddha under a canopy. The 60 m wide chamber is naturally well-
decorated with clusters of dagger-like stalactites hanging from the
ceiling which serve as a reminder of the precariousness of our
existence. The cave floor is strewn with large boulders and slopes
steeply downwards to the south and west but appears to be
completely blocked at the far end.

    Two nearby springs, Nam Rue Takhean and Nam Rue Luang, are
probably the current outflows of the stream that formed Tham Ngam.

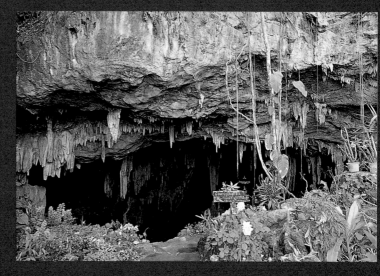

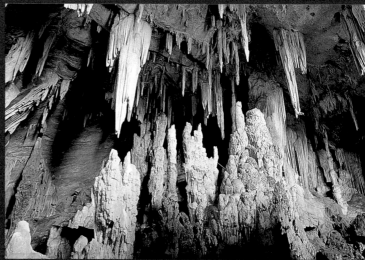

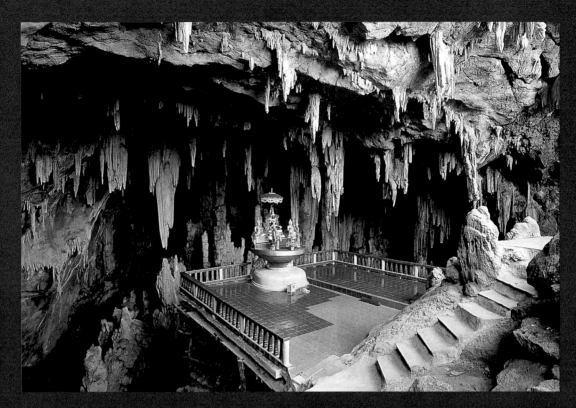

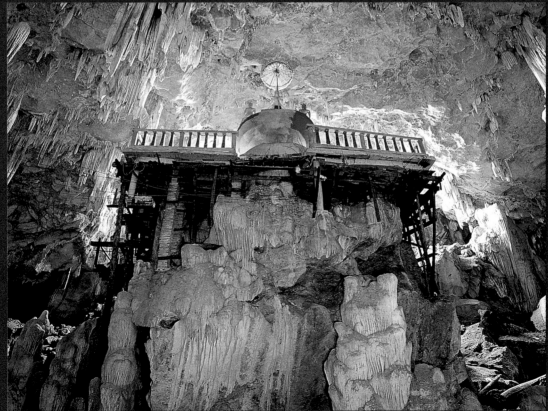

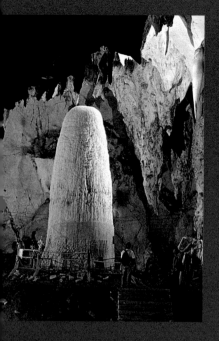

Tham Muang On   ถ้ำเมืองออน

A well-known sacred site and tourist attraction 30 km NE of Chiang Mai. The cave has been open to visitors since 1978 and is jointly managed by the temple and local people. A small admission fee is charged to maintain the lighting system and infrastructure. A newly constructed road takes visitors to within 100 m of the cave but it is still a steep climb up a flight of steps to the cave itself. A gated entrance leads down to a short upper level with a small shrine, then down a further 30 m to the main level. The most remarkable natural feature of the cave is an outstanding 15 m stalagmite, perfectly symmetrical and completely covered with glittering white crystals. This stalagmite is said to have been built by a naga to house a hair of the Buddha who gave it to the naga when he visited the cave in person 2,500 years ago (see page 126). A large reclining Buddha and a seated Buddha in a natural alcove high in the wall further add sanctity to the main chamber. A level dry passage continues behind for 150 m with a few small shrines but no other active speleothems. Close to the end of the passage there are some crinoid fossils and a band of sedimentary rock which resembles a piece of wood.

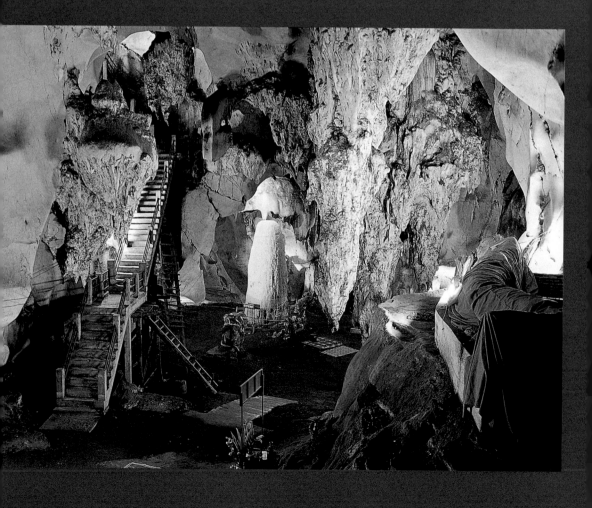

Tham Borichinda   ถ้ำบอริจินดา

Just inside Doi Inthanon National Park, 4 km NE of the main entrance gate. A rough track is drivable for the first couple of km up to a stream, after which it is a hot half-hour walk through degraded deciduous forest to reach the cave entrance on the other side of the hill. The cave itself remains hidden behind a corner of the cliff until you are nearly on top of it.

A steep slope leads down to the imposing entrance chamber, a vast hall decorated with gigantic flowstone and stalagmites. The numerous cave windows, multi-layered floor and enormous re-eroded speleothems indicate that Borichinda is a very old cave. At the back of the entrance chamber a massive section of roof has collapsed, beyond which there is another smaller chamber and a very high vertical rift.

The natural unevenness of the floor has been further accentuated by treasure seekers looking for hidden gold believed to have been left by the Japanese during World War II. The gold remains elusive but weapons dating from that period have been found.

In 1933 an archaeologist reported that the cave was a locally venerated shrine but no trace of this remains today.

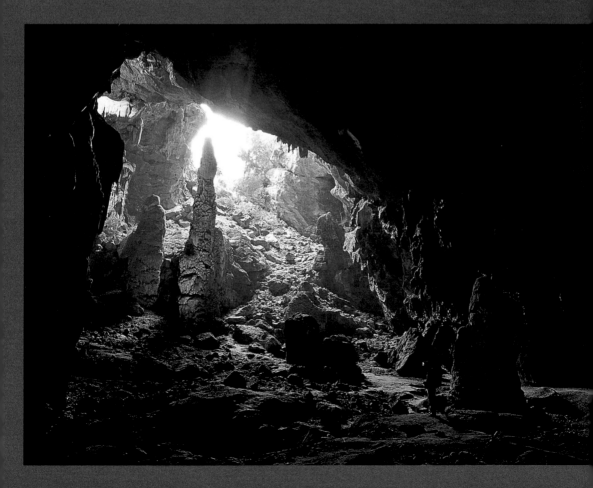

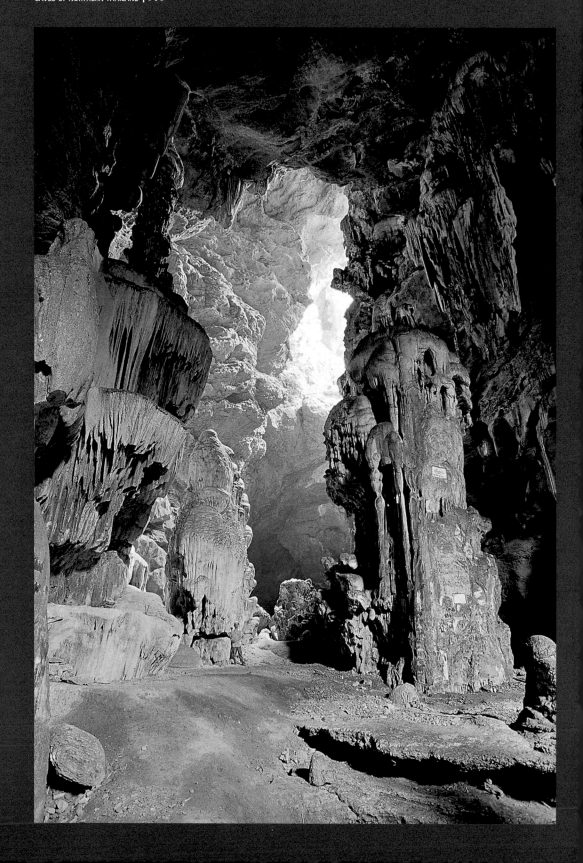

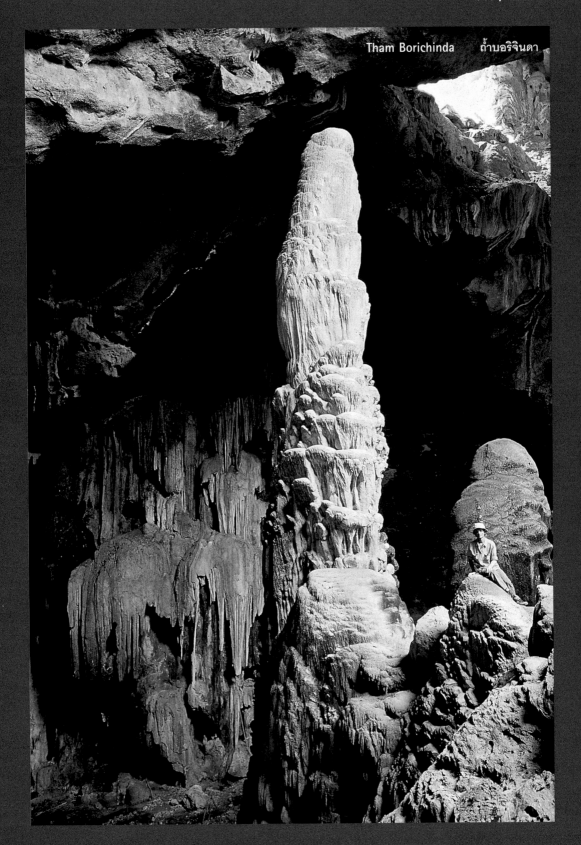

Tham Borichinda   ถ้ำบอริจินดา

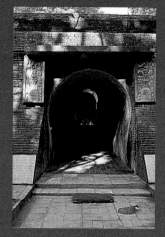

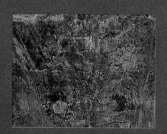

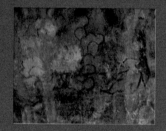

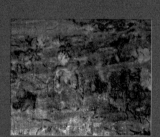

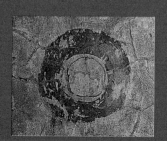

Wat Umong   วัดอุโมงค์

An ancient temple on the outskirts of Chiang Mai Town with the only example of a man-made cave in Northern Thailand. The huge brick platform with a network of subterranean passages was constructed between 1380 and 1450 AD. The inner sections of the passages are decorated with beautiful paintings of flowers and birds which were sensitively restored in 2003. The paintings and even the cave itself are probably a stylistic representation of "Pa Hin Maphan", a mythical forest where celestial beings reside. The cave were allegedly built for Phra Jan, an esoteric monk who had the habit of wandering off into the forest when the King required his council (see page 120).

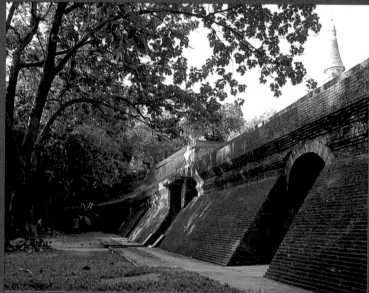

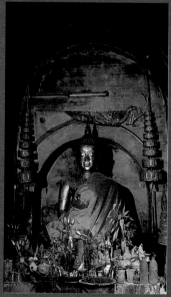

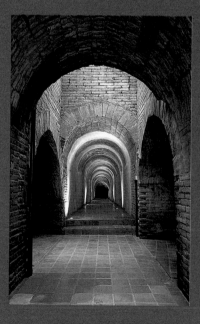

Tham Russi ถ้ำฤาษี ในวัดพระบาทตากผ้า
(Wat Phra Phuttha Bat Tak Pha)

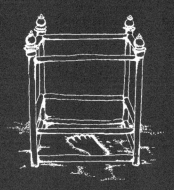

A small underground cavern hollowed out from laterite bedrock - the only underground cave in this book not made from limestone. The entrance is easily located by the ornate covering built to protect this venerated shrine. Steps lead down from floor level to a square chamber dominated by a shrine with a statue of the hermit Wa Suthep. The cave is roughly square, about 4 x 4 m and 3 m high with a large hole in the roof but it is difficult to estimate the natural size as it has almost certainly been artificially enlarged by generations of worshippers.

The temple is particularly significant in Buddhists mythology as it has no less than seven miraculous footprints carved into the laterite bed-rock. The footprints are believed to have been made by the Buddha himself, his disciple Ananda, hermit Wa Suthep, a seven-year old child saint and Kru Ba Srivichai, a former abbot of the monastery and one of the most famous monks in Northern Thailand. The Buddha's print is the largest at 2.5 m long and even the hermit's is 43 cm, whereas the print of Kru Ba is the most realistic at 30 cm. A rock nearby is supposed to bear an imprint of the Buddha's robes which were laid out to dry.

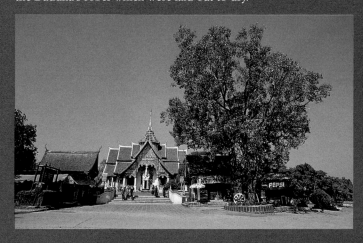

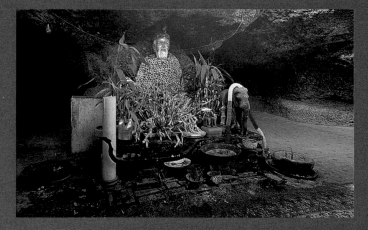

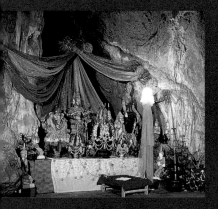

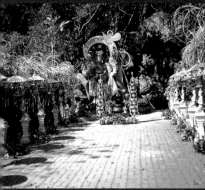

Tham Tep Ni Mit   ถ้ำเทพนิมิตร

An extraordinary cave in an isolated limestone outcrop close to Tham Erawan, the only cave in Northern Thailand dominated by Hindu deities. Tep Ni Mit means "built by God" in Thai.

An avenue flanked with angels (Thervadas) leads to a smooth black rock shaped like a shiva-lung in front of a statue of Ganesha, the elephant-headed god of creativity and prophecy. A stairway behind the statue descends into a single irregularly-shaped chamber with an eclectic mixture of religious imagery. The whole cave is dominated by a large mauve statue of Shiva, the Hindu god of energy and fertility, holding a trident. Several other smaller shrines contain Buddhist and Christian icons in addition to the Hindu ones. A natural cave window high in the ceiling infuses the scene with an air of drama and mystery.

At the time of our visit, colourful drapery and a magnificent display of fresh flowers completed the transformation of the natural cavern into a place of worship.

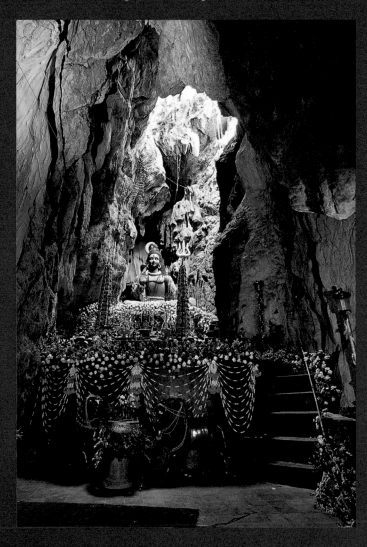

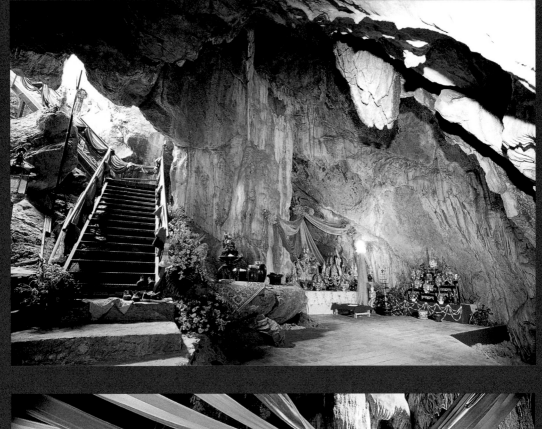
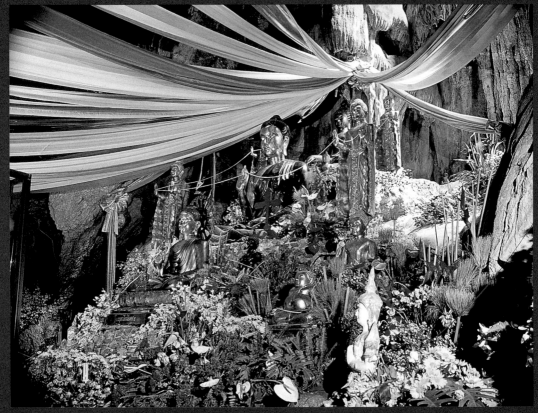

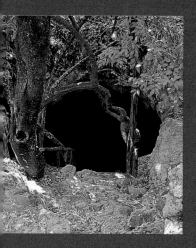

Tham Erawan ถ้ำเอราวัณ

Recently developed as a tourist attraction with a brand new tarmac road leading to within a few hundred metres of the cave entrance. In the dry season, the cool air inside the cave is a welcome relief after the parched landscape outside. A complex interconnected series of chambers and short passages with some fanastically-shaped rock formations, including one supposed to resemble a three-headed elephant (Erawan). The section of the cave developed for visitors extends for about 200 m. A lighting system has not yet been installed but a guide with lamp can be provided by the nearby ranger station and there is an informative signboard with a map at the cave entrance.

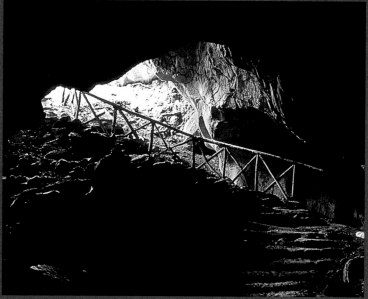

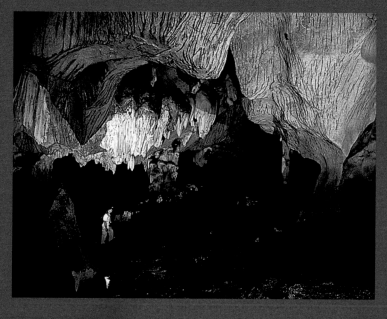

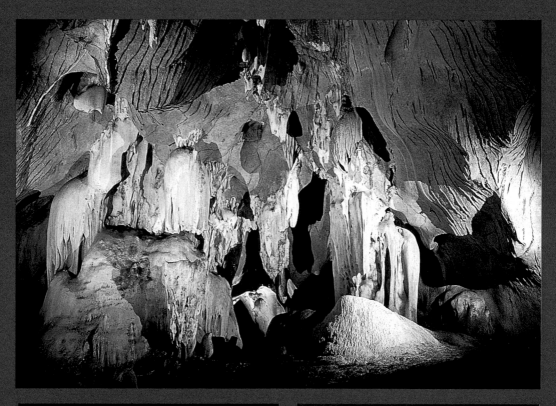

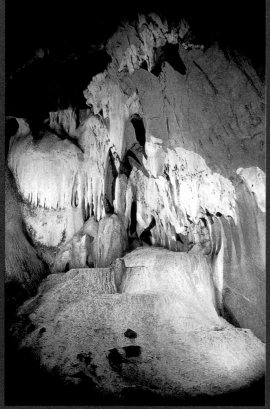

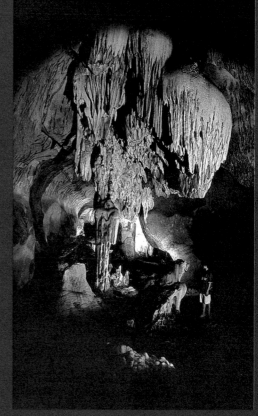

Luang Pha Wiang   หลวงผาเวียง

The largest of a series of caves south of Ban Hong. A steep flight of steps leads up through deciduous/bamboo forest to the edge of a collapsed doline 50 m across and divided by a magnificent natural arch. A concrete pathway leads down through the imposing cave entrance to a series of sparsely decorated chambers. After 150 m a short detour leads to an upper level gallery with a good view of the rest of the cavern. The main path skirts around a fine group of active stalactites and stalagmites to a point where the passage suddenly narrows and takes an unexpected U-turn before entering another sizeable chamber with a large colony of bats and an unpleasant atmosphere. The main passage is said to carry on for a considerable distance beyond.

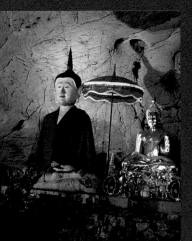

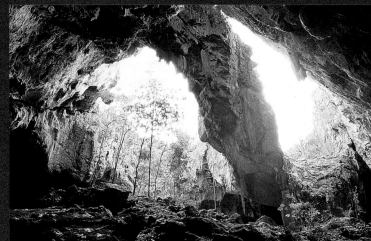

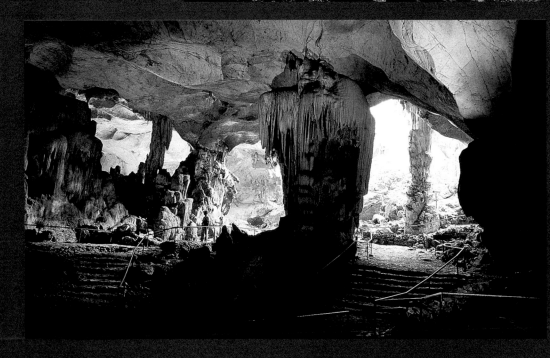

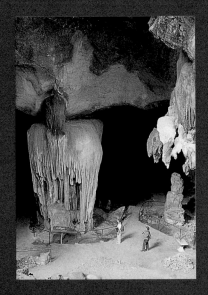
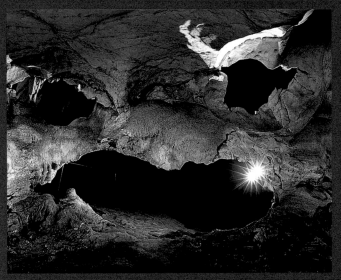
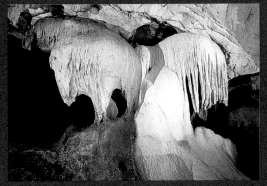
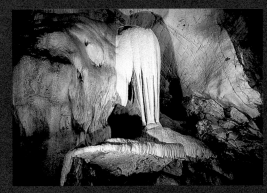
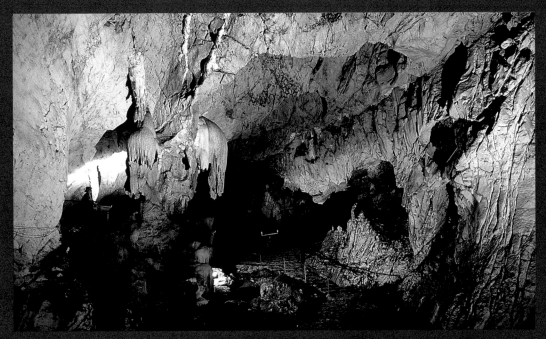

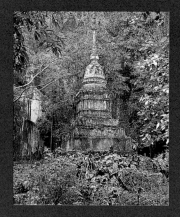

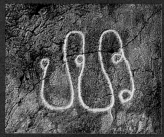

Tham Pha Thai   ถ้ำผาไทย

One of the most renowned caves in Northern Thailand. King Rama VII visited the cave in 1926 and inscribed his initials near the entrance. A flight of steps from the car park leads to a semicircular amphitheatre of cliffs enclosing a patch of lush forest.

The vast entrance chamber is lit from inside by a natural window high in the ceiling, forming a dramatic setting for a large gilded Buddha and a white stalagmite 20 m high called "the stone parasol". A path northwest from the entrance leads to Tham Chom, a beautiful but much smaller cave reputed to have been a bandit's hideout.

The spacious 350 m long main passage beyond the entrance chamber has been extensively developed for visitors with a concrete path and a sensitively hidden lighting system in harmony with the natural surroundings. Only a few of the formations are still active but many of the older inactive ones have been re-eroded into fantastic shapes.

Several of the side galleries have been partially filled with ash from an ancient volcanic eruption 9 million years ago, which indicates that Pha Thai was formed before that date and is one of the oldest known caves in the world. Although the main cave is well ventilated by roof holes, some lower levels contain one ofs the highest concentrations of CO_2 (6.5%) yet measured in Thailand.

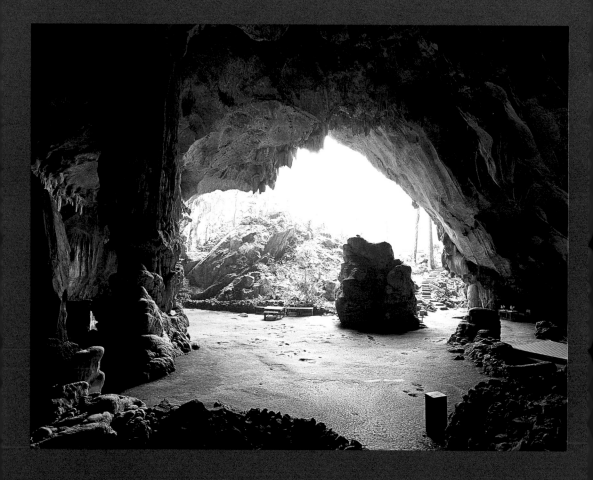

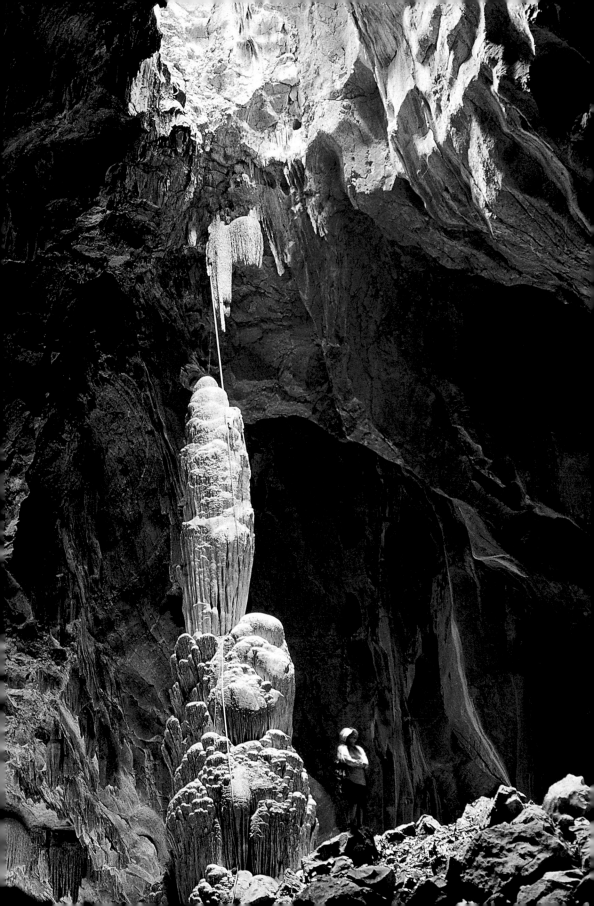

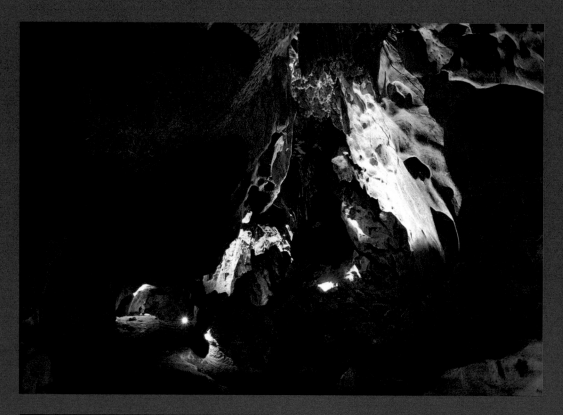

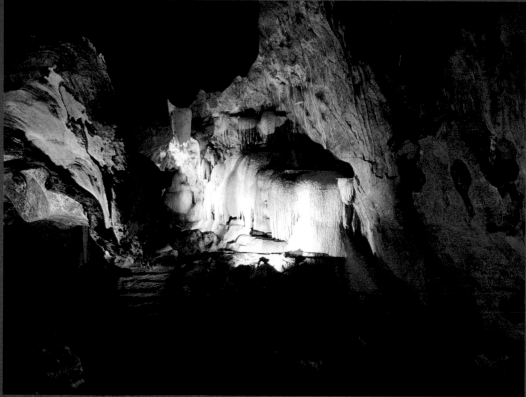

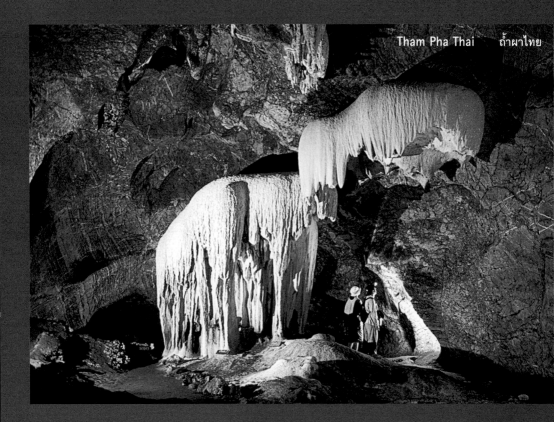

Tham Pha Thai   ถ้ำผาไทย

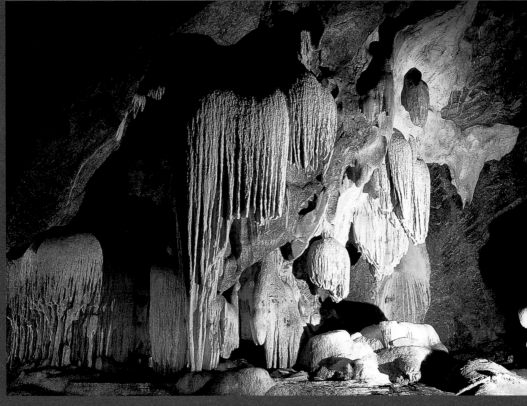

Pra Tu Pha Overhang   ประตูผา

Pra Tu Pha houses the most important collection of prehistoric rock art in Thailand. The site consists of a 300 m long overhang at the base of a narrow limestone ridge which marks the border between Mae Mo and Ngao districts. The site was only discovered in 1988 by an army officer on training exercise. The paintings are divided into 7 distinct groups, containing a staggering 1872 individual paintings many of which are overlain on top of each other. They were painted using red ochre and the vast majority are handprints of which there are examples of all styles known from Thailand. There are also numerous animal species represented including turtle, peacock, mountain goat, deer, monkey, elephant, dog and cattle. Some paintings seem to depict species not currently found in Thailand and may represent extinct species. The most famous group shows a band of hunters capturing a herd of wild cattle.

Pra Tu Pha is also a major archaeological site with some of the best preserved organic remains yet discovered in Thailand. Human hairs, rice husks, fragments of cloth, bamboo matting and baskets have been unearthed and dated at over 3,000 years old. Pieces of red ochre found in the same layers suggest that the rock paintings are of a similar period. The depiction of domesticated cattle and dogs also suggests a similar date since there is no record of animal domestication before 4,000 BP in Northern Thailand.

The gap where the road cuts through the mountain is a famous site in Thai history as it was here that the local hero Phra Ya Mulek (Iron Hand) put up an ill-fated resistance to the Burmese in 1732 AD.

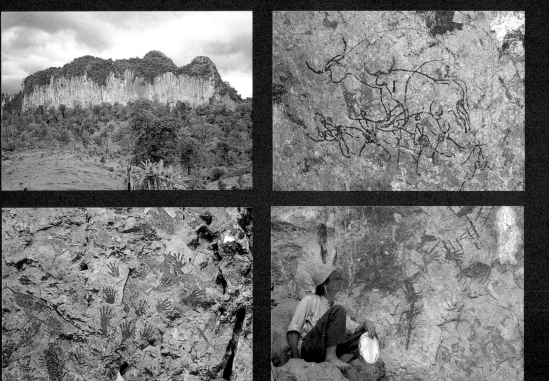

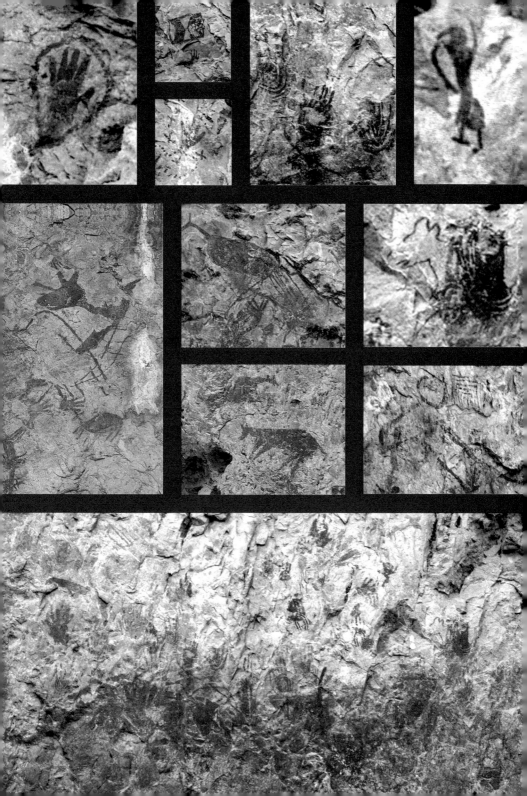

Tham Luang   ถ้ำหลวง

The third longest cave in Northern Thailand (6,220 m), first explored by a French team in 1987. An impressive entrance chamber 80 m long leads to easy walking along a sandy seasonal stream bed for the first kilometre. The stream bed disappears into an impenetrable mass of rocks but a 30 m scramble up the slippery slope on the right opens out into a huge chamber known as "float city", half filled with gigantic boulders. Any one without proper equipment and experience should turn back here. Great care is needed to weave a way through the unstable boulder fields to reach a small 1 m gap at the far side, leading back down to the stream bed. 30 m further on is another large chamber ("screen city") with vast banks of pale sand (see next page). The stream passage continues for another 5 km beyond but is only accessible during the dry season. The passage is 3-6 m wide x 1.2-10 m high with a level sandy floor and no serious obstacles until the final chamber, making for extremely easily walking but rapidly becoming tedious especially since you have to return the same way. There are very few speleothems or running water and neither the active levels nor the outflow have been located.

A sign board at the entrance has a sketch map of the first 1 km andthe visitor centre has a detailed map of the entire cave. The entrance is only 3 km from Burma and local legend has long maintained that the cave passes under the border. In reality, the passage swings south after a couple of km and its closest approach is 1 km from the border.

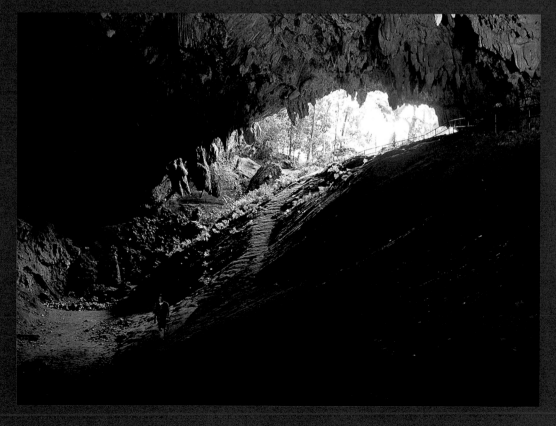

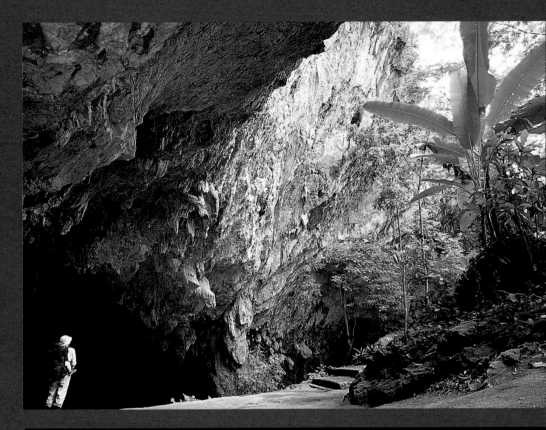

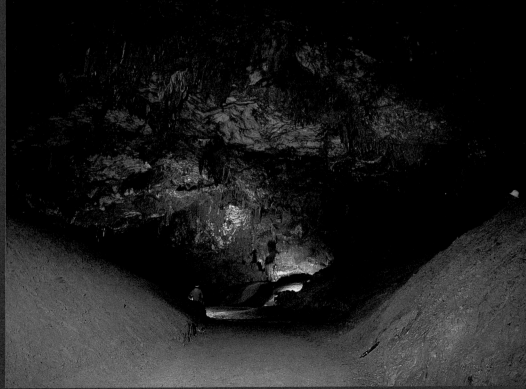

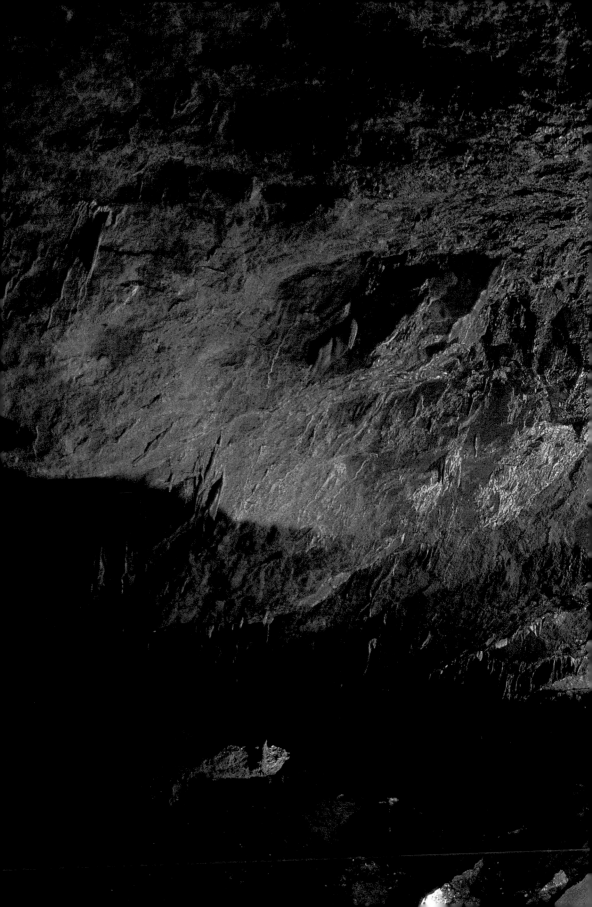

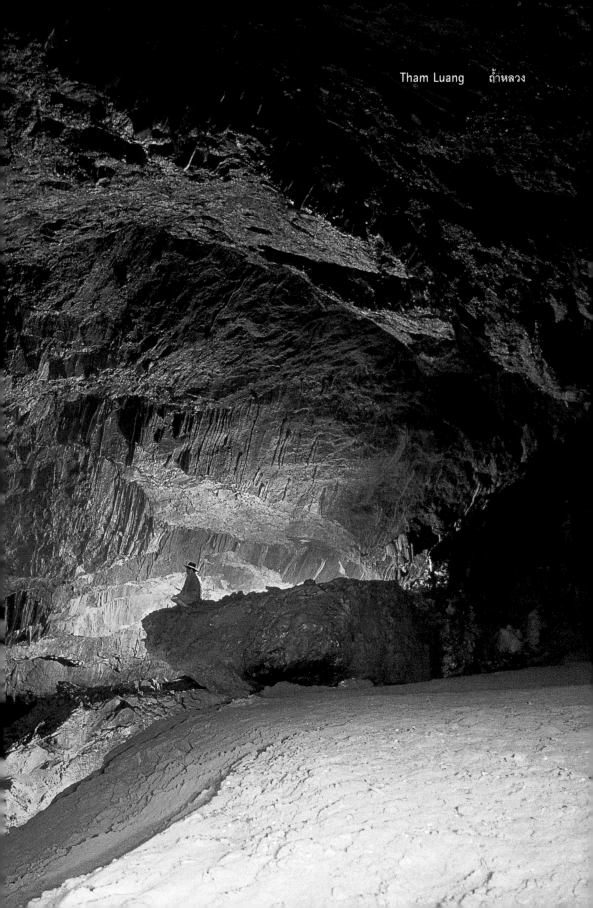

Tham Luang ถ้ำหลวง

Tham Phaya Nag   ถ้ำพญานาค

15 minutes up the hill on the south side of the forest park at Tham Luang. A series of shallow caves in the base of a 15 m cliff, sheltered by clumps of bamboo. The inner sections of the largest cave are full of rock fall and broken columns which completely block off the back of the cave, making further exploration impossible.

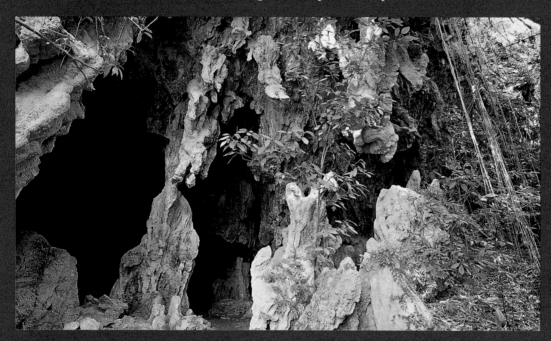

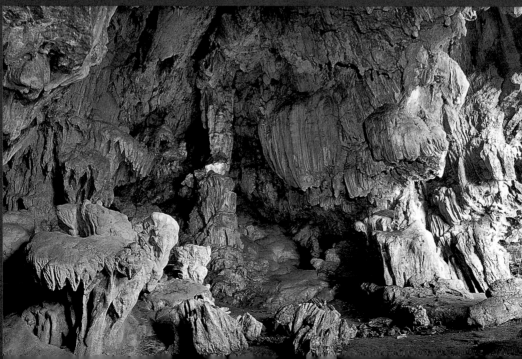

Tham Phra ถ้ำพระ

In the southwest corner of Tham Luang Forest Park, reached by a paved path winding through an attractively landscaped garden-forest with an natural amphitheatre of cliffs as a dramatic backdrop. A red devil-like guardian ('Phra Ya Yom') with a trident in one hand and a skull in the other protects the entrance, where a large mauve Buddha sits in meditation. The cave itself consists of a single dry chamber 40 x 25 x 15 m and several short side passages without any features of interest.

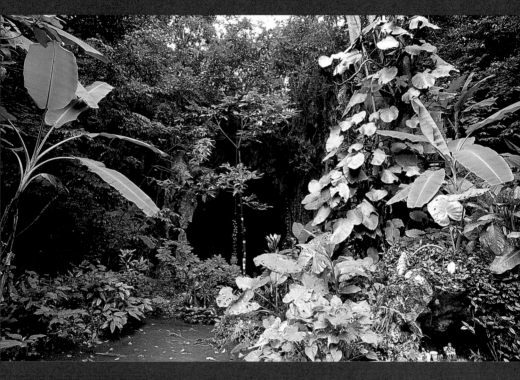

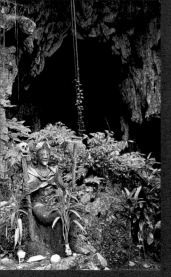

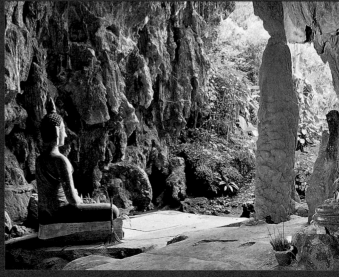

Tham Liang Pha   ถ้ำเลียงผา

Close to Tham Pra Ya Nag, a magnificent amphitheatre of cliffs
enclosing a haven of lush evergreen forest. The far end descends into
a vast steeply sloping chamber 75 m long and 50 m deep with 2
separate entrances joining together at the back of the cave and a very
high up window through which palms can be glimpsed. In the late
afternoon, rays of sunlight pierce right down to the far end of the
cave, allowing a lush growth of ferns. A semi-permanent muddy
wallow at the bottom is frequented by wild animals such as serow
(mountain goat), after which the cave is named.
(see page 335 for archaeology of this site).

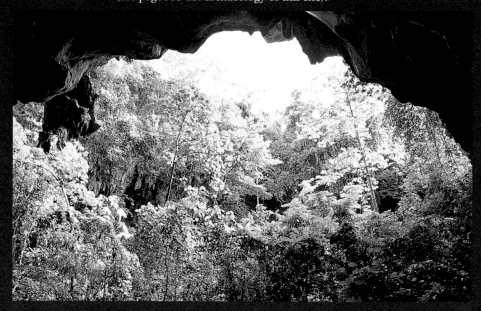

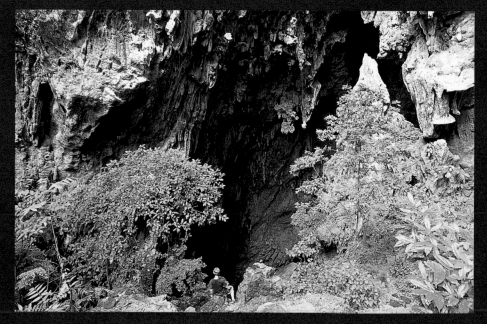

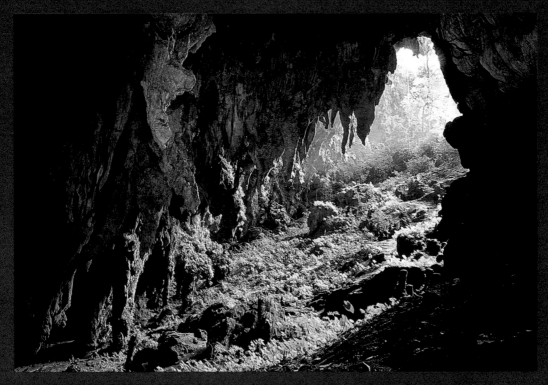

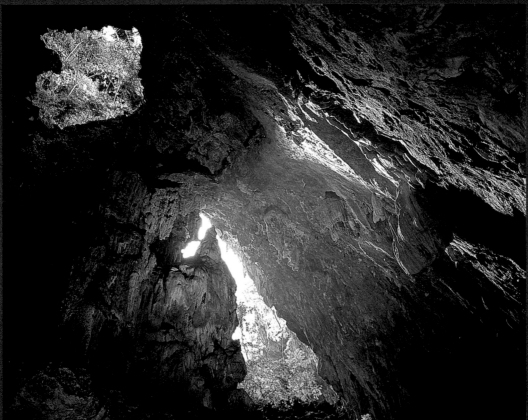

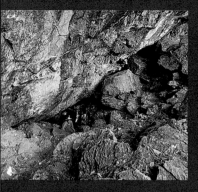

Tham Sai Thong (Tham Luek)　ถ้ำทรายทอง

Set in a picturesque forest park behind a large pond fed by a clear spring which has been turned into a popular picnic area. The cave entrance is 15 m up the hillside and obscured by large boulders. A short descent of 5-6 m leads to a level pass-age 4-10 m wide x 3-15 m high with a bed of golden sand. The sandy sections make for easy walking but are frequently interrupted by lengthy scrambles over huge unstable rockfalls. After 900 m the passage is completely blocked. Tham Sai Thong seems to be the former route of the stream that emerges at the spring but does not seems to connect at present. The spring is probably the outflow for Tham Luang (page 326).

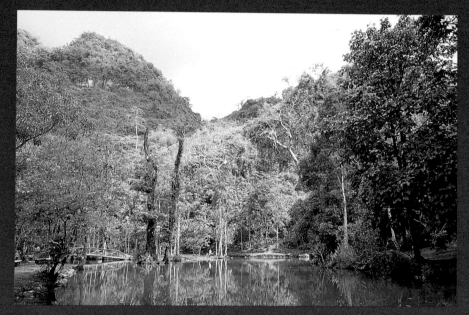

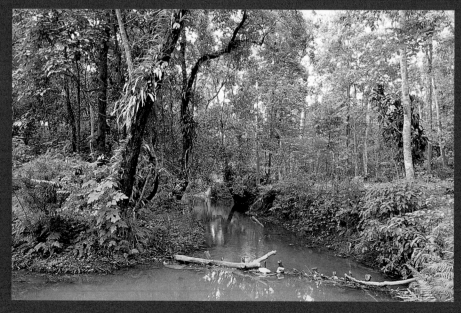

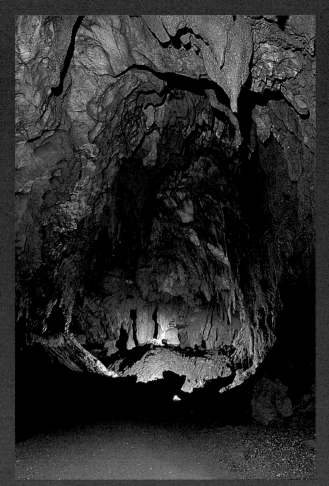

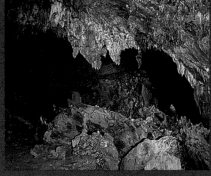

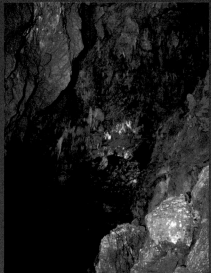

Liang Pha Overhang   เพิงผาเลียงผา

A natural rock shelter on the north side of Tham Liang Pha (page 332). The floor sediments are packed with thousands of shells and stone tools, a sure indication that the area was used by prehistoric humans as a habitation site over a considerable period. No detailed archeological excavation has yet been carried out at this potentially important site.

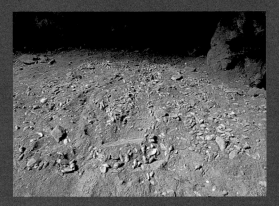

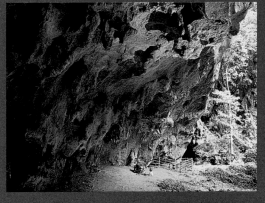

Tham Phum   ถ้ำปุ่ม

One of the most sacred caves in Thailand, said to have been visited
not only by Buddha Gotama himself but also by three previous
Buddhas and prophesied to be visited by the future Buddha Ariyametta as well (see page 126).

A naga staircase leads to a tiled platform at the cave entrance
with a beautiful view out over the plain. A graceful vase-shaped
passage houses a group of statues of the Buddha accompanied by nine
disciples in orange robes. Further on a reclining Buddha and two
seated Buddhas on brick pedestals have been placed in natural alcoves
shaped like nagas. The main passage meanders on for another 120 m,
ending at a shrine and group of decaying Buddha statues which are a
potent reminder of the impermanence of worldly things. The whole
cave has a concrete floor and electric lighting.

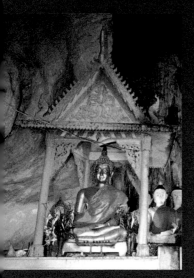
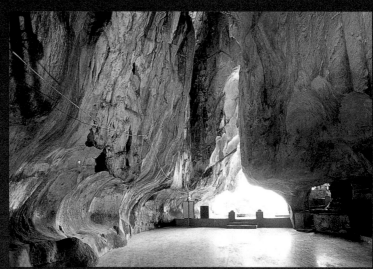
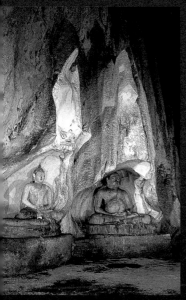
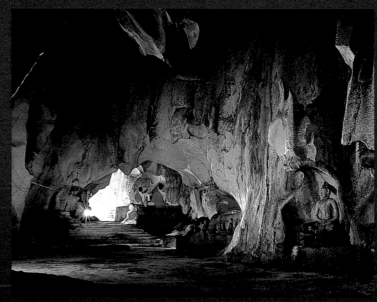

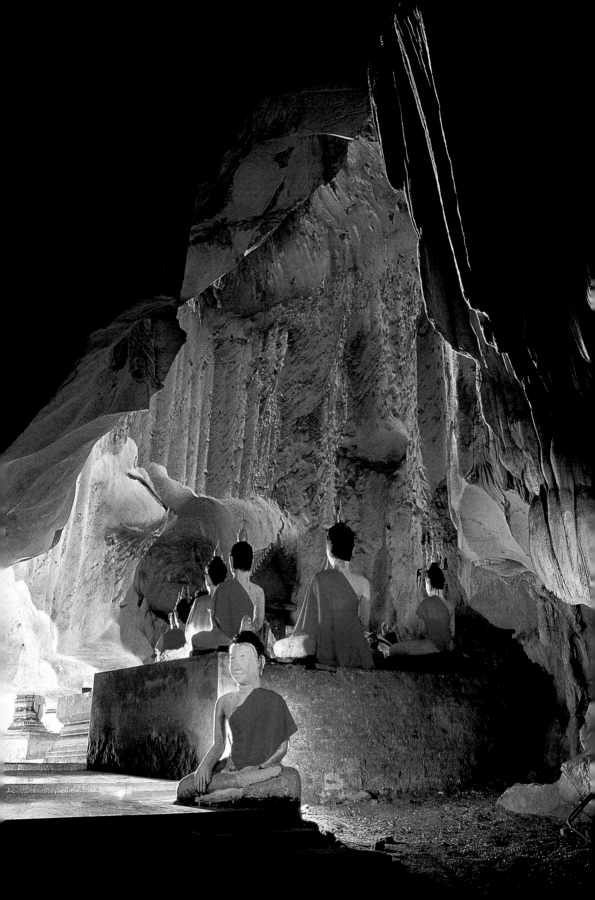

Tham Pla (Monkey Cave)   ถ้ำปลา

A sacred spring issuing from the base of a cliff in Wat Tham Phum-Tham Pla with no known access into the cave itself. The fish in the pool are said to be descendants of grilled fish that were offered to the Buddha Gotama when he visited the area and were miraculously brought back to life (page 126). Also known as Monkey Cave due to the troupes of monkeys which inhabit the temple grounds. The steps behind the pool lead up to Tham Plaew Plong Fah (page 340).

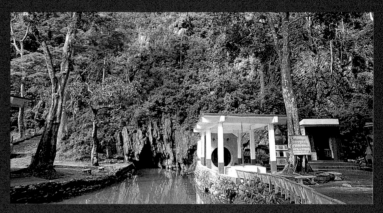

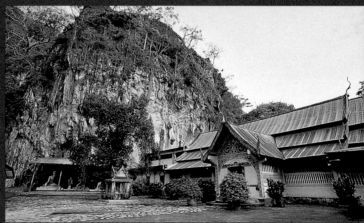

Tham Ukaew   ถ้ำอู่แก้ว

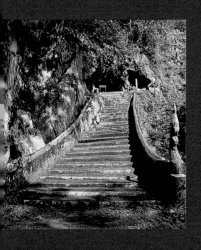

50 m NW of the main temple a flight of naga steps leads up to a platform with a reclining Buddha. After 40 m the cave bends sharply right and appears to stop at a statue of a hermit but small hole on the left leads to the inner part of the cave. The next section is fairly torturous with several crawls less than 50 cm. high. After 150 m the cave opens out into a more spacious passage 8-9 m wide x 5-6 m high, decorated with columns and suspended gour pools. 500 m from the entrance an irregularly shaped chamber has several short passages leading off it, all of which appear to be blocked. A couple of these cul-de-sacs house a marvellous collection of clay figurines, representing the Buddha surrounded by the multitude of humanity going about their daily lives.

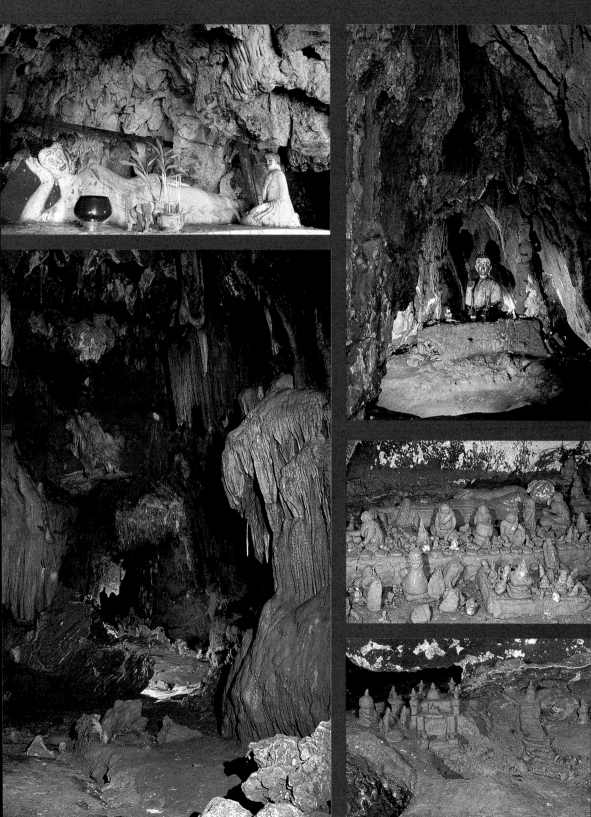

Tham Plaew Plong Fah   ถ้ำเปลวปล่องฟ้า

Behind the pool in Wat Tham Pla-Tham Phum at the top of a long
flight of steps guarded by stone lions. Approached through a
spectacular natural canyon formed by the collapse of the roof of a
former extension of the cave. The cave itself consists of a single high
rift chamber with a Buddha in front of a golden chedi naturally lit by
a window high up in the ceiling. A short cu-de-sac on the right houses
another shrine with a golden Buddha accompanied by a hermit.

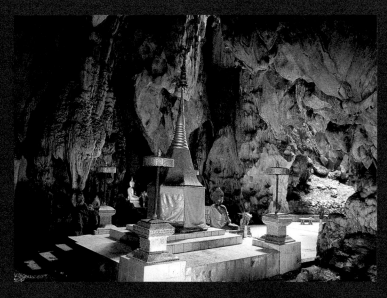

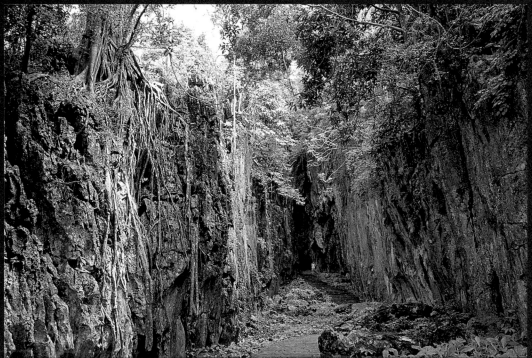

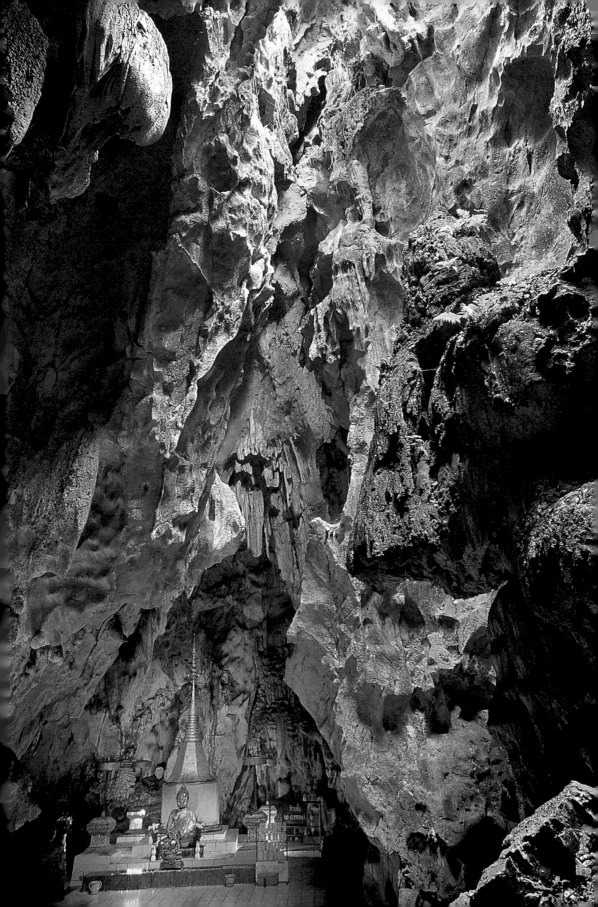

Tham Sao Hin   ถ้ำเสาหิน

A popular tourist spot in a picturesque setting beside of Sao Hin
Reservoir, accessible by boat. The lower entrance is next to the jetty
and contains a small shrine of a Buddha on the back of an elephant.

A squeeze up through a maze of columns leads to the second level
which is connected to a third level by a bamboo ladder. Both of these
levels display such an impressive array of stalactites, stalagmites and
columns that some sections of the cave have been completely enclosed
by them. From the third level you can exit the cave through a small
window and return to the jetty along the edge of the cliff.

Alternatively, a small hole in the southeast corner of this level (easily
located by the blast of air coming from it) leads down a steep muddy
slope to a completely dark inner chamber with a large bat colony.

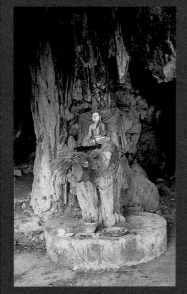

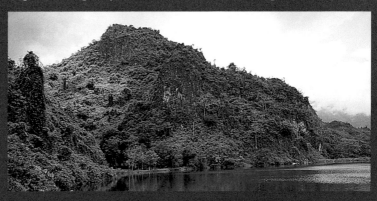

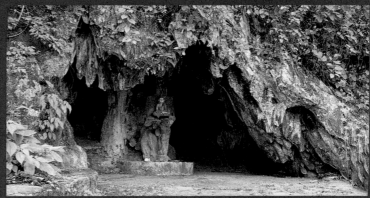

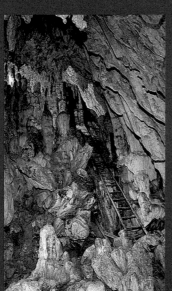

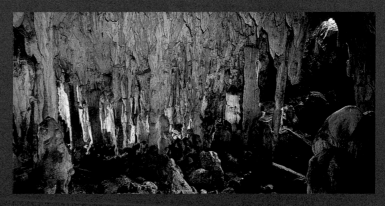

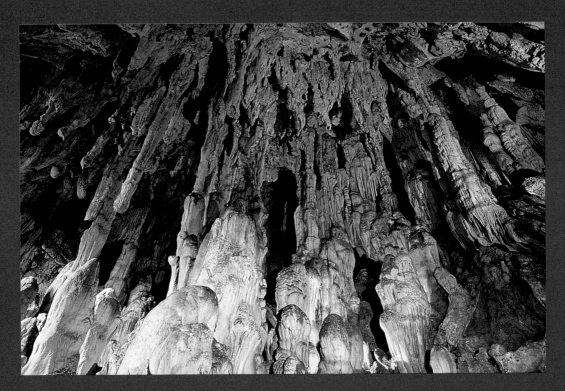

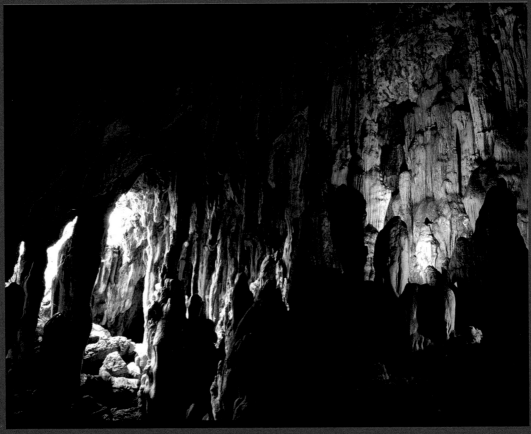

Tham Pha Dtak   ถ้ำผาแตก

A pair of vertical rifts in an isolated lime-stone tower in the middle of a temple on the outskirts of Mac Sai town. Each of the rifts has a concrete floor and a small shrine with a golden Buddha at the end.

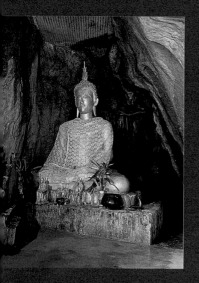

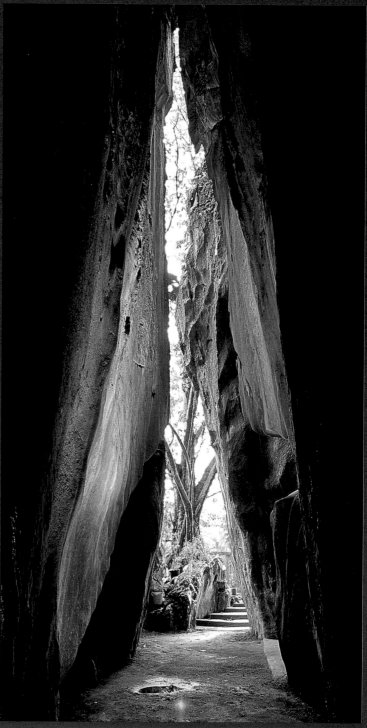

Tham Pha Rua ถ้ำผาเรือ

A less well-known cave in an isolated limestone tower 300 m south of
Tham Pha Dtak. The entrance is reached by a steep climb up a flight
of 215 steps. The main chamber has three openings which light up a
glowing golden chedi and a large reclining Buddha statue beneath a
graceful canopy, surrounded by minature figures. Red lanterns
suspended on poles lend a Chinese touch to the scene. A level passage
behind the chedi leads to a small exit on the other side of the
mountain after 100 m.

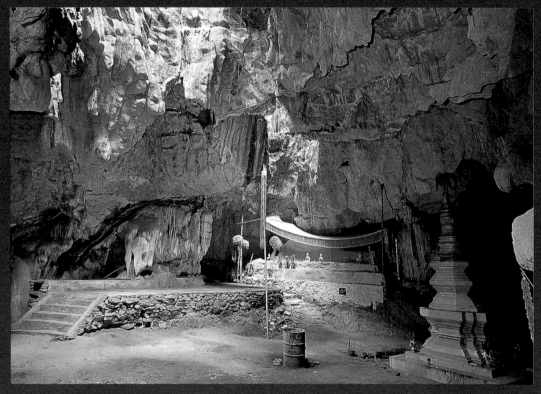

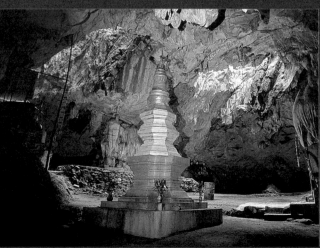

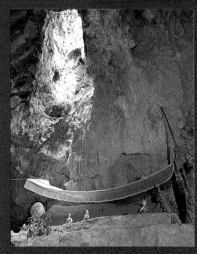

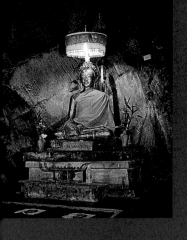

Tham Pha Chom   ถ้ำผาจม

Wat Pha Chom has two small dry caves, one of which is used by monks for meditation and closed to the general public. The other cave is electrically lit and tiled throughout. A narrow entrance opens into two small chambers on different levels separated by a short passage. The first chamber is dominated by a golden chedi surrounded by numerous shrines. The second chamber has a more aesthetic feel with a single Buddha image under a golden canopy and a painting of a fasting Buddha on the wall. Steps lead up to an upper exit with a terrace overlooking the valley.

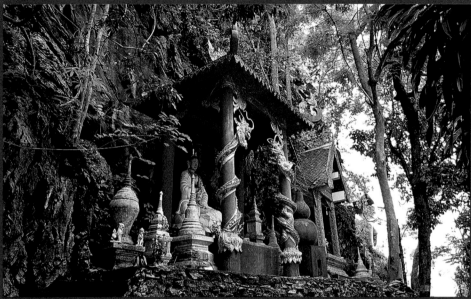

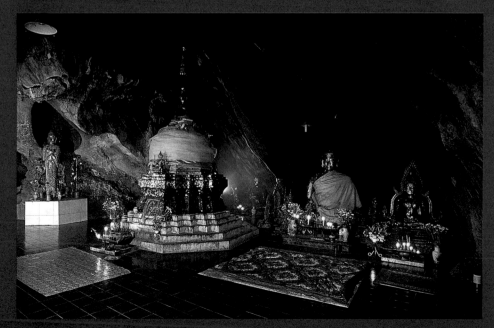

Tham Phra Bam Pen Bun   ถ้ำพระบำเพ็ญบุญ

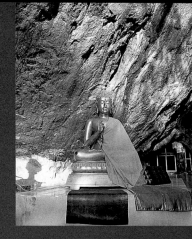

Two separate caves on different levels in an isolated limestone
tower visible east of the main Phayao-Chiang Rai road.
The lower cave is at the base of an overhanging cliff which
provides shelter for a Buddha in saffron robes. The cave entrance
is flanked by concrete monkeys and a locked gate. The cave itself
is spotlessly clean with white tiles covering every available surface
and netting over the window to keep out bats. The overhang
outside has shells and fragments of bone buried in the floor
sediments which suggests than it was a prehistoric habitation site.

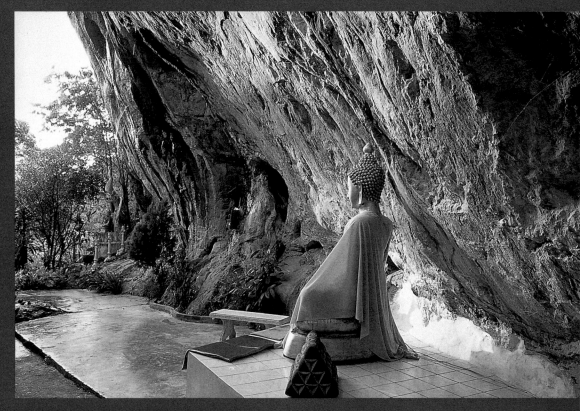

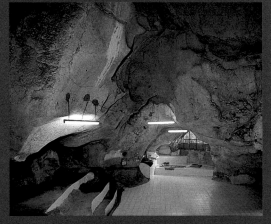

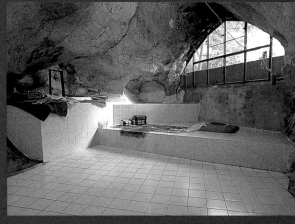

Tham Nam Lod   ถ้ำน้ำลอด

A through stream cave on the edge of Doi Luang National Park 10
km southwest of Phan Town. The outflow is set amongst lush semi-
evergreen forest next to a rocky stream which is a popular picnic spot
for local people in the hot season. The main passage is 4-6 m wide x
3-6 m high with very few active spelaeothems. A straight forward
walk along the rocky stream bed re-emerges on the other side of the
mountain 115 m further on. Flood debris in niches several metres
above the floor indicate that the cave is flooded almost to the roof
after heavy rains.

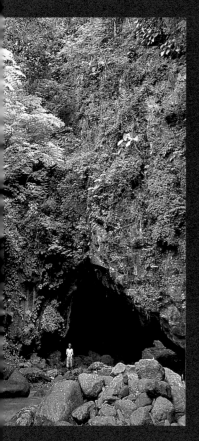
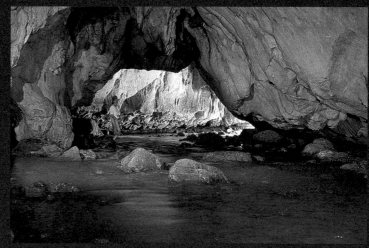
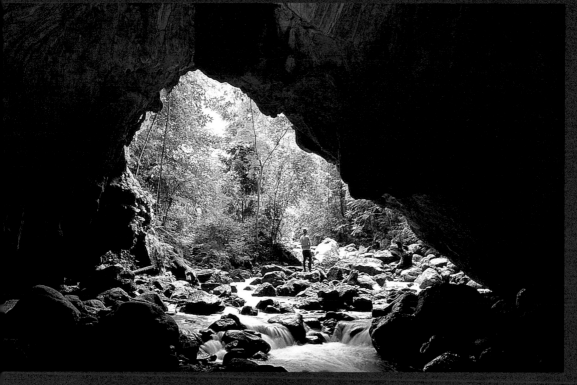

Tham Pha Ja Rui   ถ้ำผาจรุย

A series of small caves in a prominent isolated tower behind Wat Pha Ja Rui. The first cave is a series of high rifts with spectacular tree roots dangling down 20 m from the ceiling and a colony of bats. Fragments of bone and stone tools suggest that this was a prehistoric habitation site. A second cave has a small gated entrance leading into a gloomy chamber with a multi-layered shrine and an overpowering stench of bats. A flight of steps leads to another small cave close by. The entrance is flanked by stone lions and numerous white stupas which contain the ashes of deceased people.

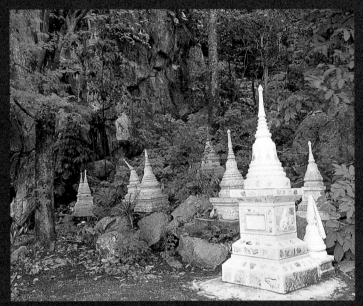
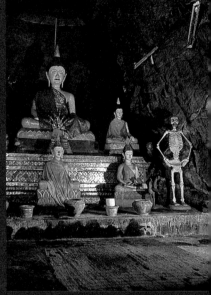
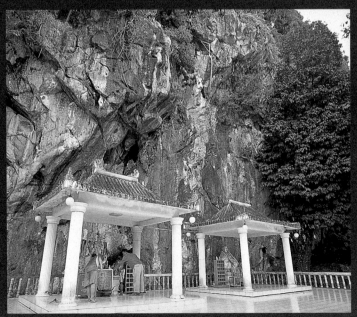
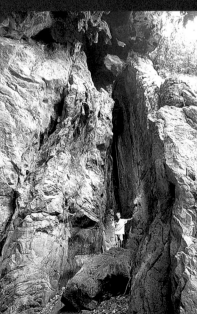

Tham Pha Daeng ถ้ำผาแดง

Newly developed cave in Phu Sang National Park, 20 km east of
Chiang Kham. The cave entrance is close to the river behind a
picturesque campsite. A 200 m concrete path connects the three main
chambers, the largest of which is 30 m wide. Most of the speleothems
are long since inactive and have been eroded into bizarre shapes. Rare
gypsum speleothems can be seen in the middle chamber. The inner
sections support a colony of bats whose roosting spots have peppered
the walls and ceiling with brown marks.

Tham Nam Lod is only 30 m away, reached by a flight of steps.
A narrow entrance opens into a seasonally flooded passage leading to
a stream and pond after 50 m.

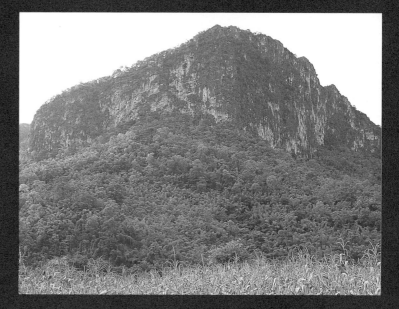

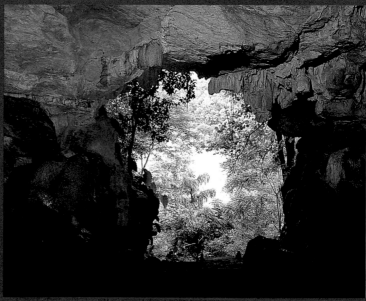

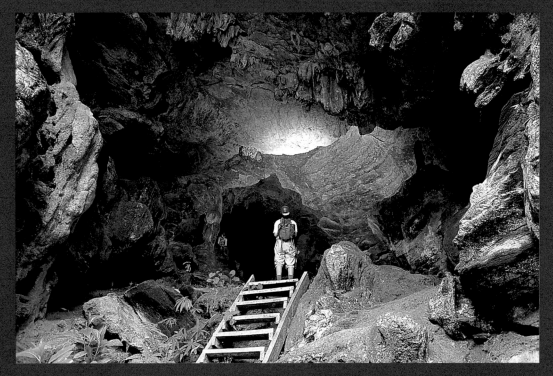

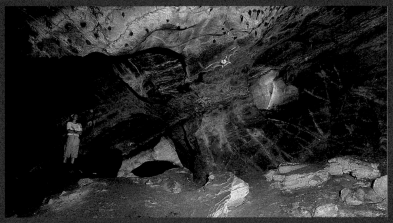

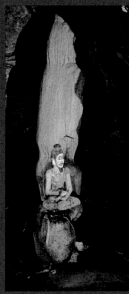

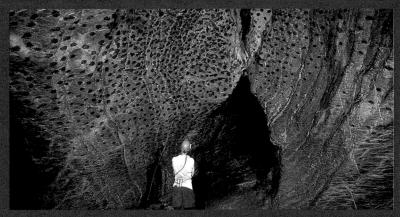

Tham Phra (Phayao)   ถ้ำพระ

A couple of tall narrow rifts one above the other at the base of an impressive white cliff close to Tham Sakoen on the border between Phayao and Nan provinces. The lower level opens into a large man-made platform under a sheltering overhang and is inhabited by a solitary monk. A tall narrow passage extends behind for 30 m but has no interesting features. A wooden ladder leads to the upper level, but at the time of our visit it was in such an advanced stage of decay that we decided not to risk our lives.

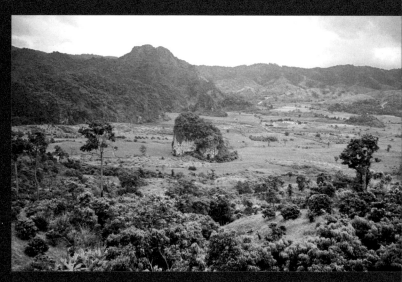

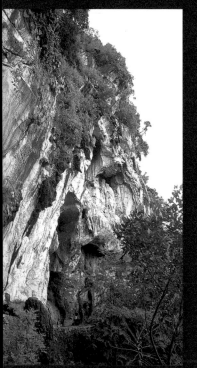

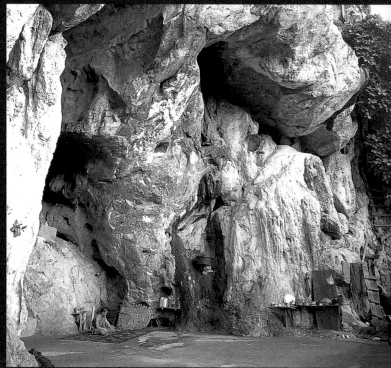

Tham Phra (Nan)   ถ้ำพระ

Situated in an isolated limestone tower at the back of a small
tranquil temple with a charming atmosphere. The cave entrance is at
the base of an imposing white cliff and has been completely sealed
up except for an ornately carved wooden door. The two small
chambers have been extensively refurbished and are electrically lit.
A 20 m passage on the right leads to a cave window high up in the
cliff with a fine view over the surrounding countryside. Another
passage on the left with sleeping platforms continues for 50 m to a
narrow crawl which was not explored.

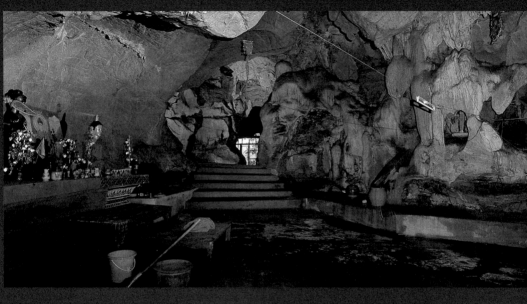

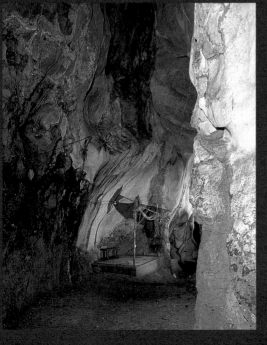

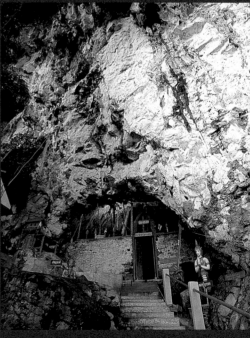

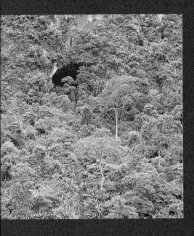

Tham Sakoen   ถ้ำสะเกิน

One of the most impressive caves in the region, giving its name to a whole national park. The 45 m wide entrance is set 150 m above the valley floor and is visible throughout the area. A large colony of bats produces copious quantities of guano which are collected by the locals and ferried directly down the mountainside on ropes. There are at least five entrances leading into a series of four large chambers separated by short steeply sloping passages. Collapsed floors, large rock falls and eroded speleothems indicate the great age of this cave.

A small hole at the top of the final chamber leads to a marvellous view and a breath of fresh air after the overpowering stench of bat guano. Unfortunately, the only way down is to retrace your steps back through the cave.

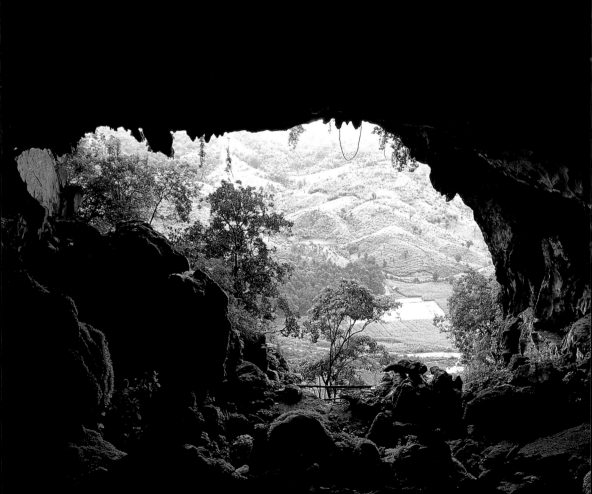

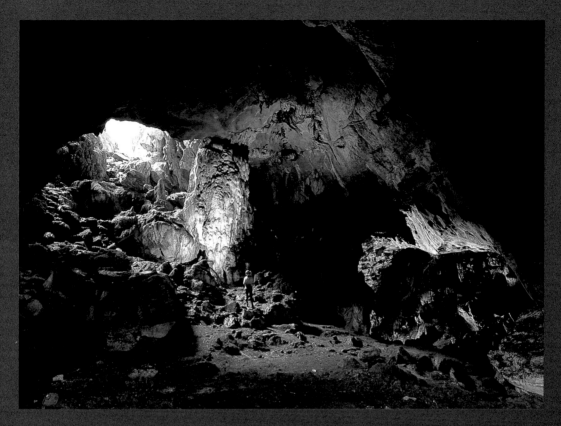

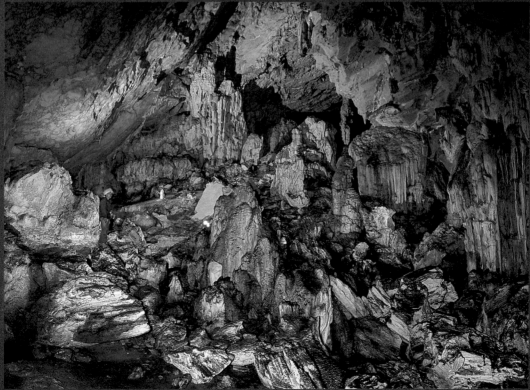

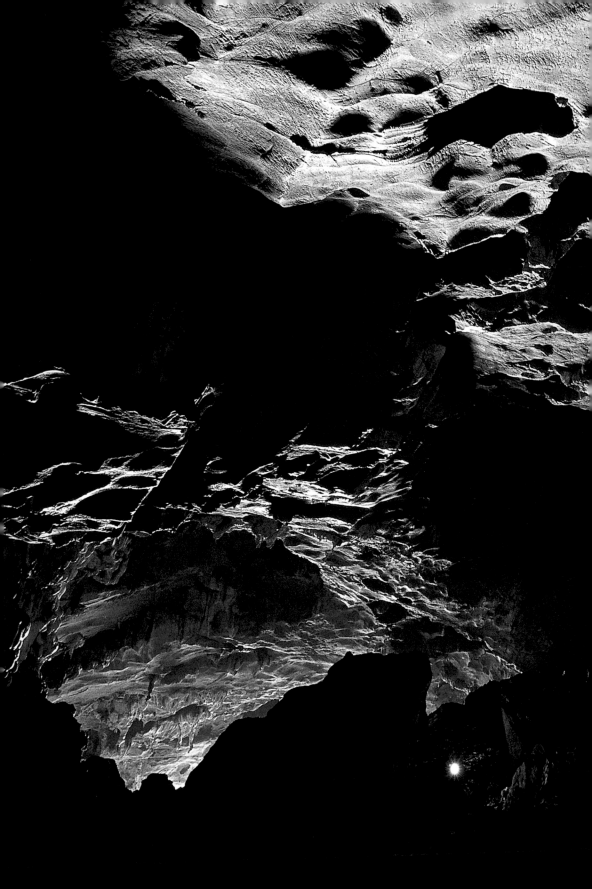

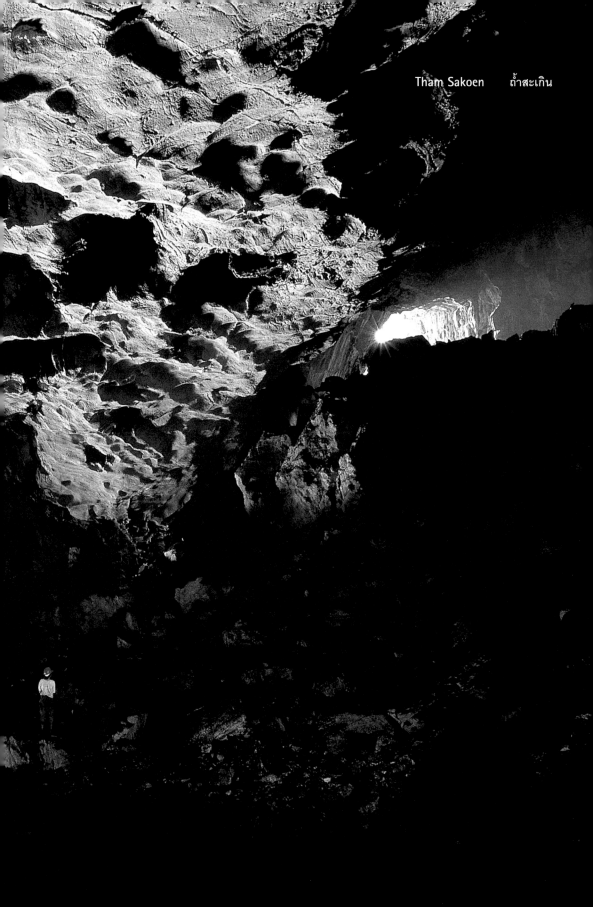

Tham Sakoen   ถ้ำสะเกิน

Tham Pha Kong ถ้ำผาฆ้อง

An easily accessible cave in Doi Phu Kha National Park, 6 km southwest of park headquarters. A lovely 3 km trail winds down through lush vegetation and fields to a stream. The small cave entrance is 30 m up the other side of the valley. A series of small seasonally active chambers leads to a permanent spring and a shallow stream which disappears down a hole 20 m further on. Good examples of cave breccia can be seen in several places (see page 29). The inner parts of the cave become progressively narrower and were not explored due to high CO_2 levels. The cave is said to emit a booming sound like a bell ("kong") which is probably caused by pressure differences between the various chambers.

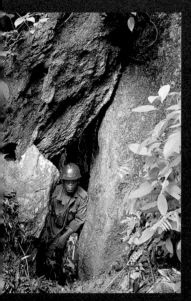

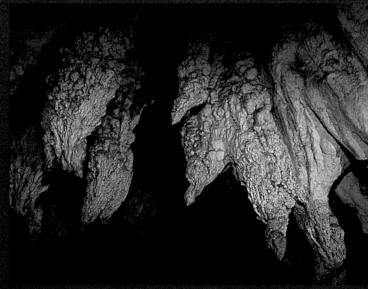

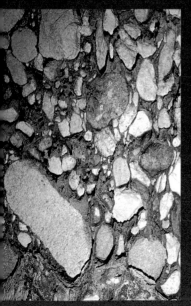

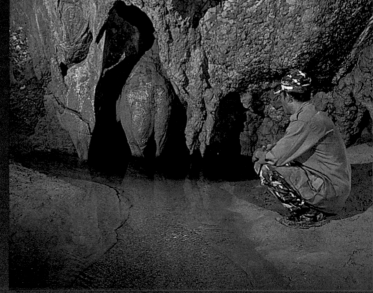

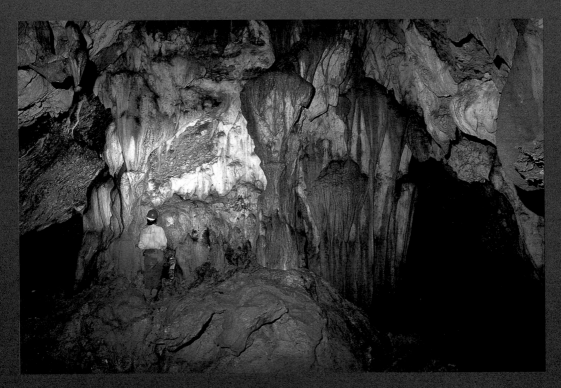

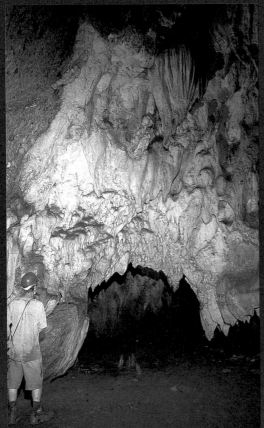

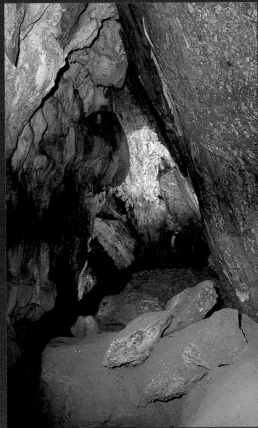

Pha Tub Caves   ถ้ำผาตูบ

A cluster of at least 16 small caves in two prominent limestone towers
just west of the main road 10 km north of Nan. The area has been
developed into a district park with an extensive network of paths
winding between huge boulders entwined with lianas and tree roots.
Tham Bor Nam Tip houses a seated Buddha under a purple canopy in
a high narrow chamber lit from above by a cave window. A rimstone
pool nearby with permanent water is believed to have special powers
but is littered with plastic garbage. Tham Phra (T. Pha Sawan) is a
series of vertical rifts with two large bells at the entrance leading to a
platform with two Buddhas. Tham Chedi Kaew has a remarkable
concreted structure resembling a chedi.

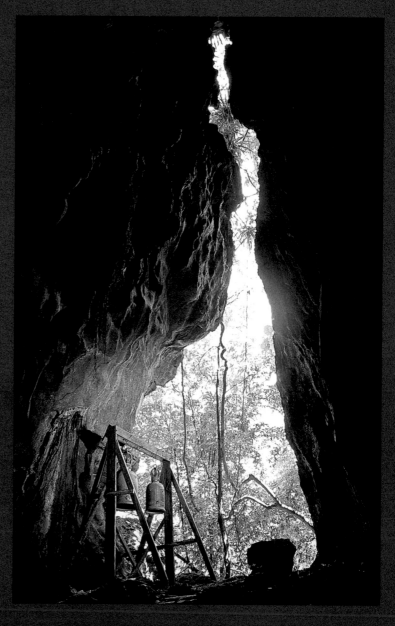

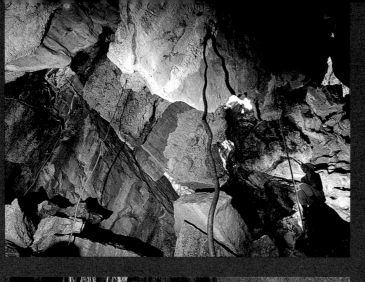
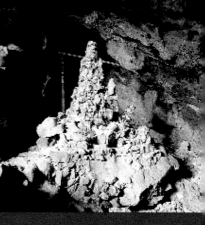
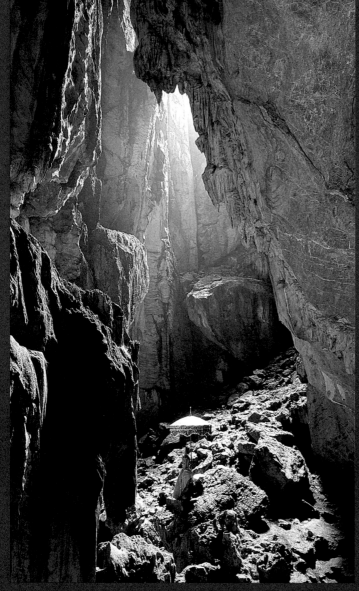
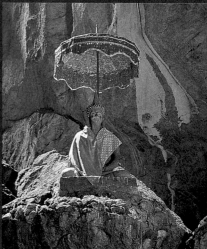
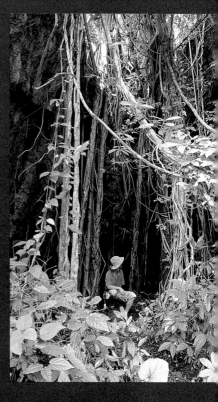

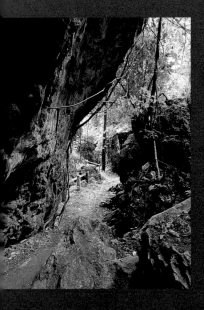

Tham Pha Nang Koi   ผานางคอย

A well-known tourist cave close to the main Phrae-Nan road 4 km of
Rong Kwang, hosting a local festival during Songkran festival in April.
A steep flight of steps between towering boulders leads up to the cave
entrance at the base of a cliff. The main passage is dry and dusty with
very few speleothems or other natural decoration. An easy 200 m walk
along a level trail with electric lights leads to the main attraction-a
natural rock formation which resembles a recumbent lady holding a
baby. The upper cave entrance 30 m further on provides dramatic
backlighting for a seated Buddha at the head of vast tiled platform.
From here you can follow an attractive and well-maintained trail
around the side of the mountain back to your starting point.

According to local legend, the lady is Princess Aranyani, daughter
of the King of San Wi who fell in love with a commoner and fled to
this cave with him to escape the king's wrath. One day her lover went
looking for food but was killed by the King's soldiers and the princess
turned to stone waiting for him to return.

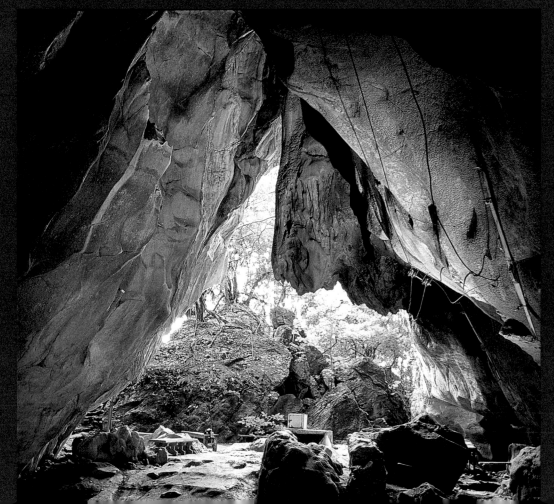

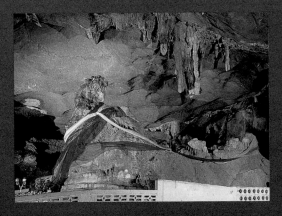

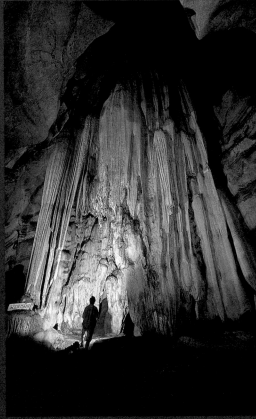

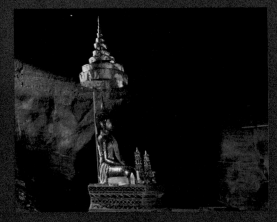

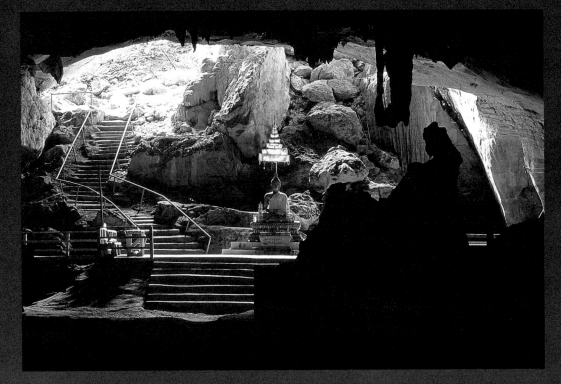

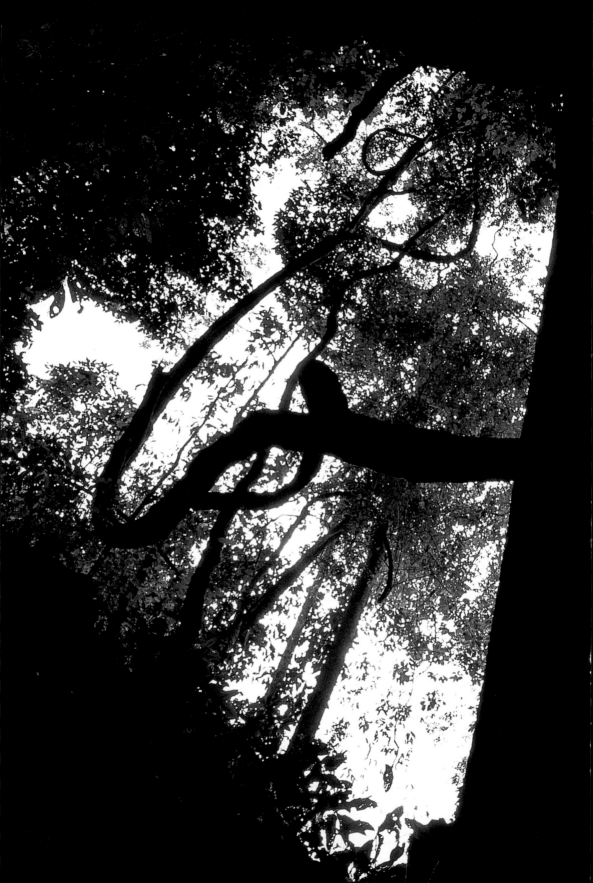

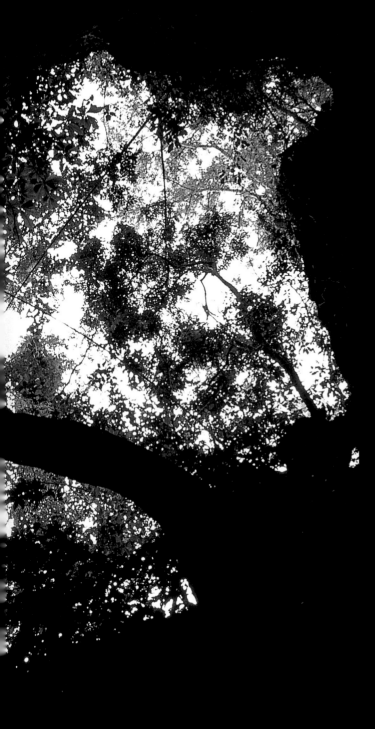

APPENDIX

ADDITIONAL INFORMATION

English Name   Thai Name   Page No.

Latitude, Longitude (to nearest minute) Length (m)
Other Names (English & Thai)

Access how to get there, when to go, level of difficulty.

Access information for sensitive caves is ommitted.

References letters & numbers in parentheses are the cave
identification codes for particular references.

PANG MA PHA

Pha Daeng   ผาแดง   **179**

19°36 'N, 98°06 'E   1,100 m

Pha Dhaeng, Nam Pha Daeng

Follow the dirt road leading north from the check point
on highway 1095 about 1 km west of the Nam Khong
Bridge. 3 km south of Ban Sale take a footpath leading up
through fields to the base of prominent red cliffs to the
west. The cave is about 15 minutes walk from the road.
There are no real difficulties during the dry season. A couple of sections are very low and completely impassable
during the wet season.

Dunkley 1995 (MH66); Dilokvanit 2000 (STD 609, 610);
Siripornpibul 2000 (map); Spies 2000

Pha Daeng Outflow (Pi Man) ผาแดงผีแมน **184**

19°36 'N, 98°06 'E   10 m est.

Pha Daeng Resurgence (Pi Man)

30 m south of Tham Pha Daeng outflow

Dunkley 1995 (MH108); Dilokvanit 2000 (STD 19);
Shoocondej 2000 (site 1, map)

Pha Daeng Scaffolds (Pi Man) เก้าชิ้น **184**

19°36 'N, 98°06 'E   10 m

Pha Daeng 7 pieces, Pha Daeng coffin cave 1

Less than 100 m north of Tham Pha Daeng inflow, a steep
scramble up along the base of the cliffs from the exit
window.

Dunkley 1995 (MH142); Dilokvanit 2000 (STD 13);
Shoocondej 2000 (site 2, map); Spies 2000

Long Yak   โลงยักษ์   **186**

19°36 'N, 98°06 'E   50 m est.

Long Yuk, Pha Daeng coffin cave 1

100 m south and 50 m higher thanTham Pha Daeng
inflow, at the top of a steep scree slope.

Dunkley 1995 (MH139); Dilokvanit 2000 (STD 15);
Shoocondej 2000 (site 3, map); Spies 2000

Chan Long   ชันโลง   **186**

19°36 'N, 98°06 'E   10 m

Lum Karn (Pi Man), Ram Karn, Pha Daeng coffin
cave 2, รำคาญ

100 m south of Tham Srisophon, 300 m south of Tham
Pha Daeng inflow.

Dilokvanit 2000 (STD17); Shoocondej 2000 (site 15);
Spies 2000

Spirit (Chester Gorman)   ผี   **187**

19°36 'N, 98°06 'E   70 m est.

Tham Piman, Tham Phi

Sensitive archaeological site of international significance.

Gorman 1972; Higham 1989, 2002; Higham & Thosarat
1998; Dunkley & Brush 1986; Dunkley 1995 (MH64);
Dilokvanit 2000 (STD 76, 107); Shoocondej 2000 (site 19,
map); Spies 2000

Srisophon   ศรีโสภณ   **187**

19°36 'N, 98°06 'E   100 m

Sri Saporn, Srisuphone, Boromathet

250 m south of Tham Pha Daeng inflow, follow the trail
to Tham Long Yak and keep going along the base of the
cliffs for another 150 m.

Dunkley 1995 (MH140); Dilokvanit 2000 (STD 16);
Shoocondej 2000 (site 4, map); Spies 2000

Pang Kham   ปางคาม   **188**

19°41 'N, 98°13 'E   1,370 m

Nam Pang Kham

The upper entrance window is 500 m south of Ban Pang
Kham at the foot of the cliffs. A recently upgraded road
leads from the "Mae Lana" turnoff on highway 1095 all
the way to Ban Pang Kham, a distance of 23 km. A good
walking path leads to the cave from the village. The
outflow entrance is 1km northeast of Ban Nam Jang,
along a steep path leading to the base of the cliffs. The
road to Ban Nam Jang is impassable in the wet season, as
is the cave itself. A long rope is needed to descend the
vertical 40 m north entrance since the ladder is now too
dangerous.

Dunkley & Brush 1986 (map); Kiernan 1990 (map);
Dunkley 1995 (MH74); Dilokvanit 2000 (STD 601, 602);
Siripornpibul 2000 (map); Spies 2000

Pang Kham 1 (Pi Man) ปางคาม 1   **190**

19°04 'N, 98°12 'E

Tham Hpi, Pang Kham coffin cave 1, Pang Kham Piman
West, Tham Wak, ผีแมนน้ำจาง

Sensitive archaeological site in the vicinity of Ban Nam
Jang. There are several other important archeological sites
in this area.

Pha Phuak   ผาเผือก   **192**

19°39 'N, 98°13 'E   > 680 m

Pha Puak, Pha Puek

One km west of the road to Pang Kham, 13.5 km from
highway 1095. A steep 4WD track, driveable in the
dry season, descends to a Lahu village on a ridge in front
of white cliffs. A good walking trail leads down to the
cave entrance 200 m further on. The gigantic entrance
chamber is full of dangerously unstable rockfalls which
become extremely slippery after rain. The inner part of
the cave has vertical sections and there have been reports
of bad air.

Dunkley & Brush (map); Boland 1989; Kiernan 1990
(map); Dunkley 1995 (MH76); Cosslett 1999; Dilokvanit
2000 (STD 604); Siripornpibul 2000 (map)

Banyan Valley   ผีแมนปุงฮุง   196

19°38 'N, 98°12 'E   100 m

Mae Hung (Piman), Boong Hoong (Piman), ไม้ฮุง, บ้านไม้ฮุง

50 m north of the Pung Hung stream inflow, a large dry chamber used to access the main Pung Hung cave.

Signs of Gorman's original archaeological excavation in 1972 can still be seen although all the interesting artifacts have long since been removed. There are 3 separate excavation pits 90-130cm deep. A unique example of a triangular stone tool shaped like a spear head was found at this site - see photograph in Higham (1998).

Gorman 1972; Higham 1989, 2002; Higham & Thosarat 1998; Reynolds 1990; Dunkley 1995 (MH65); Dilokvanit 2000 (STD 87, 674); Shoocondej 2000 (site 30, map)

Pung Hung   ปุงฮุง   196

19°38 'N, 98°12 'E   4,442 m

Boong Hoong, Long Snake Cave

Geologically sensitive site, only suitable for experienced caving parties.

Dunkley 1995 (MH160); Boland 1989, 1992; Dilokvanit 2000 (STD 675, map)

Jabo   ผาจะโบ่   200

19°35 'N, 98°13 'E   30 m est.

Jabo Cliff (Pi Man)

In the cliffs 1 km north of Jabo village, on the road to Pang Kham 4 km from highway 1095. Shortly before the army barracks, a footpath cuts through fields for 300 m to the base of the cliff. From here, it is a steep scramble with no clearly defined trail over thickly vegetated rocks to the cave entrance 70 m above.

Dunkley 1995 (MH125); Dilokvanit 2000 (STD 25, 680); Shoocondej 2000 (site 38)

Mae Lana   แม่ละนา   202

19°34 'N, 98°13 'E   >12,600 m

Nam Mae Lana, น้ำแม่ละนา

The inflow entrance is located at the southwest end of a large closed depression 3 km south of Ban Mae Lana. A rough 4WD track leads from the village to within 100 m of the cave entrance, but is currently only driveable as far as the northern end of the depression, even in the dry season. From here, it is a 20 minute walk down to the cave entrance at the bottom of the depression.

The outflow entrance is about 45 minutes walk north of highway 1095, about 500 m west of the Ban Luk Khao Lam turnoff. Head down the pine-covered ridge on the right slightly before the viewpoint. A camp site has been made close to the entrance next to the stream and would make a good base for exploring the cave at leisure. Guides can be hired from Ban Mae Lana.

Dunkley & Brush 1986 (map); Boland 1989, 1992; Dunkley 1995 (MH42); Anderson 1999; Dilokvanit 2000 (STD 628, 629); Spies 2000

Pet   เพชร   206

19°35 'N, 98°13 'E   100 m est.

Diamond Cave

On the north side of the same doline as Tham Mae Lana, just off the track leading down to the main cave 2.5 km

south of Ban Mae Lana, 300 m past the sign to Tham Pakarang.

Dilokvanit 2000 (STD 622); Spies 2000

Kai Mook   ไข่มุก   208

19°35 'N, 98°13 'E   40 m est.

Pearl Cave

2 km south of Mae Lana Village, on the northern edge of the same doline as Tham Mae Lana. 300 m before Tham Pakarang a signpost next to the 4WD track points to an indistinct path leading to the cave entrance hidden behind enormous boulders 150 m to the west.

Dilokvanit 2000 (STD 621); Spies 2000

Pakarang   ประการัง   208

19°34 'N, 98°13 'E   150 m est. +762 m

Coral Cave

2.5 km south of Mae Lana Village, on the north east side of the same doline as Tham Mae Lana. In the dry season, 4WD vehicles can be driven all the way to the "Tham Pakarang" signpost. From here, a walking path leads south for 100 m to the inconspicuous cave entrance.

Dilokvanit 2000 (STD 624); Spies 2000

Bor Krai   บ่อไคร้   210

19°33 'N, 98°14 'E   50 m est.

Extremely sensitive archaeological site, not suitable for visitors or film crews.

Dunkley 1995 (MH217); Dilokvanit 2000 (STD 42, 102); Shoocondej 2000 (site 36, map); Spies 2000

Lang Jan   ลางจันทร์   212

19°32 'N, 98°15 'E

Tham Axi, ช้อน

The most easily visited coffin caves in Pang Ma Pha, clearly visible in the cliffs close to the main road 2 km west of Soppong. A sign next to the road points to a steep path leading up to the base of the cliffs after 150 m.

Deharveng 1986 (NAL2, map); Dunkley 1995 (MH39); Dilokvanit 2000 (STD 53, 99, 100, 101); Shoocondej 2000 (sites 14, 34, 35, map); Spies 2000

Ya Pa Nae 1   ยาป่าแหน1   213

19°35 'N, 98°14 'E   50 m est.

Ping Yar (Piman), ผีแมนปิ้งยา

1.5 km along a walking path leading south from Ban Ya Pa Nae, on the dirt road between Mae Lana and Tham Lod Villages. The inconspicuous entrance is hidden behind boulders less than 50 m west of the path.

Dunkley 1995 (MH56); Dilokvanit 2000 (STD 612); Shoocondej 2000 (site 41, map)

Fossil   ฟอสซิล   214

19°35 'N, 98°15 'E   750 m est.

South of the dirt track between Ban Tham Lod and Ban Huai Haeng. 4 km from Ban Tham Nua, take a side track on the left for 500 m to a group of huts on a ridge. From here, a walking trail leads southwest around the sides of two dolines, crosses over the saddle between two prominent limestone outcrops then descends through thick forest to a third doline 1km from the huts. The cave entrance is 30 m south of the path close to the northeast corner of the doline. The first part of the cave up to the fossils is generally straightforward although there are some low crawls for short sections. A rope is necessary to explore the lower passages beyond the fossils. Several sections can have high CO_2 levels, especially the right-hand branch and the lower levels. Estimated total length 750 m, of which 441 m surveyed. Depth 35 m.

Dunkley 1997 (MH258); Dilokvanit 2000 (STD 614)

Christmas   คริสต์มาส   216
19°35 'N, 98°16 'E   300 m est.
Xmas Cave

South of the dirt track between Ban Tham Lod and Ban Huai Haeng. 3.3 km from Ban Tham Nua, head down the pine-covered ridge southwest to a limestone outcrop after 600 m. The cave entrance is at the head of a blind gully 100 m to the south of the saddle. Alternatively, follow the path along valley leading north west from Tham Hued (p. 999). Take the northern branch at the valley junction and the cave is close to the path slightly over 1 km from the road. The entrance chamber requires some scrambling over loose rocks and the inner passage needs short (5 m) rope or bamboo pole to get down to the stream bed.

Dunkley 1995 (MH109); Dilokvanit 2000 (STD 618)

Hued   ปลาตาบอด   220
19°35 'N, 98°16 'E   1,365 m
Tham Hud, Tham Pla, Blind Fish Cave, ฮืด

1.5 km WNW of Ban Tham Nua, 100 m west of the dirt track to Ban Huai Haeng on the edge of a small doline where the road cuts between two rocky outcrops. WARNING! do not consider visiting the inner sections of this cave unless you are certain you know what you are doing because the levels of CO_2 are so high that you could die in the attempt.

Deharveng 1986, 1988 (NAL13, map); Dunkley 1995 (MH33); Dilokvanit 2000 (STD 620); Spies 2000

Nam Hu   น้ำฮู   220
19°32ûN, 98°14ûE   800 m est.
แม่ฮู, Tham Nam Hoo, Tham Mae Hu, Toxic Wind Cave

20 m north of the main road 3 km west of Soppong market, shortly after Lang Jan Cliffs.

Deharveng 1987, 1988 (NAL3, map); Dunkley 1995 (MH10); Dilokvanit 2000 (STD 653); Siripornpibul 2000 (map); Spies 2000

Nong Pha Cham   หนองผาแจ่ม   221
19°29 'N, 98°14 'E   60 m est.
Close to the 4WD track between Soppong and Guet Sam Sip, 500 m south west of Nong Pha Cham village.

Dunkley 1995 (MH91); Dilokvanit 2000 (STD 668); Spies 2000

Lod   ลอด   222
19°34 'N, 98°17 'E   1,600 m
Tham Nam Lod, Tham Lot, น้ำลอด

Close to Ban Tham Lod, 9 km north of Soppong on a surfaced road, signposted from highway 1095 slightly northeast of the market. The road ends at Tham Lod Nature Education Centre, where there is a large car park and several food stalls. An excellent trail leads from here to the cave inflow entrance, 300 m to the south. There is no entrance charge or artificial lighting system installed

anywhere in the cave. Guides with lanterns can be hired from the car park and bamboo rafts are available at the cave entrance. The RFD have recently announced their intention to charge visitors 40 to 200 baht per person in the near future. The cave is open to visitors throughout the year, although at the height of the wet season the bridge across to the north chamber often gets washed away and it is not possible to wade or take a raft down the river to the outflow.

Tour Magazine (1985); Clarac & Pagau-Clarac 1995; Deharveng (1986); Dunkley 1983, 1985, 1995 (MH1); Dunkley & Brush 1986 (map); Dilokvanit 2000 (STD 632, 84, map); Shoocondej 2000 (site 27, map); Spies 2000

Lod Overhang   เพิงผาถ้ำลอด   229
19°34 'N, 98°17 'E   7 x 20 m est.
Pha Ba Mai, Forestry Cliff

In the grounds of Tham Lod Nature Education Centre, at the base of cliffs 100 m SSW of the car park. A signboard at the site gives background information on the excavation findings. Visitors are requested to stay outside of the roped area - this is for your own safety as the flimsy boards cover a 4 m deep pit!

Shoocondej 2003

Long Yaow   ล่องยาว   230
19°34 'N, 98°19 'E   516 m
Tham Doi Yaow

2.5 km south west of Muang Phaem Village, accessible part of the way on a 4WD track in the dry season. Follow along the main stream past Tham Chedi until you reach a major side valley on the right after about 1.5 km. The path crosses the main stream and follows a tributary stream for 200 m before veering south east into a dry side valley. The small cave entrance is at the base of a low cliff on the south side of the valley after 300 m. A safe and easy cave, no special equipment needed other than torches.

Dunkley & Brush 1986 (map); Dunkley 1995 (MH19); Dilokvanit 2000 (STD 637); Spies 2000

Phra (Muang Phaem)   พระ   232
19°34 'N, 98°19 'E   100 m est.
Huai Rai 2, ถ้ำห้วยราย 2

Close to Tham Long Yaow, 2 km south east of Muang Phaem village. The inflow is 200 m northwest of Tham Long Yaow, at the head of a dry gully. The outflow is on the north side of the ridge, close to the main stream. After Tham Chedi continue east along the main valley through rice fields for 200 m. The cave entrance is on the south side of the stream. The inner section of the cave is only passable at the height of the dry season. The cave is best visted by raft, which can sometimes be hired from Muang Phaem Village.

Dunkley & Brush 1986; Dunkley 1995 (MH21); Dilokvanit 2000 (STD 635, 636)

Chedi (Muang Phaem)   เจดีย์   233
19°34 'N, 98°19 'E   20 m est.
Jedi Cave, Pagoda Cave

On the way to Tham Long Yaow, 1.5 km southeast of Muang Phaem Village. A steep path on the north of the main track climbs up to the base of a brick staircase leading to the cave. Easily visited at any time of year.

Dunkley 1995 (MH185); Dilokvanit 2000 (STD 634)

Talu (Muang Phaem)   ทะลุ   **234**
19°34 'N, 98°19 'E   285 m

ถ้ำห้วยราย, Huai Rai 1

The inflow is at the head of a blind gully 200 m northeast of Tham Long Yaow, 2.5 km southeast of Muang Phaem Village. The outflow is on the north side of the mountain, close to the main stream. From Tham Chedi, continue east along the main valley for 400 m. The cave is subject to flooding in the wet season.

Dunkley & Brush 1986 (map); Dunkley 1995 (MH20); Dilokvanit 2000 (STD 638, 639)

Pha Mon   ผามอน   **236**
19°31 'N, 98°18 'E   4,050 m

The whole cave is currently under the protection of the Thai Military and closed to the general public. A wooden and bamboo walkway leads into the cave for the first 1km from the inflow entrance and electric lighting has been installed but it is unclear whether these facilities will be maintained in the future.

Deharveng 1986 (NAL17, map); Dunkley & Brush 1986; Dunkley 1995 (MH4); Dilokvanit 2000 (STD 657); Siripornpibul 2000 (map); Spies 2000; Tourism Authority of Thailand 1999

Ban Rai Overhang   เพิงผาบ้านไร่   **242**
19°30 'N, 98°11 'E   80 m est.

Ban Rai (Pi Man), Guet Nam Lang (Piman), Spirit Cave 1, Cave, NL, กึ๊ดน้ำลาง

2 km west of Soppong, take a side road south off highway 1095 for 1km to Ban Rai Village. A good path follows down the Nam Lang River for 2.5 km to a group of huts 500 m above the final sink. From here, the site is a steep 150 m climb up boulder-strewn slopes on the south side of the valley. A sign board at the site explains the history and archaeology of the site. This is an extremely important archaeological site so please stay out of the marked off areas as you may inadvertently cause irreparable damage to the fragile remains.

Dunkley & Brush 1986 (NL1, map); Dunkley 1995 (MH7); Dilokvanit 2000 (STD 52, 77, map); Shoocondej 2000 (site 20, map); Spies 2000

Face Cave   ผีแมนหน้าคน   **246**
19°33 'N, 98°07 'E   25 m est.

Wilderness 3 (Pi Man), ผีแมนน้ำของ 3

At the base of cliffs on the eastern side of Doi Makaeng, 1.5 km northwest of the Nam Khong Bridge on highway 1095. The quickest route is to head directly uphill behind Wilderness Lodge, about 1 km from the main road. The steep 300 m climb takes about 1 hour. A short rope is useful to safely negotiate the 6 m cliff face below the cave entrance. There are several other coffin caves nearby which contain coffin fragments, but they are all in poor condition and none of them have the unique face carvings.

Dunkley 1995 (MH151); Dilokvanit 2000 (STD 39); Spies 2000

Pha Tao Daeng   ผาเต่าแดง   **247**
19°31 'N, 98°08 'E

Tao Daeng Overhang, Dtao Daeng (Pi Man), Red Tortoise Cliff, Nam Khong Rock Painting, เพิงผาผีแมนน้ำของลาง, ภาพเขียนสีน้ำของ

A steep scramble along the base of towering cliffs 600 m north of Tham Nam Lang, 200 m beyond Tham Pha Nam Lang. The site is not easy to find but the bamboo ladders left by honey collectors serve as a useful pointer. The rock paintings are 2 m above the ground behind an enormous boulder.

Dilokvanit 2000 (STD 48, 81); Shoocondej 2000 (site 24, map)

Pha Nam Lang   ผีแมนน้ำของ   **247**
19°31 'N, 98°09 'E   20 m est.

Nam Lang Cliff Cave, Piman Nam Lang, Piman Nam Khong 3

A rough scramble along the base of the cliffs 400 m north of Tham Nam Lang. The main entrance is 20 m up the vertical cliff face and clearly visible from a distance but cannot be entered. A steep rocky gully to the south leads to an inconspicuous side entrance window less than 1m diameter.

Dunkley 1995 (MH58); Dilokvanit 2000 (STD 50, 83); Shoocondej 2000 (site 26, map)

Nam Lang   น้ำลาง   **248**
19°31 'N, 98°08 'E   8,550 m

3 km southeast of the Nam Khong Bridge on highway 1095. Follow the path down the river from behind the RFD station, crossing and recrossing the river several times before reaching the junction between the Nam Khong and Nam Lang rivers. The cave entrance is at the base of enormous cliffs 1km to the east, at the top of a long pile of boulders above the stream resurgence.

Tham Nam Lang is best visited during March or April - at other times of year the water is either dangerously high or too cold for comfort. The inner part of the cave can only be visited by well-equipped parties and may require an overnight camp deep inside the cave.

Dunkley 1983; Dunkley & Brush 1986 (map); Spies 1987; Kottelat 1988; Boland 1989, 1992; Dunkley 1995 (MH57); Dilokvanit 2000 (STD 650)

Susa   ซู่ซ่า   **252**
19°28 'N, 98°08 'E   2,617 m

A tiring 6 km walk east from Ban Mae Suriya, or a much longer but pleasanter 12 km walk down the river from the Nam Khong Bridge further north on highway 1095. A better option is to take a two-day rafting or kayaking trip from the bridge down to the Pai river and on to Nam Tok Mae Surin National Park headquarters, 7 km north of Mae Hong Son Town. The cave is at the foot of the cliff face 500 m east of where the Susa Stream tumbles over a spectacular 300 m wide travertine delta into the Nam Khong River.

Inexperienced cavers are advised not to enter the stream passage. The lower part of the passage is technically not difficult but can be dangerous due to the high CO_2 levels. The spray from the waterfalls contains rock chips which could cause infection or serious eye damage. The upper waterfall defeated several professional caving expeditions, until it was finally scaled in 1990.

Dunkley 1985; Dunkley & Brush 1986 (map); Kottelat 1988; Boland 1990 (map); 1992 (map); Dunkley 1995 (MH59); Dilokvanit 2000 (STD 717); Spies 2000

Seua   เสือ   **254**
19°28 'N, 98°07 'E   3,100 m (inc. T. Lom)

Tiger Cave, Huai Seua Neung, Tiger Creek Cave1, Huai Pha Tung

At the head of the Huai Pha Tung stream, 200 m west of Than Lom. The lower cave can be reached by following up the stream to where it emerges from a maze of tree roots and boulders into a small pool, but a much easier dry entrance is close to a path 60 m higher up hill.

Dunkley & Brush 1986; Smart 1994 (map); Dunkley 1995 (MH61); Dilokvanit 2000 (STD 649)

Lom ลม 254

19°28 'N, 98°07 'E 3,100 m (inc. T. Seua)

Wind Cave, Tham Huai Seua Song, Tiger Creek Cave 2

From the Susa waterfalls, follow down the west bank of the Nam Khong River for about 1.5 km. At the large permanent camp turn west up the Huai Pha Tung stream for 500 m. The cave is hidden behind a large boulder pile entangled with trees just south of the stream.

Dunkley & Brush 1986; Smart 1994 (Map); Dunkley 1995 (MH62); Dilokvanit 2000 (STD 663)

MAE HONG SON

Woa วัว 257

19°32 'N, 98°05 'E

Tham Woa, Tham Sunnata, Cow Cave

Behind Wat Wua, 2 km west of Mae Suriya village on highway 1095 between Soppong and Mae Hong Son.

Dunkley 1995 (MH134); Dilokvanit 2000 (STD 648)

Pla ปลา 258

19°26 'N, 97°59 'E < 10 m

Tham Pla, Tham Plah

17 km north of Mae Hong Son Town, just off the main road to Pai. The spring is a 200 m walk from the car park and food stalls.

Clarac & Pagau-Clarac 1985; Dunkley & Brush 1986; Deharveng 1988; Dunkley 1995 (MH96); Suwannabun 1981

Khang Khao ค้างคาว 259

19°25 'N, 97°55 'E 1,800 m

Pha Daeng, Ru Hoa Koa

16 km north of Mae Hong Son Town on highway 1095, 500 m before Tham Pla. The cave entrance is 50 m south of the main road behind a group of Buddha statues under black canopies. There are no difficult sections requiring special equipment for at least 1 km, although some passages are rather low. Not suitable during the wet season as the entire cave is subject to flooding and you could easily get trapped. The main danger of this cave is the slow build-up of CO_2 levels the further into the cave you go, which is more difficult to detect than a sudden change because your body adjusts to the lack of oxygen. Take several lighters and turn back as soon as anyone in your party starts to feel uncomfortable.

Deharveng 1987, 1988 (Ru Hua Koa, map); Dunkley and Brush 1986 (map); Kiernan 1990; Dunkley 1995 (MH99)

Pha Phung ผาผึ้ง 262

19°25 'N, 97°59 'E < 20 m

200 m east of highway 1095 to Pai, 15 km north of Mae Hong Son Town and 2 km southwest of Tham Pla. There is a clear footpath made by honey gatherers in the dry season, but it becomes very overgrown after the rains.

Jak Dtor พระบาทคู่ 265

19°26 'N, 97°58 'E 483 m

Jak Tor, Pra Bat Ku, Yak Tok 1

On the northwest side of Doi Khao Chang Mob, the mountain behind Tham Pla. 1km west of Tham Pla, a side road on the right leads to Ban Mok Jum Pae. After 500 m, cross over the stream on a footbridge, 100 m downstream of a small dam. A footpath on the other side of the stream heads northeast back towards the mountain then climbs up 150 m along the base of cliffs. Tham Jak Dtor is at the bottom of the northernmost of a series of three dolines on the right of the path, about 1.5 km from the bridge. The cave entrance is an active forest monastery which hosts spiritual retreats for groups of up to 30 people. The inner section of the cave can be quite dangerous as you have to scramble over steep slopes of loose boulders coated with mud and guano. Climbing gear and plenty of experience would be needed to explore the full length of the cave. Surveyed depth: 65 m.

Deharveng 1988 (Yak Tok 1, map); Dunkley 1995 (MH174); Pradit (undated)

Pha Phung ผาผึ้ง 266

18°04 'N, 97°57 'E < 20 m

1.6 km east of Ban Nong Haeng, about 25 km south of Khun Yuam on the main road between Mae Hong Son and Mae Sariang. A good track leads from the village past a hot spring to a river crossing. The site is at the base of cliffs on the opposite site of the river.

Mae Ga Noi แม่กาน้อย 267

18°34 'N, 97°57 'E < 100 m

Mae La Ka

In remote country north east of highway 108, 96.5 km south of Mae Hong Son and 67.5 km north of Mae Sariang. Close to km post 170 take a rough track heading up and along the ridge for 3.5 km, probably once driveable but currently only so for the first 100 m. Just before a slight dip in the main ridge head west down a side ridge for 100 m then north along the side of steep valley for another 100 m to the small cave entrance halfway up the hillside.

Mae Hu แม่ฮุ 268

18°24 'N, 98°00 'E 80 m est.

About 2 km east of Tham Kaew Komol. At the abandoned visitor centre, continue straight on towards the mining site until you reach the Mae Hu Stream. The caves are in a small forest temple 100 m above the stream on the east side of the road.

Crystal Mae La Noi 2 กลางเหมือง 268

18°23 'N, 98°00 'E > 30 m

Recently discovered cave, no information available. This cave is situated in an active mining site and is extremely dangerous to visit because ongoing blasting by the mine may have affected the structural intregriy of the cave.

Kaew Komol แก้วโกมล 270

18°24 'N, 98°00 'E 140 m

Crystal Mae La Noi, Calcite Cave, คริสตัลแม่ลาน้อย

4 km east of Mae La Noi Town, with a surfaced road right up to the cave entrance where there is a small car park but no food stalls. From the centre of town follow signs to

Ban Huai Ma Fai. At the enormous abandoned visitor centre turn left for 200 m then turn right straight up the hill for another 400 m. The entire cave is electrically lit and all visitors must be accompanied by an RFD guide, for which there is no charge.

Dunkley 1995 (MH243); Spies (map, undated); RFD pamplet

CHIANG MAI

Chiang Dao   เชียงดาว   274
19°23 'N, 98°56 'E   5,190 m
Luang Chiang Dao

At the base of Doi Chiang Dao, 5 km west of Chiang Dao Town. The cave is inside the temple at the centre of Ban Tham. A small charge is levied at the cave entrance to cover maintenance costs for the electric lighting system. Guides with lanterns can be hired inside the cave for anyone wishing to explore the unlit sections.

Anon. 1972; Kaczmarek 1972; Penth 1974; Kusch 1975; Windecker 1976; Bhamarabutr 1980; Suwannabun1981; Deharveng & Gouze 1983; Dunkley 1983; Clarac & Pagau-Clarac 1985; Deharveng 1986 (map); Price 1986; Dunkley 1995 (CM1); Munier 1998; Thikakhayano (date?)

Ki Nok   ขึ้นก   280
19°23 'N, 98°56 'E   695 m
Grotte Superieure

In the cliffs above Tham Chiang Dao, reached by a steep trail from the back of the temple complex 100 m north-west of the main cave entrance.

Deharveng & Gouze 1983; Deharveng 1986 (CHD3, map); Dunkley 1995 (CM2)

Pak Sung   ปากสูง   282
19°24 'N, 98°56 'E   20 m est.
Pok Soong

1 km northwest of Tham Chiang Dao, a driveable track leads up to the temple buildings just in front of the cave. The cave is securely locked with a gate but the key can be obtained from the monks.

Deharveng 1986 (CHD17, map); Dunkley 1995 (CM21)

Pak Pbiang   ปากเปียง   282
19°24 'N, 98°56 'E
Pak Piong

A few hundred metres northwest of Tham Pbak Sung, on the other side of the same rock outcrop.

Deharveng 1986 (CHD16, map); Dunkley 1995 (CM20)

White Buddha Cave   282
19°24 'N, 98°56 'E   110 m
200 m NW of Tham Pak Sung, within the temple grounds
Deharveng 1986 (CHD15, map); Dunkley 1995 (CM22)

Ki Mi   ขี้หมี   284
19°22 'N, 98°44 'E   400 m est.
Barami

On the western edge of Chiang Dao Wildlife Sanctuary, 4 km south of Muang Kong. A rough 4WD track leading from the west side of town stops about half way. From

here there is a good trail through scenic surroundings to a resting point by a stream 50 m below the cave.

Tham Ki Mi presents no real difficulties although it could be subject to flooding after heavy rains.

Travellers Guide June 1982; Clarac & Pagau-Clarac 1985 (Barami); Dunkley 1995 (CM91, Barami)

Klaeb Yai   แกลบใหญ่   288
19°34 'N, 99°04 'E   2,190 m
Ngam, Pha Chan

On the eastern edge of Chiang Dao National Park in the vicinity of Ban Tham Klaeb on the main road between Chiang Dao and Fang, about 25 km north of Chiang Dao Town. 1 km north of Ban Tham Klaeb, turn northwest on a surfaced road for about 1 km. The upstream entrance window is hidden in a dense patch of forest at the base of cliffs less than 100 m south of the road. The 20 m scramble down from the window to the stream level requires some care as many of the boulders are loose and slippery. The stream passage itself is generally straight-forward, although there are some rather low sections. The whole cave could be subject to flooding after heavy rains.

Deharveng 1986 (PHD6, map); Dunkley 1995 (CM35)

Long   โลง   290
19°34 'N, 99°04 'E   50 m est.

Follow the nature trail starting from the upstream entrance window of Tham Klaeb Yai. The trail starts with a steep climb up the cliffs above Tham Klaeb Yai, becoming increasingly fainter further uphill. The impressive entrance is set high up in a cliff face about 1.5 km to the southwest of the road. Huai Ja Kan sub-station near the turn off from the main road can provide guides if necessary.

Klaeb   แกลบ   291
19°33 'N, 99°04 'E   176 m
50 m north of Ban Tham Klaeb on the main road between Chiang Dao and Fang, about 25 km north of Chiang Dao Town.

Deharveng 1986 (PHD9, map); Dunkley 1995 (CM33)

Ngung Chang   งวงช้าง   291
19°3 'N, 99°04 'E   > 25 m
Nguang Chang

200 m northwest of Ban Tham Klaeb on the main road between Chiang Dao and Fang, about 25 km north of Chiang Dao Town. A track from the main road leads right up to the cave entrance inside a small temple complex.

Watson & Loof 1967; Dunkley 1995 (CM34)

Klang Muang 1   กลางเมือง 1   292
19°42 'N, 98°54 'E   > 300 m (1,000 m reported)

About 2 km south east of Muang Na in the northern part of Chiang Dao National Park. A rough track through fields leads around the base of the cliffs to within about 200 m of the caves, although it is still a steep climb up to the entrance. The cave itself presents no real difficulties, although care is needed to negotiate the numerous loose rockfalls and breakdown sections on the floor. The bamboo ladder up to the entrance needs to be critically examined before venturing forth as a sudden collapse could be fatal. Guides are available from the RFD ranger station 1 km south west of Muang Na Town.

RFD pamplet (Chiang Dao National Park)

Klang Muang 2   กลางเมือง 2   293

19°42 'N, 98°54 'E   < 100 m

At the base of the cliff immediately below Tham Klang Muang 1.

Naresuan   นเรศวร   294

19°43 'N, 98°55 'E   100 m est.

Chang-Pha Hok, Jang-Pha Hok, แจ้งป่าหก

On the northern edge of Chiang Dao National Park, about 1 km from the Burmese border. A track signposted "Muang Na Caves" leads northwards from near the bridge just east of Muang Na Town. After about 1 km turn right on a dirt track without a signpost, winding through fields to a forest temple in front of the cave entrance a few km further on.

RFD pamplet (Chiang Dao National Park, as 'Tham Chang-Pha Hok')

Luek   ลึก   296

19°43 'N, 98°55 'E   > 400 m

The inconspicuous entrance is at the western end of a small closed depression, a few hundred metres north of Tham Naresuan. Care is needed to descend the steep slippery slope down to the main passage, which is level and easy going for at least 300 m. The stream bed is flooded for several months during the rainy season.

Guides are available from the RFD ranger station 1km southwest of Muang Na Town.

Mai   ใหม่   298

19°43 'N, 98°55 'E

50 m above Tham Nam at the eastern end of a small closed depression, a few hundred metres north of Tham Naresuan.

Nam   น้ำ   298

19°43 'N, 98°55 'E

100 m east of Tham Luek, on the opposite side of a small closed depression, a few hundred metres north of Tham Naresuan. A vertical drop of at least 15 m leads down to the main level. A temporary bamboo ladder has been constructed by the monks to negotiate this difficulty.

Huai Bon (Wat)   วัดห้วยบอน   299

19°57 'N, 99°12 'E   < 20 m

Huai Bon

10 km northwest of Fang, near the eastern edge of Fang National Park close to the 4WD track leading to Doi Pha Hom Pok. From the by-pass, take the road signposted to Ban Huai Mayom, which turns into a dirt track after a few km. About 6 km from the ring road, turn left just after the border of the national park. The cave is halfway up the cliffs at the back of the temple complex, reached by a steep staircase.

Deharveng 1986 (FAN1, map); Dunkley 1995 (CM32); RFD pamplet (Fang National Park)

Huai Bon (Haeng)   ห้วยบอน   300

19°57 'N, 99°12 'E   > 300 m

The lower entrance window is 200 m southwest of Wat Huai Bon, close to the water tanks 50 m above the temple. From the upper entrance, a path leads back around the side of the mountain to your starting point.

RFD pamplet (Fang National Park)

Huai Bon (Nam)   น้ำห้วยบอน   300

19°57 'N, 99°12 'E   > 200 m

About 1 km southwest of Wat Huai Bon. From the upper exit of Tham Huai Bon Haeng a faint path cuts through thick vegetat-ion to the inconspicuous cave entrance less than 100 m away.

Tab Tao   ตับเต่า   302

19°42 'N, 99°06 'E   > 1,500 m

Tub Tao, Tap Tao

2 km west of the main road between Chiang Dao and Fang, 46 km north of Chiang Mai and 34 km south of Fang. A small surfaced road close to km post 120 leads through the village into the temple grounds. Much of the cave is electrically lit, for which a small donation is expected. Do not be tempted to explore any unlit passages as many inner sections have high CO_2 levels.

Bock 1884; Anon. 1895; Deharveng & Gouze 1983; Clarac & Pagau-Clarac 1985; Deharveng 1986 (PHD1); White 1988; Dunkley 1995 (CM28)

Ngam   งาม   306

19°42 'N, 99°05 'E   150 m est.

2 km south-west of Tham Tab Tao, reached via a small surfaced road branching off to the south just before Tab Tao Temple. The cave entrance is 10 minutes walk up a flight of steps from the road head.

Deharveng 1986 (PHD4, map); Dunkley 1995 (CM29)

Muang On   เมืองออน   308

18°46 'N, 99°13 'E   335 m

Maung On, Muang Oon

About 30 km east of Chiang Mai Town, 17 km beyond San Kamphaeng and well-signposted from the ring road. A newly constructed surfaced road winds up the mountainside from Ban On Luai to a car park and food stalls. From here, a flight of steps leads to the cave entrance about 100 m above. A small entrance charge is levied and there is electric lighting throughout the cave.

Deharveng 1986 (ONL1, map); Dunkley 1995 (CM53); Mahidol 1989

Borichinda   บอริจินดา   309

18°31 'N, 98°41 'E   > 100 m

Just inside Doi Inthanon National Park, 70 km southwest of Chiang Mai. Less than 1 km after the entrance gate a 4WD track on the right is driveable in the dry season for the first couple of km up to a rocky stream. From here, it is about half an hour walk to the cave entrance on the other side of the hill. There is no artifical lighting or other facilities. Guides are available from the national park office.

Sarasin 1933; Kusch 1975; Clarac & Pagau-Clarac 1985; Dunkley 1995 (CM52)

Wat Umong   312

18°47 'N, 98°56 'E

On the western edge of Chiang Mai Town, close to the base of Doi Suthep about 1 km south of the university.

Penth 1974 (as Wat Umong Thera Jan)

LAMPHUN

Russi ฤาษี **313**

18°25 'N, 98°52 'E <10 m

Wat Phra Phuttha Bat Tak Pha ในวัดพระบาทตากผ้า

20 km southwest of Lamphun, about 1 km east of highway 106 to Amphoe Li. Follow the signs for Wat Phra Phuttha Bat Tak Pha.

Bidyalankarana 1935; Dunkley 1995 (LP2); Munier 1998

Tep Ni Mit เทพนิมิตร **314**

18°22 'N, 98°52 'E < 50 m

A few hundred metres north of the surfaced road to Tham Erawan, about 8 km from the main road and 3 km before the RFD ranger post.

Erawan เอราวัณ **316**

18°21 'N, 98°52 'E > 200 m

About 30 km southwest of Lamphun, east of highway 106 to Amphoe Li. The cave is well-signposted from the main road but unfortunately none of the signs are in English. Shortly before Ban Hong turn left on a newly surfaced road for 4.5 km, then turn right for another 7 km to the road head just past the ranger station. A good footpath leads to the cave entrance a few hundred metres to the south. A lighting system has not yet been installed but a guide with torch can be provided by the nearby ranger station and there is an informative signboard with a map at the cave entrance.

Luang Pha Wiang หลวงผาเวียง **318**

18°16 'N, 98°51 'E > 200 m

East of highway 106 between Lamphun and Amphoe Li. 15 km south of Ban Hong, turn onto a surfaced road past the mining site to a car park at the base of a limestone hill 2.8 km from the main road. A 10 min climb up a flight of steps leads to the cave entrance. Guides with torches are available from the car park. A lighting system might be installed in the near future.

LAMPANG

Pha Thai ผาไทย **320**

18°33 'N, 99°5 'E 1,185 m

On the edge of the newly designated Tham Pha Thai National Park, 2 km west of highway 1, 67 km northeast of Lampang and 17 km southwest of Ngao. The cave entrance is a 10 min. walk up a flight of steps from the car park and visitor information centre. The main sections of the cave are electrically lit and guides are available free of charge. Do not try to explore any unlit sections as some of the inner passages have extremely high CO_2 levels.

Kusch 1975; Suwannabun1981; Clarac & Pagau-Clarac 1985; Deharveng 1986 (PHT1); Deharveng & Gouze 1983; Dunkley 1995 (LA1); RFD pamphlet (Tham Pha Thai National Park)

Pra Tu Pha ประตูผา **324**

18°26 'N, 99°43 'E < 10 m

Cliff Gate

Close to highway 1 where it passes through a narrow gap in the limestone ridge, 55 km northeast of Lampang. From the car park behind the shrines walk east along the road for 50 m then follow the footpath south along the base of the cliffs. The rock paintings are in 6 distinct

groups 200 to 500 m south of the road. Information leaflets in Thai, English and French are available from the car park and there is an excellent little book in Thai detailing the history and significance of the site.

Suwannabun1981; Dunkley 1995 (LA28)

CHIANG RAI

Luang (Mae Sai) หลวง **326**

20°25 'N, 99°51 'E 6,220 m

Nam Cham, Tham Mai, Big Cave, ใหม่

3 km west of highway 110, two signposted turnoffs 5.5 and 7.5 km south of Mae Sai both lead to the cave. The northern route is shorter but unsurfaced. The southern route is surfaced all the way, turning north slightly before Khun Nam Nang Non RFD Station, then west after about 1km for another 500 m into the headquarters of Tham Luang Forest Park. The cave is well signposted in the southeast corner of the park, less than 100m from the car park. There is a visitor centre with a detailed map of the cave but not much else.

Deharveng 1986 (map), 1988 (map); Dunkley 1995 (CR3)

Pra Ya Nag พญานาค **330**

20°24 'N, 99°52 'E 20 m est.

Tham Na, Naga Cave

250 m south of Tham Luang Forest Park headquarters, about 15 mins. up the hill on a good path starting behind the visitor centre. At the junction after 200 m, keep straight ahead and the cave is less than 100 m up the hill. The left branch of the path leads to Tham Liang Pha.

Deharveng 1988 ('Tham Na'); Dunkley 1995 (CR7)

Phra (Mae Sai) พระ **331**

20°24 'N, 99°52 'E 50 m est.

In the southeast corner of Tham Luang Forest Park headquarters, about 50 m from the car park.

Deharveng 1988 (as 'Tham Pha'); Dunkley 1995 (CR6)

Liang Pha Overhang เลียงผา **332**

20°24 'N, 99°52 ' E 75 m

Mountain Goat Cliff

300 m southwest of Tham Luang Forest Park headquarters, about 15 mins up the hill on a good path starting behind the visitor centre. After about 200 m, turn left down into the doline in front of the cave. The path straight on leads to Tham Pra Ya Nag. Depth: 50 m est.

Deharveng 1988 (map); Dunkley 1995 (CR08)

Sai Thong ทรายทอง **334**

20°23 'N, 99°51 'E 920 m

Luek, ลึก

1 km south of Tham Luang, at the foot of Doi Nang Non. Turn off the main road 7.5 km south of Mae Sai and continue straight west through Ban Jong to Khun Nam Nang Non RFD Station. The cave entrance is on the far side of the pond, 5 minutes walk from the car park and food stalls. The main stream passage is straight-forward for the first few hundred metres, but becomes progressively more arduous with lots of scrambling over loose rockfalls.

Deharveng 1988 (map); Tourist Magazine Nov 1981; Dunkley 1995 (CR1)

Phum ปุ่ม 336

20°21 'N, 99°51 'E 192 m

Between Tham Pla and Tham Luang, 3 km west of the main road 10 km south of Mae Sai. A twisting road with many junctions leads to the well-known temple close to the mouth of a steep-sided valley cutting through the mountains. The cave entrance is at the top of a flight of steps just behind the temple.

Suwannabun1981; Deharveng 1988 (map); Munier 1998

Pla (Mae Sai) ปลา 338

20°20 'N, 99°51 'E

Tham Plaa, Tham Plah, Monkey Cave, Wat Tham Phum-Tham Plaa, ลิง,วัดถ้ำ ปุ่มถ้ำปลา

1.5 km west of highway 110 near km post 879, 12 km south of Mae Sai. A surfaced road leads through Ban Huai Khrai to a large car park at the entrance of Wat Tham Phum-Tham Pla. In addition to Tham Pla, there are two other caves within the same temple grounds - Tham Plaew Plong Fah and Tham Ukaew. Tham Phum is 2 km to the north (see above).

Suwannabun1981; Clarac & Pagau-Clarac 1985; Dunkley 1995 (CR13); Munier 1998

Ukaew อู่แก้ว 339

20°20 'N, 99°51 'E 530 m

Phu Kaeo, Kou Khan, Ku Khan

Within the grounds of Wat Tham Pla-Tham Phum, 50 m northwest of the main entrance. A tight squeeze has to be negotiated to penetrate into the inner section of the cave. Some of the breakdown sections in the floor are unstable and have pockets of bad air.

Deharveng 1986 (as 'Tham Kou Khan', map); Dunkley 1995 (CR10); Munier 1998

Plaew Plong Fah เปลวปล่องฟ้า 340

20°20 'N, 99°51 'E 70 m

Within the grounds of Wat Tham Phum-Tham Pla, at the top of a long flight of steps behind the spring.

Tourist Magazine Nov. 1981; Dunkley 1995 (CR15); Munier 1998

Sao Hin เสาหิน 342

20°19 'N, 99°51 'E 80 m est.

Sao Praya Nak, Praya Nak, Rock Column Cave, King of Naga Column Cave, เสาพญานาค

1 km south of Wat Tham Phum-Tham Pla, on the far side of a reservoir which can be reached on foot from the temple, or from the main road at the check point near km post 878, 13 km south of Mae Sai. The cave entrance itself can only be reached by boat, departing from in front of the food stalls on demand. The obvious cave entrance in the cliffs to the northwest of the reservoir is just a shallow overhang and not worth the steep scramble up to it except for the view.

Dunkley 1995 (CR28)

Pha Dtak (Mae Sai) ผาแตก 344

20°25 'N, 99°52 'E < 30 m

1.5 km west of Mae Sai Town, turn off the main road just after the local police station a few hundred metres before Immigration. Cross over the canal and continue straight on towards a prominent rock outcrop on the edge of town.

The cave is in the grounds of the small temple at the base of the outcrop.

Pha Rua ผาเรือ 345

20°25 'N, 99°52 'E 100 m est.

Pha Lua, Pha Rae

2 km southwest of Mae Sai Town. From Tham Pha Dtak, turn south on a dirt track for 300 m to the next limestone outcrop. The cave entrance is at the top of a long flight of steps on the east side of the hill.

Pha Chom ผาจม 346

20°26 'N, 99°52 'E 40 m est.

Wat Pha Chom

In Wat Pha Chom, on the outskirts of Mae Sai Town 2 km west of town centre. At the international border post, turn west along the river for about 1.5 km, then south along a small side valley. The extensive temple complex is on the right after a few hundred metres. This is probably the same cave referred to as 'Tham Wat Phra Doi Wao' in Deharveng (1988) and Dunkley (1995). Doi Wao Temple is much closer to the town centre and does not appear to have any caves.

Deharveng 1988 (as Wat Phra Doi Wao); Dunkley 1995 (CR22); Munier 1998

Phra Ban Pen Bun พระบำเพ็ญบุญ 347

19°42 'N, 99°45 'E

East of highway 1 between Phayao and Chiang Rai, a turnoff 18 km north of Phan winds through rice fields and a series of villages for 6 km before reaching Wat Tham Phra Ban Pen Bun. The main cave is at the base of a cliff face about 100 m from the parking.

Nam Lod น้ำลอด 348

19°32 'N, 99°40 'E 115 m

Talu

Southwest of Phan Town on the border of Doi Luang National Park. Turn off the main street next to the town hall, heading due west on a surfaced road for 6.5 km before turning south on another surfaced road for a further 6.5 km. A dirt track on the right leads after 1 km to a car park and picnic spot in front of the cave entrance. Several other caves are reported in the vicinity of Ban Tham Village, a couple of km further west at turning 6.5 km from Phan. The village itself is named after a shallow overhang which used to be occupied by a hermit. The floor of the overhang is littered with bones and pieces of pottery and is almost certainly a prehistoric habitation site. A few curious 'donut'-shaped stones which are probably also of prehistoric origin have been collected by the locals and placed on the shrine.

Dunkley 1995 (CR31)

Pha Jarui ผาจรุย 349

19°32 'N, 100°02 'E 25 m est.

Pha Cha Lui, Phra, Russi, Lom

In remote country between Phan and Thoen, 2.8 km north of Ban Pai Ngae on highway 1128. The caves are behind Wat Pha Ja Rui in the southeast corner of a isolated limestone outcrop about 100 m west of the road.

Dunkley 1995 (CR20, 37, 38 and 39)

PHAYAO

Pha Daeng (Phayao) ผาแดง 350

19°30 'N, 100°27ûE

In Phu Sang National Park, about 25 km east of Chiang
Kham. From the centre of town, take road 1210 as far as
Phra Tat Doi Kam, then turn right on a dirt track for 14
km to an RFD station. Continue straight on for another
1km to Ban Pang Tham, then follow a rough track on the
right which winds through fields for about one km to a
picnic site beside the river. Tham Pha Daeng is at the top
of a short flight of steps on the south bank of the river.
Tham Nam Lod is at the foot of the same hill less than
30 m away.

Dunkley 1995 (PY03)

Nam Lod (Phayao) น้ำลอด 351

19°30 'N, 100°27 'E

30 m below the entrance to Tham Pha Daeng.

Phra (Phayao) พระ 352

19°21 'N, 100°32 'E

On the southern edge of a prominent limestone outcrop
about 3 km west of Tham Sa Koen. From the main road ,
turn north at the sign to Tham Sa Koen National Park
Headquarters. After about 2 km, turn left down a rough
track just after the bridge, heading straight towards the
limestone cliffs 1km further on.

NAN

Phra (Nan) พระ 353

19°23 'N, 100°36 'E >50 m

At the southern edge of an isolated limestone tower about
1 km north of highway 1148, about half way between
Song Khwae and Sakoen. The cave is at the back of a
small temple on the outskirts of Ban Pha Lak.

Sakoen สะเกิน 354

19°22 'N, 100°32 'E 500 m est.

Sagurn, Luang, Khaeng Khao, Khon Khao, Kaeng Khao,
ค้างคาว, หลวง

In remote country on the border between Phayao and
Nan provinces, north of highway 1148 between Chiang
Kham and Tha Wang Pha. At Ban Sakoen close to km
post 64, turn north for 5 km, passing through the village,
then over a bridge. 2 km south of Tham Sakoen National
Park headquarters, follow an indistinct footpath up along
the base of the cliffs to the obvious entrance 150 m above
the plain.

Dunkley 1995 (PY4), 1998 (NA35); RFD pamplet (Tham
Sakoen National Park)

Pha Kong ผาม้อง 358

19°11 'N, 101°03 'E >100 m

In Doi Phu Kha National Park, about 75 km northeast of
Nan. Take highway 1256 east from Pua town, following
signs to 'Doi Phu Kha National Park'. About 6 km before
park headquarters, turn south on a steep track down to
Ban Pha Rai. A well-developed 3 km trail leads from the
back of the school to the cave entrance close to the bottom of the valley.

Dunkley 1995 (NA19 or 20); RFD pamplet (Doi Phu Kha
National Park)

Pha Tub ผาตูบ 360

18°51 'N, 100°44 'E

Pha Tup

A cluster of at least 20 small caves in two prominent
limestone towers just west of highway 1080, about 10 km
north of Nan Town. The area has been developed into a
district park with an extensive system of surfaced trails.
There is a detailed map near the car park. See also Tham
Bor Nam Tip, Tham Phra, Tham Pha Sawan and Tham
Chedi Kaew.

Suwannabun 1981; Clarac & Pagau-Clarac 1985; Dunkley
1995 (NA4, 5, 6, 21-34); RFD pamplet

Bo Nam Tip บ่อน้ำทิพย์ 360

18°51 'N, 100°44 'E 100 m est.

In Tham Pha Tub District Park, on the west side of the
limestone outcrops, about 1.4 km from the car park. The
final 300 m is a steep scramble up an indistinct trail along
the base of the cliffs. There are at least two possible small
entrance windows hidden amongst the boulders but the
large window above the Buddha statue is inaccesible. The
sacred pond seems almost dried up and is littered with
plastic.

Suwannabun 1981; Clarac & Pagau-Clarac 1985; Dunkley
1995 (NA6)

Chedi Kaew เจดีย์แก้ว 360

18°51 'N, 100°44 'E 60 m est.

Chae Di Khao

In Tham Pha Tub District Park (p. 999), less than a hundred metres from the car park.

Dunkley 1995 (NA21)

Pha Sawan ผาสวรรค์ 360

18°51 'N, 100°44 'E 20 m est.

In Tham Pha Tub District Park, about 250 m west of the
car park on the north side of the main trail between the
two rock outcrops.

Dunkley 1995 (NA23)

Phra (Pha Tub) พระ 360

18°51 'N, 100°44 'E 120 m est.

Tham Kong, Pha Suwan

In Tham Pha Tub District Park, about 300 m west of the
car park.

Suwannabun 1981; Dunkley 1995 (NA5, 23, 24)

PHRAE

Pha Nang Koi ผานางคอย 362

18°22 'N, 100°21 'E 200 m est.

Phra Nang, Pha Nang Khoi, Princess Cave

Slightly north of the highway 101 near km post 168, about
4 km east of Rong Kwang, 34 km from Phrae and 84 km
from Nan. The signposted turn off leads after 800m to a
car park at the base of a flight of steps up to the cave
entrance. The main passage is electrically lit throughout.
A charity fair is held in the cave during Songkran festival
in April.

Suwannabun 1981; Dunkley 1995 (PA1 & 9)

GLOSSARY

active cave a cave which has a stream flowing through it or is beneath the water table and is still forming by the solution of bedrock.

adaption an inherited characteristic of an organism in structure, function or behaviour which makes it better able to survive and reproduce in a particular environment.

aggressive wate water capable of dissolving limestone, not saturated with calcium carbonate ($CaCO_3$)

allogenic water or sediment which has a source on nonkarst rocks (compare autogenic).

anastomoses tiny network of tubes and openings found in joints, faults and bedding planes.

anthodite tufts of crystalline aragonite radiating from a common centre, resulting in a spiky appearance. Needles of aragonite and small bristles growing on other speleothems are also called anthodites.

aquifer a water-bearing layer of rock, sand or gravel that yields water to a well or spring.

aragonite a mineral of calcium carbonate, chemically identical to calcite but of a different crystal structure and higher specific gravity. Aragonite crystals are orthorhombic in form, whereas calcite crystals are rhomohedral.

arthropod invertebrate animals belonging to the phyllum Arthropoda including insects, crustaceans, spiders, millipedes and centipedes. All arthro pods have jointed limbs and external skele tons. They are by far the most common group of animals inhabiting caves.

autogenic water or sediment that is derived from lime-stone or other karstic rocks (compare allogenic).

azurite a blue copper compound which in rare caes can form speleothems, such as the blue stalac-tite in Tham Pha Mon. Azurite is a recrystal-ized form of malachite, Cu_2 (CO_3)-(OH).

bacon a type of cave drapery with lighter and darker bands, resembling a strip of fatty bacon.

basalt a dark, hard igneous rock formed from volcanic lava.

bedding plane a surface in the rock that separates individual layers or beds of rock.

blind valley a valley that is closed abruptly at its lower end by a cliff or a slope facing up the valley. It may have a perennial or intermittent stream which sinks at its lower end into a cave or it may be a dry valley.

botryoid form small bead-or knob-like projections from cave walls, usually made of calcite. Cave popcorn and cave coral are examples of botryoid forms.

boxwork interlacing mineral blades projecting from cave walls. Boxwork can be composed of calcite, gypsum, limonite, silica or other minerals.

breakdown rock slabs, blocks or chips on the floor of a cave that have fallen from the walls or ceiling.

breccia angular fragments of rock and/or fossils cemented together or with a matrix of fine sediment or clay. Breccia sometimes contains fragments of bone or organic matter and canbe of great archeological importance.

calcite the commonest crystalline form of calcium carbonate ($CaCO_3$), the main constituent of limestone and most speleothems (compare aragonite).

calcium carbonate $CaCO_3$, the main mineral constituent of limestone which occurs in two chemically identical crystalline forms - calcite and aragonite.

canopy a projection of flowstone hanging above the floor of a cave, formed by erosion of sediments beneath the flowstone. Canopies look superfi-cially similar to shields but usually have pieces of sediment attached to their lower surface.

carbide calcium carbide, CaC2, a greyish mineral that reacts with water to produce calcium hydrox-ide and acetylene gas which burns with a bright white flame and is used for lighting.

carbide lamp a lamp that produces light by burning acetylene gas produced by the reaction between calcium carbide and water. Carbide lamps were the chief source of lighting in mines before the advent of electricity but they have largely been superseded by head torches.

carbonic acid $H2CO_3$, formed by the reaction between carbon dioxide and water, the most important acid involved in the process of cave formation.

cave underground space wide enough for a person to enter. The majority of caves are formed in limestone but they are also found in sandstone, basalt and other rocks.

cave breathing movement of air in and out of a cave entrance at regular intervals.

cave fill transported materials such as silt, clay sand and gravel which cover the bedrock floor or partially or wholly block some part of the cave.

cave system a network of interconnected cave passages and chambers, or a series of caves in the same area that used to be connected at one time.

cavern a large chamber within a cave or at its entrance

chert a hard and fragile rock composed of small silica crystals, often found in thin layers imbedded in limestone bedrock. Chert is usually dark brown in colour but can also be pale grey or reddish.

column a speleothem stretching from the floor to the ceiling of a cave, formed by the growth and joining of a stalactite and a stalagmite, or by the growth of either to meet the bedrock.

conduit an underground stream course completely filled with water and under hydrostatic pressure or a circular or elliptical passage inferred to have been such a stream course.

conulite calcite-lined holes drilled into sediment floors by dripping water. Sometimes the calcite lining extends above the top of the hole. Shallow, plate-like conulites are called 'bird baths' and form by splashing drips.

coral (cave) small speleothems consisting of short stalks with bulbous ends, usually occuring in patches and often smothering the surface of other speleothems. An alternative name for cave popcorn.

crawl   a constricted cave passage requiring a caver to traverse on hands and knees or to squeeze through on his back or stomach.

crystal pool   a pool of water lined with well-formed crystals, typically with little or no water flow.

curtain   carbonate deposit in a wavy or curved form, nor mally partially translucent and hanging down from the ceiling of the cave. Also called drapery.

dark zone   the part of a cave where daylight never reaches

dead cave   a non-active cave without flowing or dripping water (compare active cave).

decoration   cave features such as stalactites which are formed by secondary mineral precipitation, usually of calcite. In most cases, cave decoration has the same meaning as speleothem or cave formation.

dendritic   a branching pattern resembling a tree.

Devonian   the fourth period of the Palaeozoic Era lasting from 395-345 million years ago, between the Silurian and Carboniferous Periods. Only a small amount of limestone in Northern Thailand is currently thought to have been deposited during the Devonian Period.

dip   the slope of a bedding plane, expressed as the angle between a straight line along the bed and a horizontal line in the same direction. See strike.

dog-tooth spar   calcite with a sharp point, often found along the edge of cave drapery.

doline   a closed depression draining underground in karst landscape, of simple but variable form, eg: cylindrical, conical, bowl or dish-shaped. A doline can be anything from 2-3m to many hundred metres in diameter.

dolomite   calcium magnesium carbonate, $CaMg(CO_3)2$, similar to limestone but usually darker. Not found in Northern Thailand.

dome   a large hemispherical hollow in the roof of a cave, formed by breakdown, usually in mechanically weak rocks, which prevents bedding and joints dominating the form.

driphole   hole in the ground caused by dripping water.

dripline   a line on the ground at a cave entrance formed by drips from the rock above. Useful in cave survey to define the beginning of the cave.

dripstone   calcite deposit formed by dripping water from the ceilings or walls of the cave.

dry cave   an inactive cave with no water either flowing or dripping (compare active cave).

dry valley   a valley without a surface stream, common in karst landscapes and other areas with porous bedrocks.

eccentric   a speleothem of abnormal shape or attitude. see also Helictite ecology (cave) the study of the interaction between cave organisms and their environment.

epiphreatic   a cave formed at or just below the water table, spending time in both the phreatic and vadose zones. A common type of cave in Northern Thailand - examples include Tham Mae Lana, Tham Nam Lang & Tham Chiang Dao.

erosion   the wearing away of bedrock or sediment at the surface or in caves by mechanical or chemical actions of all moving agents, such as rivers. wind and glaciers.

eucladiolith   a type of stalactite found near cave entrances that bends towards the light, formed by the differential growth of plants on the light side only.

fault   a fracture separating two parts of a once continuous rock body, with relative movement along the line of the fault (compare joint).

fault cave   a cave developed along a fault line in the bedrock, formed by erosion of water exploiting the line of weakness or by a gap appearing between rocks physically moving apart; usually a combination of both.

fissure   narrow crack in a rock, soil or ground material.

fissure cave   a cave developed along a narrow joint in the rock, similar to a fault cave but on a smaller scale.

flagging tape   thin plastic ribbon of any colour about 5 cm wide used for marking survey stations, trails etc.

flower (cave)   a rare and very fragile speleothem made of gypsum with delicate curling forms growing from a central point.

flowstone   a very common type of speleothem deposit formed from thin films or trickles of water flowing over cave floors or walls, usually composed of calcite (see also travertine).

formation   a rock unit with distinct characteristics within a sequence of rocks. Also used to mean any mineral deposits within a cave, essentially with the same meaning as speleothem or cave decoration. A confusing term which is best avoided.

fossil   the remains or traces of animals or plants preserved in rocks or sediments.

frostwork   an anthodite speleothem composed of aragonite with a radiating, needle-like form.

gour   a deposit formed by precipitation from water flowing over the rim of a pool. An alternative name for rimstone.

gour pool   a pool held up by a rimstone dam. Rimstone-pools are frequently in tiers and usually have a curved shape with the width about half the length. An alternative name for rimstone pool.

grotto   a room in a cave of moderate dimensions but richly decorated with speleothems.

guano   a material found on the floor of bat or bird caves made up principally of their excrement but also including rock fragments, animal skeletal material and products of reactions between excretions and rock. Guano is an excellent agricultural fertilizer.

gypsum   an unusual cave mineral composed of hydrated calcium sulphate, $CaSO_4.2H_2O$, usually associated with hydrothermal activity.

helictite   an erratic form of speleothem which at one or more stages of its growth changes its axis from the vertical to give a curving or angular form. Helictites differ from anthodites which are composed of aragonite in that they have a central canal.

Holocene   the second and most recent epoch of the Quaternary Period, starting 10,000 years ago and continuing up to the present. Some authorities subdivide the epoch into three sections - Lower Holocene (10,000-7,500 BP), Middle Holocene (7,500-2,500 BP) and Late Holocene (2,500 BP up to the present).

hurricane lamp a type of pressure lamp that uses paraffin as a fuel.

hydrology the scientific study of the nature, distribution and behaviour of water.

hypogenic cave a cave that forms as a result of water rising upwards from deep within the earth, often as a result of hydrothermal activity.

igneous a type of rock formed by volcanic activity, such as basalt or granite.

inactive cave a dry cave above the water table with no stream flow and no water dripping from the ceiling. Most of the famous tourist and temple caves in Northern Thailand are inactive caves.

inclination the angle that a cave passage forms to the horizontal. Also used to refer to the angle of the bedplanes in the bedrock itself.

inflow cave a cave into which a stream enters or is known to have entered but which cannot be followed downstream to the surface.

joint a fracture in a series of bedrock units, generally at an angle to the bedding planes (compare fault).

karren the minor forms of karst as a result of solution of rock on the surface or underground.

karst a landscape where the topography is formed by the dissolving of bedrock. Karst landscapes are usually formed on limestone and are characterized by dolines, dry valleys, subterranean drainage and caves.

karst window a closed depression which has a stream flowing across its bottom, formed by cave breakdown.

laminar flow smooth flow of water or other liquids where the sides and the centre of the flow move at the same speed-ie. there is little mixing and no turbulence.

limestone a hard grey or bluish-grey sedimentary rock composed of calcium carbonate, $CaCO_3$, formed from the shells of marine animals

manganese a common metal, Mn, which forms blackish coatings over cave sediments. Manganese coatings may be the result of bacterial activity.

marble a metamorphosed form of limestone that has been subjected to intense heat and pressure.

maze cave a cave formed along a close network of frac tures and joints.

meander a curve in the course of a river or stream as a result of sideways erosion.

metamorphic a hard rock type formed by the transformation of softer rocks under intense heat and/or pressure (compare sedimentary, igneous).

micro-cave a tiny water-filled crack in limestone bedrock, the potential precursor of a true cave (compareproto-cave).

moonmilk a white, semi-liquid material which flows down cave walls or drips onto speleothems or the cave floor. Moonmilk is usually composed of the mineral hydromagnesite and may be formedby bacteria and other organic processes.

mud stalagmite a cone shaped stalagmite made from mud mixed with calcite, resembling a small volcano. Vulcanites are rare in Northern Thailand but there are good examples in Tham Pha Mon. Sometimes erroneously called "vulcanite".

network a complex pattern of repeatedly connecting passages in a cave system.

oolite see pearl, cave.

Ordovician the second period of the Palaeozoic Era lasting from 500-435 million years ago, between the Cambrian and the Silurian Periods.

outflow cave a cave from which a stream flows or formerly did so and which cannot be followedupstream to the surface.

paragenetic an unusual type of cave which is formed from the bottom upwards as the floor of the cave is progressively filled with sediment, forcing the stream upwards. Tham Klaeb Yai is a classic example of a paragenetic cave (compare phreatic, vadose).

passage a cavity which is much longer than it is wide or high and may join larger cavities or chambers.

pearl (cave) or oolite a type of speleothem found in gour pools or other shallow hollows into which water drips, formed around a grain of sand or small stone which becomes coated with concentric layers of calcite. Cave pearls are typically roughly spherical with a smooth surface and vary in size from a few mm to the size of an apple.

pendants smooth bedrock pillars hanging from the cave ceiling or from ledges. They are part of the original bedrockmaterial left behind after solution and are characteristic of phreatic and paragenetic caves.

pendulite a kind of stalactite which has been partly sub merged and the submerged part covered with dog-tooth spar to give the appear of a drum stick.

percolation water water moving mainly downwards through pores, cracks and tight fissures in the vadose zone above the water table.

permeability the property of rock or soil permitting water to pass through it. Primary permeability depends on interconnecting pores between the grains of the material. Secondary permeability depends on solutional widening of joints, bedding planes and other cavities in the rock.

Permian the last period of the Palaeozoic Era lasting from 280-225 million years ago, between the Carboniferous and Triassic Periods. Most of the limestone in Northern Thailand was deposited during this period, including Doi Chiang Dao and Pang Ma Pha.

pH a measure of the acidity of water and other liquids. A pH of less than 7.0 is called acidic whereas a pH of greater than 7.0 is called alkaline.

phreatic cave a cave which was formed beneath the water table, characterized by smooth walled passages, even shape & large scallop marks (compare vadose).

phreatic zone the zone where voids in the rock are completely filled with water, below the water table.

pillar a column-like structure made of bedrock from the roof to the floor left by the removal of the surrounding rock (compare column).

pitch a vertical or nearly vertical part of a cave for which ladders or ropes are normally used for ascent or descent.

Pleistocene the first epoch of the Quaternary Period, 1.8 million-10,000 years ago. Some authorities subdivide the epoch into three sections - Lower Pleistocene (1.8 million-730,000 BP),

Middle Pleistocene (730,000-125,000 BP) and Late Pleistocene (125,000-10,000 BP).

plunge pool a large circular hole, occuring at the foot of a waterfall or rapid, either on the surface or under ground in a cave.

polje a type of doline with a large, flat floor and steep sides.

popcorn a botryoidal form of speleothem resembling popcorn, caused by water droplets condensing onto cold surfaces. Popcorn often grows on the surface of other speleothems and is particularly common in draughty places (= Cave coral?).

Pothole a vertical or nearly vertical shaft of chimney open to the surface.

Precambrian the oldest geological period, ending with the commencement of the Cambrian Period 600 million years ago. Small areas of marble in the south of Chiang Mai Province are of Precambrian origin.

proto-cave a small cavity in the bedrock which has the realistic potential to grow and become a cave (compare micro-cave).

pseudokarst a landscape that resembles karst but which was not formed by solution.

quartz the main mineral constituent of sandstone, composed chiefly of silica.

Quaternary the most recent geological time period, starting 1.8 million years ago and divided into two epochs - the Pleistocene (1.8 million-10,000 years BP) and the Holocene (10,000 years BP up to the present).

racer snake elaphne taeniura, a harmless snake often found in caves, with a black & white striped tail, a dark eye stripe and often with a yellow band down the side of its body.

resurgence a spring where an underground stream which has a course on the surface higher up reappears at the surface.

rift a long narrow high and straight cave passage controlled by planes of weakness in the rock. Compare fissure.

rimstone a deposit formed by precipitation from water flowing over the rim of a pool. An alternative name for gour rimstone dam a wall-shaped calcite deposit that holds or used to hold a pool of water.

rimstone pool a pool held up by a rimstone dam. Rimstone pools are frequently in tiers and usually have a curved shape with the width about half the length. An alternative name for gour pool.

rising see spring rock shelter. A cave with a more or less level floor reaching only a short way into the hillside or under a fallen block so that no part is beyond the daylight. Rock shelters in Northern Thailand are often prehistoric human habitation sites and may be important for archeological research.

saturated water which has dissolved as much limestone as it can under normal conditions.

scallops oval-shaped hollows formed on cave walls and stream beds by flowing water. Scallops are usually curved on the upstream side and slightly pointed on the downstream side. Fast-moving water creates smaller scallops, whereas slow-moving water creates large ones.

scavenger an animal that eats dead remains and wastes of other animals and plants.

sediment material deposited or precipitated from water, wind or other agents.

sedimentary a class of rocks formed by the deposition of sediments, for example limestone, shale and sandstone (compare igneous, metamorphic).

shaft a vertical or nearly vertical cave passage.

shield a plate or slab of calcite that juts out from the cave wall at an angle, consisting of two parallel plates dividedby a very thin crack from which water under slight pressure enters the cave show cave (tourist cave) a cave that is open to the general public and does not need specialist equipment.

Silurian the third period of the Palaeozoic Era lasting from 435-395 million years ago, between the Ordovician and the Devonian Periods. Only a small amount of limestone in Northern Thailand is currently thought to have been deposited during the Silurian Period.

sink hole a depression in the ground caused by the solution of the underlying rock or the collapse of the roof of an underlying cavern (different meaning in USA?).

sinking stream a stream that disappears underground, usually in a depression. compare swallow hole?

siphon (syphon) a water-filled passage of inverted U-pro file which delivers a flow of water whenever the head of water upstream rises above the top of the inverted U. Sometimes used more generally to refer to any water filled passage (ie: a sump) soda straw (calcite straw) a thin hollow stalactite resembling a sipping straw. It grows from the tip by water flowing down the inside of the tube and consists of a single calcite crystal.

solubility the ease with which a solid dissolves in water.

speleogenesis the processes by which caves form. speleologist a person who studies caves.

speleology the exploration, description and scientific study of caves and related phenomena.

speleothem a secondary mineral deposit formed in caves, usually of calcite. Stalactites, stalagmites, flowstone, gour pools, cave pearls and many other cave features are all speleothems. The terms "cave formation" and "cave decoration" are often used to refer to speleothems in general.

spelunker a term usually used by non-cavers to mean a caver. Originally coined as a term to describe cavers but it fell out of favour with cavers as it was picked up by the media to describe all kinds of people who went into caves.

spongework a highly complicated system of tiny holes, tubes and interconnected cavities found in cave walls, resembling swiss cheese.

spring the place where an underground stream reaches the surface, indicating the level of the water table. A resurgence is a spring where the stream has a section above the surface further upstream before sinking into the ground.

squeeze a very tight narrow passageway (compare crawl)

stalactite a secondary deposit of calcium carbonate in a cylindrical or conical shape, hanging from the cave ceiling or ledges, resembling icicles.

stalagmite a secondary cave deposit in a cylindrical or

conical shape, rising up from the cave floor or a ledge, formed by precipitation of calcium carbonate from water drips (see also Mud Stalagmite).

straw a long, thin-walled tubular stalactite less than 2cm in diameter. It grows from the tip by water flowing down the inside of the tube and consists of a single calcite crystal. Also referred to as calcite straw or soda straw.

stream sink the point at which a surface stream disappears underground. An inflow cave is a stream sink that can be entered by people.

strike a horizontal line on a bedding plane surface at right angles to the direction of dip at any point. On level ground the strike is the direction of rock outcrop of inclined beds.

sump a place where the ceiling of a cave passage drops below the water level, leaving no air space, with the cave passage continuing under water (compare siphon).

supersaturated referring to water that has more limestone in solution than the maximum corresponding to normal conditions.

swallow hole a place where a stream sinks into the ground in limestone terrain, usually in a closed depression (an alternative name for stream sink)

swirlhole a hole in rock in a stream bed eroded by eddying water, often with sand or pebbles.

talus a sloping mass of rock, dirt and debris at the base of a drop. A pile of jumbled rocks.

Tertiary the first period of the Cenozoic Era, starting 65 million years ago and finishing 1.8 million years ago at the commencement of the Quaternary Period. The Tertiary is the time period when mammals came to prominence.

through cave a cave which may be followed from entrance to exit along a stream course or along a passage which formerlycarried a stream.

travertine calcium carbonate which is deposited from streams or other ground water sources in a series of flowstone dams. Can also refer to any flow stone or dripstone deposit.

Triassic the first period of the Mesozoic Era, 225-190 million years ago. The Triassic is the beginning of the age of the Dinosaurs and is the most recent geological period from which there are limestone deposits in northern Thailand.

troglobite (troglobyte) an animal that is fully adapted to life in total darkness and is unable to live outside the cave environment troglophile an animal that may live underground but may also be found on the surface. Troglophiles sometimes complete their entire lifecycle in a cave but they are not confined to this habitat tower (tower karst) feature of karst landscapes in which the residual hills have very steep to over hanging lower slopes. There may be alluvial plains between the karst towers and flat floored depressions within them.

trogloxene an animal which spends only part of its life cycle in caves and returns periodically to the surface for food.

tube a cave passage which is elliptical or nearly circular in cross-section and has a smooth surface.

tufa spongy or vesicular calcium carbonate deposited from spring, river or lake waters.

twilight zone the part of the cave to which at least some daylight penetrates.

uvala a complex closed depression with several lesser depressions within its rim.

vadose cave a cave formed above the water table, typically with a narrow cross-section like a canyon and a steep descent (compare phreatic).

vadose flow water flowing in free-surface streams in caves

vadose zone the zone above the water table where the spaces in the bedrock are partially filled with air and through which water descends under gravity virgin passage a cave passage or an entire cave that has not previously been entered; a new discovery.

vulcanite see Mud Stalagmite.

water table the top or highest level of ground water in a given area; below this level cave passages may be flooded (see vadose, phreatic).

window (cave) a hole in the roof of a cave through which daylight enters (see also karst window).

BIBLIOGRAPHY

Anon.
1972   Story of Chieng Dao cave.
1982   Survey of geological and mineral resources in Mae Hong Son province, northern Thailand; Krom Sapphayakon Thoranu, Bangkok.
1986   Tham Phiman (Spirit Cave) at Mae Hong Son, Thailand; Look East, pp. 48-51.
1992   Tham Luang Chiang Dao, Chiang Mai; Chiang Dao Municipality.
1999   (Management report for Tham Pha Mon - in Thai); Tourism Authority of Thailand.

Allen, J. et al. (ed.)
1977   Sunda and Sahul. Prehistoric Studies in Southeast Asia, Melanesia and Australia; Academic Press, London.

Anderson, B.
1979   Caving with Buddha; N. S. S. NEWS, 37 (2).
1987   Earliest Inhabitants of Thailand; Siam Society.
        Anderson, S.
1999   Tham Mae Lana: values and vulnerability. Paper in Pang Ma Pha cave project interim report; Faculty of Environment & Resource Studies, Mahidol University, Bangkok. Bellwoods, P.
1997   Prehistory of the Indo-Malaysian Archipelago; University of Hawaii Press, Honolulu.

Bhamarabutr, A.
1980   The Story of Chiang Mai. Bidyalankarana, H. H.Prince.
1935   The Buddha's Footprints; J. Siam Soc., XXVIII-XXIX pp. 1-1.

Bloom, A. L.
1979   Geomorphology: A systematic analysis of late cenozoic landforms; New Delhi.

Bock, C.
1884   Temples and Elephants: The narrative of a journey of Exploration through Upper Siam and Laos; Sampson Low, Marston, Searle Rivington London.

Boland, K.
1989   Under the Triangle-Thailand '88; Nargun, 21 (10): 90-95.
1990   Thailand 1990; Nargun, 23 (part 1 in no. 2 2 in no. 5).
1992a   The Exploration of Tham Susa, Nam Khong Basin, North-West Thailand; International Caver, vol 3: 36-40.
1992b   North-West Thailand: The story so far; International Caver, Vol. 5: 30-35.

Chapman, P.
1993   Caves and Cave Life, Harper Collins, London.

Charoenwongsa, P. (ed.)
1988   Muang Mae Moh Obluang and Ban Yang Thong Tai; FAD, Bangkok.

Clarac, A. & Pagau-Clarac, H.
1985   Thailande: Guide Touristique; DK Book House, Bangkok.

Dasher, G. R.
        On Station: A Complete Handbook for Surveying & Mapping Caves; National Speleological Society, Alabama, USA.

Deharveng, L. & Gouze, A.
1983   Grottes et Karsts des Environs de Chieng Mai (Thailande); Karstologia, vol. 2 (2): 55-60.
1986   Thai-Maros 85: Rapport Speleologique et Scientifique; Assoc. Pyreneenne de Speleologie, Toulouse.
1987   Expedition Thai-Maros 86: Rapport Speleologique Scientifique; Assoc. Pyreneenne de Speleologie, Toulouse.
1988   Expedition Thai '87-Thai '88: Rapport Speleologiqueet Scientifique; Assoc. Pyreneenne de Speleologie, Toulouse.
1993   Ten new species of *Troglopedetes* Absolon 1907 from Caves of Thailand (Collembola, Paronellidae). Bijdragen totde Dierkunde, 63(2), p. 103-111, SPB Academic Publishing, Holland.

Deuve, T.
1988   Coleopteres Caraboidea des Milieux souterrains de l'Asia du Sud Est. Expeditions de l'APS en Asie du Sud Est, travaux scientifiques 1 (1988) 17- 20 p. 19 (drawing of Ozaenaphaenops beetles).

Dilokwanich et al.
2000   An Exploration and Database System of Caves: Mae Hong Son Province; Faculty of Environment & Resource Studies, Mahidol University, Bangkok, 6 vols.

Dilokwanich, S.
2000   Objectives & Previous Research. Vol. 1 in "An Exploration and Database System of Caves: Mae Hong Son Province"; Faculty of Environment & Resource Studies, Mahidol University, Bangkok.

Dunkley, J. R.
1983   Chiang Dao Cave - N. Thailand; A. S. F. Newsletter, 99.
1984   Thailand Caves Catalogue; Speleological Research Council, Sydney.
1985   Karst and caves of the Nam Lang-Nam Khong Region, Northwest Thailand; Helicite, 23 (1): 3-22.
1995   The Caves of Thailand; Speleological Research Council, Sydney.
1998   Caves of Thailand - Addendum 1995-1997; Speleological Research Council, Sydney.

Dunkley, J. R. & Brush, J. B. (ed.)
1986   Caves of north-west Thailand; Speleological Research Council, Sydney. Exley, S.
1994   Caverns Measureless to Man, Cave Books, St Louis, USA.

Faniran, A. & Jeje, L. L.
1983   Humid Tropical Geomorphology; London. Farr, M.
2000   The Darkness Beckons, Diadem Books, London.

Fix, A.
        Foragers, Farmers & Traders in the Malayan Peninsula: Origins of cultural & biological diversity.

Flenley, J. R.
1984   Late Quaternary Changes of Vegetation and Climate in the Malesian Mountains; Erdwissen Schaftliche Forschung, 18. pp. 261-267.

Ford, T. D. & Cullingford, C. H. D
1976   The Science of Speleology; Academic Press, London, UK.

Ford D. & Williams, P.
1989   Karst Geomorphology and Hydrology, Chapman & Hall, London.

France, S.
1988   Slave unit design; Caves and Caving, 42, pp. 2-4.

Gillieson, D.
1996   Caves: Processes, Development & Management; Blackwell Publishers, Oxford, UK.

Gorman, C.F.
1970   Prehistoric Research in North Thailand: A Culture-Chronographic Sequence from the Late Pleistocene to the Early Recent Period; Ph.D. Dissertation, Department of Anthropology, University of Hawaii (unpubl.).
1971   The Hoabhinian and after: Subsistence Patterns in S.E.Asia during the late Pleistocene and early Recent Periods; World Archeology, vol. 2(3): 300-320.
1972   Excavations at Spirit Cave, North Thailand: some interim explanations.; Asian Perspectives, Vol. 13: 80-107.
1977   A priori models and Thai prehistory: a reconsideration of the beginnings of agriculture in Southeast Asia; In Reed, C.A. (ed.), Origins of Agriculture, pp. 322-55. Mouton, The Hague.

Grave, P. et al.
1994   Dating the Log Coffins of Northwest Thailand; paper

presented at IPPA Conference, Chiangmai 5-12 Jan 1994.

Gray, I. D; Piprell, C. & Graham, M.
1991 National Parks of Thailand; Communication Resources, Bangkok. Hair, W.
1999 Notes from the Underground; Sawasdee Magazine (Thai Airways International), 28 (4).

Hallett, H. S.
1890 A Thousand Miles on an Elephant in the Shan States; Blackwood & Sons, London, 484 pp.

Hastings, P. & Leingsakul, M.
1973 Chronology of the late Quaternary Climatic Changes in Thailand; Proceedings of the First Symposium of Geomorphology & Quarternary Geology, Bankok, pp. 24-34.

Hawkins-Salt, J. R. G.
1991 Caving in northern Thailand; Speleopod, vol. 2.

Heaney, T.
1991 Climatic and vegetational change in Southeast Asia; Climatic Change, 19: 53-60.

Henley, T.
1996? Reefs to Rainforests – Mangroves to Mountains, Dawn of Happiness Resort Co, Krabi.

Hess, A. & Koch, K. E.
1979 Geological Map of Northern Thailand 1:250,000; Federal Institute for Geosciences and Natural Resources, Hannover, Germany.

Higham, C.
1989 The Archeology of Mainland Southeast Asia from 10000 BC to the fall of Angkor; Cambridge University Press, UK.
1996 The Bronze Age of Southeast Asia; Cambridge Univeristy Press, Cambridge.
2002 Early Cultures of Mainland Southeast Asia; River Books, Bangkok.

Higham, C. & Lu, T.
1998 The origins and dispersal of rice cultivation; Antiquity, vol. 72: 867-77.

Higham, C. & Thosarat, R.
1998 Prehistoric Thailand; River Books, Bangkok.

Hill, C. & Forti, P.
1997 Cave Minerals of the World; National Speleological Society, Alabama, USA, (2nd ed.).

Howes, C.
1987 Cave Photography: A practical guide; Caving Supplies, Buxton, Derbyshire, UK, 68pp.
1989 To Photograph Darkness: the History of Underground & Flash Photography; Alan Sutton, Gloucester, UK, 22 + 330pp illus.
1997 Images Below: A manual of underground and flash photography; Wild Places Publishing, Cardiff.

Jansen, E. R.
2002 The Book of Buddhas: Ritual Symbolism used on Buddhist Statuary and Ritual Objects; New Age books, New Delhi.

Jennings, J. N.
1985 Cave and Karst Terminology; in P. G. Mattews (ed), Australian Karst Index 1985, ASF Broadway.

Johnson, A. and Earle, T.
1991 The Evolution of Human Societies; Stanford University Press, Stanford.

Jones, C.
1995 Some Cave Terminology Explanations and Thai Translations; unpubl.
1996 Management of Tourism Impact within Caves in Thailand's National Parks; Tiger Paper, 23: 4, pp. 14-17.

Judson, D. (ed.)
1995 Caving Practice & Equipment; British Caving Research Association.

Keesing, R. M.
1981 Cultural Anthropology: A Contemporary Perspective;

HRW International Editions, CBS Publishing Asia Ltd., 2nd ed.

Kiernan, K.
1986a Karst Development in the Pang Kham-Pha Daeng Area; in Dunkley & Brush, Caves of North West Thailand.
1986b Prehistoric caves of the Nam Khong-Nam Lang area; in Dunkley & Brush, Caves of North West Thailand, p. 40-49.
1987 Cave management: values and management of karst resources; Australian Ranger Bulletin, 4 (3).
1988a Geomorphology of a tropical Intermontane basin in the Sino-Burman Ranges; 26th International Geographical Congress, Sydney.
1988b Mountain Karst of North-west Thailand; Proc. International Symposium on Speleology, Liuzhou, China.
1988c Prehistoric Mountain dwellers of north-west Thailand; Asia Wise, Oct: 3.
1990 Some Limestone Caves North-East of Mae Hong Son, Northern Thailand; Nat. Hist. Bull. Siam. Soc., Vol. 38: 59-67.
1991 Geomorphological evidence for quaternary climatic change in the lower Sino-Burman ranges; Singapore J. Tropical Geog., vol. 12 (2): 112-123.
1991 Tropical mountain geomorphology and landscape evolution in north-west Thailand; SZ fur Geomorphol., vol. 35 (2): 187-206.

Kiernan, K.; Spies, J. & Dunkley, J.
1988 Prehistoric occupation and burial sites in the mountains of the Nam Khong area, northwestern Thailand; Austr. Archeol., vol. 27: 24-44.

Kijngam, A.
1988 Archeological Sites of Thailand; Division of Archeology, Fine Arts Department, vol. 1, no. 6/2531.

Kottelat, M.
1988 Two Species of cave fishes from northern Thailand in the genera Nemacheilus and Homaloptera (Osteich-thyes: Homalopteridae); Rec. Austr. Museum, vol. 40: 225-231.

Kulparadit, K. & Spies, J.
2000 Database & Management. Vol. 6 in "An Exploration and Database System of Caves: Mae Hong Son Province" (in Thai); Faculty of Environment & Resource Studies, Mahidol University, Bangkok.

Kusch von, Heinrich
1975 Hohlen in Laos, Nord-und West Thailand; Die Hohle, 26(4): 114-123.

Lekagul, B. and McNeely, J. A.
1988 Mammals of Thailand; Bangkok, Sala Karn Bhaet Co., Ltd.

Lhote, H.
1987 Oasis of Art in the Sahara; National Geographic, vol. 172, no. 2.

Loofs, H. H. E.
1967 Elements of the Megalithic Complex in Southeast Asia; Australian National University Press, Canberra.

Lowe D. & Waltham, T.
1995 A Dictionary of Karst and Caves, British Cave Research Association, UK.

Maffres, L.
1988 Expedition "Thai88"; Spelunca, 30. Maffres, L. & Parma, C.
1988 Expedition Polonaise "Chiang Dao 87"; Spelunca, 30.

Maffres, L. & Rigal, D.
1986 Le Karst de la Nam Lang; in Thai-Maros 85, Dehvarenget al., pp. 38-50.

Maleipan, V.
1992 Old Stone Age tools at Chiang Saen; Archeology Journal, vol. 4: 35-43.

Mallory, B. F. & Cargo, D. N.
1979 Physical Geology; New York, McGraw-Hill Book Co.

Manahan, S.E.
1984   Environmental Chemistry; Brooks/Cole Publishing Co., Monterey.

Matics, K. I.
2001   Gestures of the Buddha; Chulalongkorn University Press, 2nd ed.

Meredith, M. & Wooldridge, J.
1992   Giant Caves of Borneo; Tropical Press Sdn Bhd, Kuala Lumphur.

Millar, I. R. & Wilde, K. A.
1989   General Policy and Guidelines for Cave and Karst Management.; Department of Conservation, Wellington, N. Z.

Moore, G. W. & Sullivan, N.
1997   Speleology: Caves & the Cave Environment; Cave Books, St. Louis, USA.
1978   The Study of Caves; St. Louis, U. S. A.

Mouhot, H.
1864   Travels in the Central parts of Indo-China (Siam), Cambodia and Laos) during the years 1858, 1859 and 1860; John Murray, London.

Mulvaney, J. & Kamminga, J.
1999   Prehistory of Australia; Smithsonian Institution Press.

Munier, C.
1998   Sacred Rocks and Buddhist Caves in Thailand; White Lotus, Bangkok.

Nicholl, C.
1988   Borderlines: A Journey in Thailand and Burma; Viking, New York, 238 pp.

Odel, B. & Odel, C. W.
1984   Karstformer i Thailand; Grottan, 19: 3-14.

Olivier, G.
1969   Practical Anthropology; Charles C. Thomas, Illinois.

Osborne, M.
        River Road to China: The Mekong River Expedition 1866-1873; Liveright, New York.

Pearn, B. R.
1963   An Introduction to the History of Southeast Asia; Longmans of Malaya, Kuala Lumphur.

Penth, H.
1974a   A Note on the History of Wat Umong Thera Jan; J. Siam Soc., 62, pp. 269-274.
1974b   Kunst im Lan Na Tha die Glocke in der Hohle bei Chiang Dao (Art in Lan Na Thai. The bell in Chiang Dao cave); Artibus Asiae, Ascona, XXXVII pp. 61-74.
1994   A Brief History of Lan Na Civilizations of Northern Thailand; Silkworm Books.

Penth, H. (ed.)
1993   History of Phra That Doi Tung; Mae Fah Luang Foundation, Bangkok/Chiang Rai.

Pookajorn, S. et al.
1992   The Phi Tong Luang (Mlabri): A hunter-gatherer group in Thailand; O. S. Printing House, Bangkok.

Pope, G. G. et al.
1985   Taxonomy, dating and paleoenvironment: The paleoecology of the early Far Eastern hominids; Modern Quaternary Research in Southeast Asia, 9: 65-80.
1980   Palaeanthropological Investigations of the Thai-American Expedition in Northern Thailand (1978-1980): An interim report; Asain Perspectives, vol. 21(2): 147-163.

Prishanchit, S.; Santoni, M. & Pautreau, J.-P.
1988   Ob Luang. Report on survey and exploration in 1985; In Charoenwongsa, P. (ed.), Muang Mae Moh Obluang and Ban Yaang Thong Tai, FAD, Bangkok, pp. 36-68.

Prosser, J. & Grey, H.
1992   National Speleological Society Cave-Diving Manual, National Speleological Society, USA.

Putman, J. J.
1988   The Search for Modern Humans; National Geographic, 174, No. 4, pp. 439-477.

Rahula, W.
1999   What the Buddha Taught; Haw Trai Foundation, Bangkok.

Rea, G. T. (ed.)
1992   Caving Basics: A comprehensive guide for beginning cavers; National Speleological Society, Alabama, USA, 3rd ed.

Renfrew, C. and P. Bahn.
1991   Archaeology : Theory And Methods & Practice; Thames & Hudson Ltd., London.

Reynolds, T. E. G.
1990   The Hoabinhian: A Review; in Barnes, G. L. (ed.), Bibliographic Review of Far Eastern Archeology, Oxbow Books, Oxford, pp. 1-30.
1990   Problems in the Stone Age in Thailand; J. Siam Soc., vol 78 (1): 109-114.
1992   Excavations at Banyan Valley Cave, Northern Thailand: A report on the 1972 Season; Asian Perspectives, 31: 77-97.

Romero, A. (ed.)
2001   The Biology of Hypogean Fishes; Kluwer Academic Publishers.

Rowland, J. J.
1995   Photographing caves. Chapter 11 in Caving Practice and Equipment, Judson, D. (ed.); British Cave Research Association.

Santoni, M.; Pautreau, J.-P. & Prishanchit, S.
1986   Excavations at Ob Luang, Province of Chiang Mai, Thailand; in Glover, I. C & Glover, E. (ed.). Southeast AsianArcheology 1986. British Archeological Report (International Series), 561: 37-54.

Sarasin, F.
1933   Prehistorical researches in Siam; J. Siam Soc, vol. 26 (2): 171-202.

Seidenfaden, E.
1939   Antiquities on Doi Suthep; J. Siam Soc., XXXI, pp. 37-43.

Shoocondej, R.
1996   Working towards an anthropological perspective on Thai prehistory; BIPPA, vol. 14: 119-32.

Shoocondej, R. et al.
2000   Archeology. Vol. 4 in "An Exploration and Database System of Caves: Mae Hong Son Province" (in Thai); Faculty of Environment & Resource Studies, Mahidol University, Bangkok.

Siripornpibul, C.
2000   Geology & Hydrology. Vol. 2 in "An Exploration and Database System of Caves: Mae Hong Son Province" (in Thai); Faculty of Environment & Resource Studies, Mahidol University, Bangkok.

Smart, D.
1994a   (Tiger/Wind survey).
1994b   Down and out in Thailand; Descent, 119: 34-36.

Solheim, W. G.
1970   Northern Thailand, Southeast Asia, and World prehistory; AP, vol. 13: 145-62.
1972   An earlier agricultural revolution; Scientific American, vol. CCVI (4): 34-41.

Spate, A. (ed.)
1987   Cave Management; Australian Ranger Bulletin, 4 (3).

Spies, J.
1987   Beneath the Golden Triangle; Sawasdee, 16 (3): 34-41.
1997   The values and vulnerability of the caves and karst resources in Mae Hong Son Province, Thailand; Journal of Ecology, 24 (3): 41-48.
2000   Suggestions for cave management, in "An Exploration and Database System of Caves: Mae Hong Son Province, vol. 6"; Faculty of Environment & Resource Studies, Mahidol University, Bangkok.

Srisuchat, A.
1992   Rock Art of the Historic Period in Thailand: part 2 (in Thai, with English summary); Fine Arts Department, Bangkok.

Stoddard, S.
19??   An Introduction to Cave Photography; BCRA, Cave studies series No. 4.

Stone, F. D.
1983   Biogeography and Ecology of the Caves of Thailand: Preliminary Observations; N. S. S. Bull., 45(1).

Suwannabun
1981   Interesting caves of Thailand. (in Thai)

Swearer, D. K.
1974   Myth, Legend andHistory in the Northern Thai Chronicles; J. Siam Soc., 62, pp. 67-88.

Tattersall, I.
2000   Once We Were Not Alone; Scientific American, Jan, pp. 38-44.

Thikakhayano
History & Guide Book to Chiang Dao Caves (Tham Luang Chiang Dao); Santipap Printing, Chiang Mai.

Trakullertsathien, C.
2003   Welcome back; Bangkok Post, Oct. 16 (outlook).

Valli E. & Summers, D.
1990   Shadow Hunters, Sun Tree Publishing, Singapore.

Warren, W.
1985   Thailand's Beautiful caves; Sawasdee, vol 14 (10): 15-21.

Watson et al.
1997   Guidelines for Cave and Karst Protection; IUCN. Wilkens, H.; Culve, D.r & W Humphreys (Eds) 2000, Subterranean Ecosystems, Elsevier Science, Amsterdam.

Windecker, R. C. et al.
1975   Map of Chiang Dao Cave; Nat. Hist Bull. Siam Soc., vol 26: 1-10.

Wyatt, D. K.
1984   Thailand: A Short History; Yale University Press.

Yen, D. E.
1977   Hoabinhian horticulture: the evidence and questions from Northwest Thailand; in Allen, J. et al. (ed.) Sunda and Sahul. Prehistoric Studies in Southeast Asia, Melanesia and Australia. Academic Press, London.

Shutter Photography
2545   Flash เทคนิคการใช้แฟลชถ่ายภาพ. กรุงเทพฯ: สำนักพิมพ์อิมเมจโฟกัส.

กรมศิลปากร
2522   ภาพเขียนสีสมัยก่อนประวัติศาสตร์ในประเทศไทย. กรุงเทพฯ: โรงพิมพ์การศาสนา.
2530   โบราณคดีสีภาค. กรุงเทพฯ: โรงพิมพ์หัตถศิลป์.
2531   แหล่งโบราณคดีประเทศไทย เล่ม1.กรุงเทพฯ: โรงพิมพ์ชุมนุมสหกรณ์การเกษตรแห่งประเทศไทย.

คณะสิ่งแวดล้อมและทรัพยากรศาสตร์
2532   การจัดทำแผนปฏิบัติการการอนุรักษ์ธรรมชาติเฉพาะแหล่ง: ถ้ำเมืองออน อำเภอสันกำแพง จังหวัดเชียงใหม่; สำนักงานสิ่งแวดล้อมแห่งชาติ กระทรวงวิทยาศาสตร์ เทคโนโลยีและสิ่งแวดล้อม.

จอห์น สปีส์, ภาณุ บุรุษรัตนพันธุ์, วิวัฒน์ พันธุวุฒิยานนท์
2537   ถ้ำ อาณาจักรมหัศจรรย์ใต้พิภพ. สารคดี, 110(10): 88-114.

ฉันท์ สุวรรณบุณย์
2524   ถ้ำที่น่าสนใจในเมืองไทย: กรมวิชาการกระทรวงศึกษาธิการ.

ชัยพร ศิริพรไพบูลย์
2542   ลักษณะทางธรณีวิทยาและอุทกวิทยา: รายงานความหน้าหน้าครั้งที่ 2 โครงการการสำรวจและจัดทำฐานข้อมูลเกี่ยวกับถ้ำจังหวัดแม่ฮ่องสอน. หน้า29-67. คณะสิ่งแวดล้อมและทรัพยากรศาสตร์ มหาวิทยาลัยมหิดล.
2543   รวมคำศัพท์เรื่องถ้ำและคำศัพท์อื่นๆที่เกี่ยวข้อง: รายงานวิจัยฉบับสมบูรณ์ เล่มที่ 2 ธรณีวิทยาและอุทกวิทยา โครงการการสำรวจและการจัดทำฐานข้อมูลเกี่ยวกับถ้ำ จังหวัดแม่ฮ่องสอน. คณะสิ่งแวดล้อมและทรัพยากรศาสตร์ มหาวิทยาลัยมหิดล.

บริษัท จีออกราฟฟิกดีไซน์ จำกัด
2538   โครงการการศึกษาเพื่อพัฒนาและอนุรักษ์ถ้ำ กิ่งอำเภอปางมะผ้า จังหวัดแม่ฮ่องสอน.

ประเสริฐ ประดิษฐ์
2543   รอยฝ่าพระหัตถ์ ถ้ำพระพุทธบาท: เอกสารนำเที่ยวถ้ำ.

พิมุข ชาญธนะวัฒน์
2537   ถ้ำผีแมนที่แม่ฮ่องสอน แหล่งล้ำค่าทางโบราณคดี: ศิลปวัฒนธรรม. 15:11หน้า 72-74.

พิสิฐ เจริญวงศ์ สายันต์ ไพรชาญจิตร์ พาสุข ดิษฮเดช
2527   รายงานการสำรวจทางโบราณคดี เหมืองแม่เมาะ อำเภอแม่เมาะ จังหวัดลำปาง:
       ฝ่ายวิชาการ กองโบราณคดี กรมศิลปากร. รัศมี ชูทรงเดช
2542   รายงานความก้าวหน้าทางด้านโบราณคดีครั้งที่ 2 โครงการสำรวจและจัดทำฐานข้อมูลเกี่ยวกับถ้ำ จังหวัด แม่ฮ่องสอน: คณะสิ่งแวดล้อมและทรัพยากรศาสตร์ มหาวิทยาลัยมหิดล.
2546   คน วัฒนธรรม และสภาพแวดล้อมโบราณ บนพื้นที่สูงใน อำเภอปางมะผ้า จังหวัดแม่ฮ่องสอน. เอกสารประกอบการประชุมวิชาการ: ภาควิชาโบราณคดี คณะโบราณคดี มหาวิทยาลัยศิลปากร. รัศมี ชูทรงเดช,นาฏสุดา ภูมิจำนงค์ ,สุภาพร นาคบัลลังก์
2545   รายงานความก้าวหน้าครั้งที่ 3 โครงการโบราณคดีบนพื้นที่สูงในอำเภอ ปางมะผ้า จังหวัดแม่ฮ่องสอน: ภาควิชาโบราณคดี คณะโบราณคดี มหาวิทยาลัยศิลปากร. วลัยลักษณ์ ทรงศิริ และวิวรรณ แสงจันทร์
2545   แหล่งโบราณคดียุคดึกดำบรรพ์ที่ประตูผา จังหวัดลำปาง: ภาพเขียนสี พิธีกรรม 3000 ปีที่ผาศักดิ์สิทธิ์. ศิลปวัฒนธรรมฉบับพิเศษ. กรุงเทพฯ: สำนักพิมพ์มติชน. วิวัฒน์ พันธุวุฒิยานนท์
2541   ปริศนาถ้ำผีแมน ปริศนามนุษย์. สารคดี 13:156   หน้า135-150. สมหมาย วงษ์มาก และคณะ
2542   รายงานการสำรวจถ้ำผามอน บ้านผามอน ตำบลสบป่อง อำเภอปาง มะผ้า จังหวัดแม่ฮ่องสอน: การท่องเที่ยวแห่งประเทศไทย. สายันต์ ไพรชาญจิตร์ มาริแอล ซังโตนิ และเมือง ปีแอร์ ปอโทร
2528   รายงานการสำรวจชุดคันนาแหล่งโบราณคดีถ้ำฝาหลวง: ฝ่ายวิชาการ กองโบราณคดี กรมศิลปากร. สิทธิพงษ์ ติลกวนิช
2542   ทรัพยากรถ้ำ เอกสารประกอบการประชุมทางวิชาการ โครงการสำรวจ และการจัดทำฐานข้อมูลเกี่ยวกับถ้ำจังหวัดแม่ฮ่องสอนและจังหวัดกาญจนบุรี. สำนักงาน กองทุนสนับสนุนการวิจัย.
2542   รายงานความก้าวหน้าครั้งที่ 2 การสำรวจและการจัดทำฐานข้อมูล เกี่ยวกับถ้ำ จังหวัด แม่ฮ่องสอน: คณะสิ่งแวดล้อมและทรัพยากรศาสตร์ มหาวิทยาลัยมหิดล.
2543   รายงานวิจัยฉบับสมบูรณ์,เล่มที่ 1-6 โครงการสำรวจและการจัดทำฐานข้อมูล เกี่ยวกับถ้ำ จังหวัดแม่ฮ่องสอน: คณะสิ่งแวดล้อมและทรัพยากรศาสตร์ มหาวิทยาลัยมหิดล. สุรพงษ์ เลิศทัศนีย์ ธรณีวิทยาของถ้ำ เอกสารประกอบ การสอน: ภาควิชาธรณีวิทยา คณะวิทยาศาสตร์ มหาวิทยาลัยเชียงใหม่.

FURTHER READING

General (geology, ecology, management)

Ford, T. D. & Cullingford, C. H. D (1976) *The Science of Speleology;* Academic Press, London, UK.

Ford D. & Williams, P. (1989) *Karst Geomorphology and Hydrology,* Chapman & Hall, London.

Gillieson, D. (1996) *Caves: Processes, Development & Management;* Blackwell Publishers, Oxford, UK.

Hill, C. & Forti, P. (1997) *Cave Minerals of the World;* National Speleological Society, Alabama, USA; (2nd ed.)

Lowe D. & Waltham, T. (1995) *A Dictionary of Karst and Caves,* British Cave Research Association, UK.

Meredith, M. & Wooldridge, J (1992) *Giant Caves of Borneo;* Tropical Press Sdn Bhd, Kuala Lumphur.

Millar, I. R. and Wilde, K. A. (1989) *General Policy and Guidelines for Cave and Karst Management.;* Department of Conservation, Wellington, N. Z.

Moore, G. W. & Sullivan, N. (1997) *Speleology: Caves & the Cave Environment;* Cave Books, St. Louis, USA.

Moore, G. W. and Sullivan, N. (1978) *The Study of Caves;* St. Louis, U. S. A.

Watson et al. (1997) *Guidelines for Cave and Karst Protection;* IUCN.

Caves techniques and equipment

Dasher, G. R. *On Station: A Complete Handbook for Surveying & Mapping Caves;* National Speleological Society, Alabama, USA.

Howes, C. *Cave Photography: a practical guide;* Caving Supplies, Buxton, Derbyshire, UK; 68pp.

Howes, C. (1997) *Images Below: A manual of underground and flash photography;* Wild Places Publishing, Cardiff.

Judson, D. et al. (1995) *Caving Practice & Equipment;* British Caving Research Association.

Rea, G. T. (1992) *Caving Basics: A comprehensive guide for beginning cavers;* National Speleological Society, U. S. A.

Archeology & Culture

Higham,C. (2002) *Early Cultures of Mainland Southeast Asia;* River Books, Bangkok.

Higham,C. & Thosarat, R. (1998) *Prehistoric Thailand;* River Books, Bangkok.

Keesing, R. M. (1981) *Cultural Anthropology: A Contemporary Perspective;* HRW International Editions, CBS Publishing Asia Ltd.; 2nd ed.

Kijngam, A. (1988) *Archeological Sites of Thailand;* Division of Archeology, Fine Arts Department; vol. 1, no. 6/2531. (in Thai)

Matics, K. I. (2001) *Gestures of the Buddha;* Chulalongkorn University Press, Bangkok.

Munier, C. (1998) *Sacred Rocks and Buddhist Caves in Thailand;* White Lotus, Bangkok.

Srisuchat, A. (1992) *Rock Art of the Historic Period in Thailand 2;* Fine Arts Department, Bangkok. (in Thai with English summary).

Caves in Northern Thailand

Deharveng, L. et al. (1986 & 1987) *Thai-Maros 85 & 86: Rapport Speleologique et Scientifique;* Assoc. Pyreneenne de Speleologie, Toulouse, France. (2 separate vols).

Deharveng, L. et al. (1988) *Expedition Thai 87-Thai 88:* Rapport Speleologique et Scientifique; Assoc. Pyreneenne de Speleologie, Toulouse, France.

Dilokwanich et al. (2000) *Exploration and Database System of Caves: Mae Hong Son Province;* Faculty of Environment & Resource Studies, Mahidol University, Bangkok. 6 vols.

Dunkley, J. R. (1985) *Karst and caves of the Nam Lang-Nam Khong Region, Northwast Thailand;* Helicite; 23 (1): 3-22

Dunkley, J. R. (1995) *The Caves of Thailand;* Speleological Research Council, Sydney.

Dunkley, J. R. & Brush, J. B. (1986) *Caves of north-west Thailand;* Speleological Research Council, Sydney.

Spies, J. (1997) Values and vulnerability of the caves and karst resources in Mae Hong Son Province, Thailand; *Journal of Ecology;* 24 (3): 41-48.

Suwannabun (1981) *Interesting caves in Thailand;* Education Department, Bangkok. (in Thai).

INDEX

Piman Pingya *see* Ya Pa Nae 1 213
Piman Ban Rai *see* Ban Rai Overhang 242
Piman Boong Hoong *see* Banyan Valley 196
Piman Chan Long *see* Chan Long 187
Piman Dtao Daeng *see* Pha Dtao Daeng 247
Piman Lum Karn *see* Chan Long 187
Piman Mae Hung *see* Banyan Valley 196
Piman Pang Kham 1 *see* Pang Kham 1 (Pi Man) 190
Piman Pha Daeng 1 *see* Long Yak 186
pine, log coffins 108
pinnacle 21
Pla "fish" in Thai. Also spelt Plaa, Plah
Pla (Mae Hong Son) 69, 142, **258,** 370
Pla (Mai Sai) 119, 122, 124-126, **338-341,** 374
Pla (Pang Ma Pha) *see* Hued 220
Plaa *see* Pla
Plaew Plong Fah 123, 126, **340-341,** 374
Plah *see* Pla
plant domestication 88, 91
Pleistocene 78, 80, 83, 87, 88, 106
Pleo Plong Fah *see* Plaew Plong Fah 340
Plong Sawan (Chiang Dao) 121, 136
Pok Piang *see* Pbak Pbiang 282
Pok Soong *see* Pbak Sung 282
polished stone tools 88, 90, 98
polje 20
pollination 147
pollution 151
Pong Saen Pik 367
Pong Tham, Ban 366
pool, sacred 141, 360
poolspar 40
popcorn 42
Pope's Pit Viper 71
porcupine 74
posture, of Buddha 128
Potamon doichiangdao 64, 274
Potamon namlang 64
Potassium - dating 46, 63
pottery 90, 98, 102, 106, 109
Pra Bat Ku *see* Jak Dtor 264
Pra Tu Pha 13, 59, **98-101, 324-5,** 373
Pra Ya Nag 330, 373
Prachinburi 125
Prachuap Kiri Khan 79
Precambrian 15
prehistory 83
pressure 53
primary burial 102
Princess Cave *see* Pha Nang Koi 362
proto-cave 15, 22, 24
pseudokarst 33
Pua 375
Puak, Pha *see* Pha Phuak 192
Puek, Pha *see* Pha Phuak 192
Pung Hung *also* spelt Boong Hoong 12, 20, 27, **196-199,** 367
Pu-tai 12, 124, 126, 131, 336 *see also* Ariyametta, Metteya

quarrying 151
quartz 16, 17
quartzite 16, 33

Racer Snake 57, 70, 164
radioactivity, dating 92
Rai, Ban (Pang Ma Pha) *see* Ban Rai Overhang 242
Rai, Ban Pha 375
rainpit 21
Raja Guha, Nepal 118
Ram Karn *see* Chan Long 187
Rama III, King 130
Rama VII, King 12, 320
Ratchaburi 119, 151
Reasoning (Buddha posture) 129
Reclining (Buddha posture) 128
Recreation Opportunity Spectrum 156
Red Tortoise Cliff *see* Pha Dtao Daeng 247
redissolved flowstone 49
Reed Flute Cave (China) 144
reflectivity 175
relic, of Buddha 125
religion 115
reptile 70
respiration 16
rhinoceros 80
Rhinolophus luctus 73
rice husks 146
rillenkarren 21
rimstone dam 40
ripple karren 21
Rishi Va Suthep 132
ritual use of caves 141
road building 151
robe, Buddha 125
rock art 83, 94, **95-101,** 115, 141, 243, 247, 262, 290, 299, 312, 324-325
Rock Column Cave *see* Sao Hin 342
rockmill 27
Romania 33, 58
Rong Kwang District 375
roof pocket 24
Rousettus 73
Ru Hua Koa *see* Khang Khao 259
Rua, Pha 345
rubbish 149
Rue Luang, Nam 306
Rue Takhaen, Nam 24, 306
rundkarren 21
Russi = 'hermit' in Thai. Tham Russi is a very common name for small caves and rock shelters. 131, 140
Russi (Lamphun) 132, **313,** 373
Russi (Phan-Thoeng) *see* Pha Jarui 349
Russi (Chiang Dao) *see* Pbak Sung 282, 140
Russi (Mae Hong Son) *see* Pla 258, Khang Khao 259, Jak Dtor 264
Russi, Wa Suthep 132, 313

Sa Morakot 125
sabre-toothed cat 80
sacred pool 141, 360
Sagurn *see* Sakoen 354
Sahara 95
Sai Thong **334-335,** 373
Sakka 134

Sakoen 32, 47, 142, **354-357,** 375
Sale, Ban 366
salt (sodium cloride or halite) 17, 19, 33
saltpetre 142
Salween River 10
San Kamphaeng 372
San People (South Africa) 116
San Wi 136, 362
sandstone 16, 33
Sao Hin 136, **342-343,** 374
Sao Praya Nak *see* Sao Hin 342
Saptaparna Guha Cave (Nepal) 118
Saraburi 119
Sarika Cave 118
Satorkopuszta (Hungary) 32
saturated 17
saturation point 34
Saturday (Buddha posture) 128
Sawan, Pha 360
Scaffolds (Pha Daeng) *see* Pha Daeng scaffolds 184
scallop 25, 26
Schistura oedipus 38, **68,** 202, 220, 221, 236, 248
scientific ethics 113
scorpion 63
Scutigera 56, 65
sea lily (crinoid fosil) 49
sea level change 79
sea water 19
secondary burial 102
sediment 19, 45, 56, 139, 166
serow (mountain goat) 74, 91, 100
serpent (naga) 134
Seua 12, 30, 53, **254-255,** 369
sewage 151
shaft 112
shaman 115
Shang Dynasty 146
shape, of caves 30
shell 106
shelter 140
shield 43
Shiva 117, 314
Shoochondej, Rasmi 91, 104 *see also* Silpakorn University
shrimp 64
signboard 153
silica 48
Silpakorn University 10, 13, 104, 155, 188, 229, 242
siltation 150
Silurian 10, 15
Singha 133, 134
skeleton 102, 106, 137
slave units (photography) 174
Slovenia 19
snail 62
snake 70, 164
sodium chloride 19
soft-shelled turtle 71
soil erosion 21
solubility 16
solution doline 20, 22
solution of limestone 16
solution rate 18
Song Khwae 375
Songkran Festival 12, 362, 375
Soppong 20, 367, 368, 369
South Africa 82, 95, 115, 146, 155

Pang Ma Pha

Mae Hong Son Province

1: 100,000

0   1   2 km

Legend

★ cave (in this book)

★ cave (not in book)

◉ village

— river

— road

< 400 m

400-599 m

600-799 m

800-999 m

1000-1199 m

1200-1399 m

>1400 m

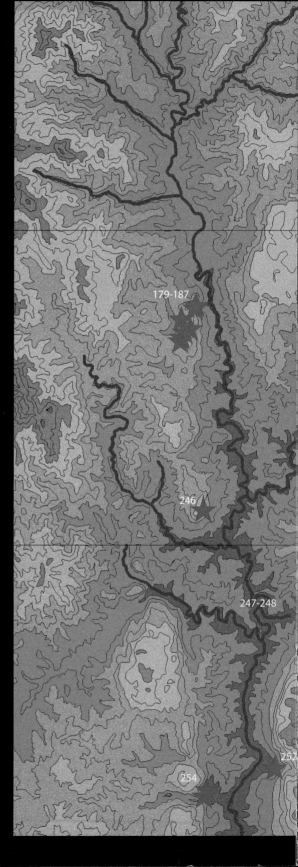